CRUDE OIL FOULING

CRUDE OIL FOULING

Deposit Characterization, Measurements, and Modeling

Edited by

FRANCESCO COLETTI, PhD
Hexxcell Ltd, London, UK

GEOFFREY F. HEWITT, PhD
Imperial College London, UK

ELSEVIER

Amsterdam • Boston • Heidelberg • London
New York • Oxford • Paris • San Diego
San Francisco • Singapore • Sydney • Tokyo

Gulf Professional Publishing is an imprint of Elsevier

G P P

Gulf Professional Publishing is an imprint of Elsevier
225 Wyman Street, Waltham, MA 02451, USA
The Boulevard, Langford Lane, Kidlington, Oxford, OX5 1GB, UK

ISBN: 978-0-12-801256-7

Library of Congress Cataloging-in-Publication Data
A catalog record for this book is available from the Library of Congress

British Library Cataloguing in Publication Data
A catalogue record for this book is available from the British Library

For information on all Gulf Professional Publishing visit our website at http://store.elsevier.com

Typeset by TNQ Books and Journals www.tnq.co.in

This book has been manufactured using Print On Demand technology.

CONTENTS

LIST OF CONTRIBUTORS

Dr. John Chew
Lecturer in Chemical Engineering, University of Bath, UK

Dr. Francesco Coletti
Chief Technology Officer, Hexxcell Ltd, UK

Prof. Barry D. Crittenden
Professor of Chemical Engineering, University of Bath, UK

Dr. Andrew J. Haslam
Research Fellow, Imperial College London, UK

Prof. Geoffrey F. Hewitt
Professor of Chemical Engineering, Imperial College London, UK

Prof. George Jackson
Professor of Chemical Physics, Imperial College London, UK

Dr. Guadalupe Jimenez-Serratos
Research Associate, Imperial College London, UK

Dr. Himanshu M. Joshi
Principal Heat Transfer Engineer, Shell Global Solutions (US), Inc., USA

Prof. Sergei G. Kazarian
Professor of Physical Chemistry, Imperial College London, UK

Prof. Sandro Macchietto
Professor of Process Systems Engineering, Imperial College London, UK

Prof. Omar K. Matar
Professor of Fluid Mechanics, Imperial College London, UK

Dr. Marcos Millan-Agorio
Reader in Chemical Engineering, Imperial College London, UK

Prof. Erich A. Müller
Professor of Thermodynamics, Imperial College London, UK

Dr. Khairul Rostani
CO2 Management R&D, PETRONAS Research Sdn Bhd, Malaysia

Dr. Daniele Sileri
Senior Business Consultant, Misys, Hong Kong

Dr. Feng H. Tay
Head of R&D, EliteTNS, Singapore

Dr. Silvia Venditti
R&D Engineer, Resource Centre for Environmental Technologies, Luxembourg

Dr. Junfeng Yang
Research Associate, Imperial College London, UK

Dr. Mengyan Yang
Research Associate, University of Bath, UK

PREFACE

Nearly every gallon (or liter!) of fuel, which is used in transport vehicles (cars, trucks, trains, airplanes, etc.), is derived from oil that is extracted in the crude state from oil wells and is then processed in an oil refinery to yield the required transport fuel (gasoline, diesel, kerosene, etc.). The first task in the refinery is to separate these useful fuels from the crude oil by the process of *distillation* in which the components of the oil are separated on the basis of their volatility. Of course, the components of crude oil, which are of volatility too low to allow them to be used as fuels, are also separated in the distillation process and can be either directly used (for example, as lubricating or fuel oils) or can be chemically converted to more volatile materials.

In the distillation of crude oils, the separated components are hot and it is important to use this heat in a series of heat exchangers (the "crude preheat train") in which heat is transferred from the separated streams into the incoming crude oil. Unfortunately, these heat exchangers are subject to *fouling*, which reduces the amount of heat they exchange and increases the amount of extra heat that needs to be added (usually through an oil-fired furnace immediately preceding the distillation column). Fouling the exchangers also leads to higher pressure losses (and hence an increased pumping power requirement) and, of course, the exchangers need to be periodically cleaned. Altogether, fouling in the crude preheat train leads to an enormous cost penalty (of the order of 1 billion dollars per annum in the United States alone).

The economic penalties associated with crude oil fouling led the UK Engineering and Physical Sciences Research Council (EPSRC) to finance a research program on crude oil fouling (CROF), which was started in 2006. This three-year program was carried out as a collaboration among Imperial College (London), Bath University, and Cambridge University. Industrial companies representing about 70% of the world's refining capacity also participated through a special Working Party. The work comprised experimental measurements of fouling, analytical modeling, and the development of techniques for the analysis and characterization of crude oils and deposits.

As a result of this research project, a step change has been possible in the characterization and modeling of crude oil fouling and it is clearly important to capture the impact of the work in a published form. Although a number of articles have been published on various aspects of the work, it was considered important to create a unified and structured presentation of the subject, and this is the objective of the present book.

Although the work described here is focused on crude oil fouling in refinery heat exchangers, most of the experimental and modeling techniques can be either directly applied or readily adapted to address to other types of crude oil fouling. We believe that engineers and researchers involved with waxing in pipelines, fouling of offshore equipment or other downstream units will also find the science and applications contained in the book useful.

Though considerable progress has been made in the work summarized here, it seems safe to predict that crude oil fouling will remain an important focus for research for many years to come!

G.F. Hewitt & F. Coletti

NOMENCLATURE

ROMANS

a	Curvature of annulus (m)
a	Material-specific constant (–)
$a_1,..a_n$	Various fouling model constants
A	Area (m^2)
A	Pre-exponential factor for chemical reaction fouling (m^2 K J^{-1})
A_a	Pre-exponential factor for aging model (s^{-1})
A_{flow}	Cross-sectional area for flow (m^2)
B_c	Baffle cut (%)
Bi_f	Biot number for fouling (–)
C	Phase volume fraction or species concentration (–)
Ca	Capillary number (–)
C_d	Discharge coefficient (–)
C_f	Friction factor (–)
c	Constant of integration (–)
c_b	Concentration in the bulk (kg m^{-3})
c_f	Foulant concentration (kg m^{-3})
c_p	Concentration of precursors (kg m^{-3})
c_p	Specific heat capacity at constant pressure (J kg^{-1} K^{-1})
c_s	Concentration at the surface (kg m^{-3})
c_r	Concentration of reactants (kg m^{-3})
D	Diameter (m)
D_d	Deposit diffusion coefficient (m^2 s^{-1})
D_H	Hydraulic diameter of annulus (m)
D_i	Inner diameter (m)
D_o	Outer diameter (m)
D_{otl}	Tube bundle diameter (m)
D_S	Shell inner diameter (m)
D_s	Inverse of diffusion coefficient (s m^{-2})
d	Asphaltene phase mole fraction (–)
d_t	Inner diameter of nozzle throat (m)
d_{tube}	Inner diameter of tube connected to nozzle (m)
dir	Direction of flow (–)
E_a	Activation energy ageing reaction (J mol^{-1})
E_f	Activation energy fouling reaction (J mol^{-1})
e	Roughness (m)
e	Mathematical constant 2.71828 (–)
e^0	Initial roughness value (m)
e^∞	Final roughness value (m)
G	Free energy (kJ mol^{-1})

H	Enthalpy (kJ mol^{-1})
H	Channel height (m)
H_D	Heptane dilution value (–)
ΔH_p	Precipitation enthalpy (kJ mol^{-1})
h	Distance between nozzle and deposit (m)
h_0	Distance between nozzle and substrate (m)
h_{id}	Heat transfer coefficient for ideal cross-flow (W m^{-2} K^{-1})
I_N	Insolubility number (–)
J	Heat transfer coefficient correction factor (–)
J	Mass flux from the oil phase to fouling phase (kg m^{-2})
K	Equilibrium coefficient (–)
K_0	Equilibrium constant at precipitation temperature (–)
k	Order of reaction
k_1	Growth rate constant (s^{-1})
k_2	Removal rate constant (s^{-1})
k_a	Roughness dynamics model constant (–)
k_f	Thermal conductivity of deposit (W m^{-1} K^{-1})
k_m	Mass transfer coefficient (m s^{-1})
k_r	Rate constant of energy barrier (m s^{-1})
k_t	Ageing rate (s^{-1})
l_e	Length of element (m)
l_{eff}	Effective length of tube connect to nozzle (m)
L	Tube length (m)
L_{bc}	Central baffle spacing (m)
L_{bi}	Inlet baffle spacing (m)
L_{bo}	Outlet baffle spacing (m)
L_{tp}	Tube pitch (m)
L_X	Number of moles present for each phase (mol)
M	Mass (kg)
$\dot{M}$	Mass flow rate (kg s^{-1})
$\dot{M}_d$	Mass rate of deposition (kg s^{-1})
$\dot{M}_r$	Mass rate of fouling removal (kg s^{-1})
MW	Molecular weight (kg mol^{-1})
M_c	Characteristic value of mobility (s kg^{-1})
$\dot{m}$	Mass flux (kg m^{-2} s^{-1})
m_g	Measured gauging mass flow rate (kg s^{-1})
$\mathbf{n}$	Normal vector (–)
N	Total number of measurements (–)
N_b	Number of baffles (–)
N_p	Number of tube–side pass per shell (–)
N_S	Number of shells per unit (–)
N_{SS}	Number of sealing strip pairs (–)
N_t	Total number of tubes (–)
Nu	Nusselt number (–)
NV	Number of measurements (–)
P, p	Pressure (Pa)
P	Wetted perimeter (m)

$\boldsymbol{P}$	Probability (–)
Pe	Peclet number (–)
$\boldsymbol{\Delta P_{ij}}$	Pressure drop between station *i* and *j* (Pa)
Pr	Prandtl number (–)
$\boldsymbol{P_t}$	Tube-side inner perimeter (m)
$\boldsymbol{P_x}$	Pitch ratio (–)
q''	Heat flux (W m^{-2})
Q	Heat duty (W)
r	Radial coordinate (m)
$\underset{\sim}{r}$	Reaction rate (mol m^{-3} s^{-1})
$\tilde{r}$	Dimensionless radial coordinate (–)
r_d	Density ratio (–)
r_k	Conductivity ratio (–)
r_i	Reaction i (–)
r_v	Viscosity ratio (–)
R	Radius (m)
R_f	Fouling resistance (m^2 K W^{-1})
$\overline{R_f}$	Average fouling resistance (m^2 K W^{-1})
R_g	Universal gas constant (J kg^{-1} mol^{-1})
Re	Reynolds number (–)
$\mathbf{Re_a}$	Reynolds number in the annulus (–)
S	Heat transfer surface (m^2)
S	Entropy (kJ mol^{-1} K^{-1})
S_{BN}	Solubility blending number (–)
T	Temperature (K)
T_b	Bulk temperature (K)
T_f	Film temperature (K)
$\widehat{T}_f$	Modified film temperature (K)
T_{in}	Inlet flow temperature (K)
T_p	Precipitation temperature (K)
T_r	Reduced temperature (–)
T_s	Surface temperature (K)
T_{s0}	Initial surface temperature (K)
T_{st}	Surface temperature at time t (K)
T_w	Wall surface temperature (K)
T_{wb}	Thermocouple temperature (K)
T_{wm}	Thermocouple temperature (K)
T_{ws}	Thermocouple temperature (K)
ΔT_e	Temperature increase of element (K)
t	Time (s)
$t_{0.5}$	Time when θ is half its maximum value (s)
t_{age}	Ageing time (s)
U	Overall heat transfer coefficient (W m^{-2} K^{-1})
$\boldsymbol{u}$	Velocity vector (–)
u	Mean fluid velocity in the tube (m s^{-1})
$\overline{V}_{bulk}$	Average mean velocity in the bulk (m s^{-1})
V	Volume (m^3)

$\dot{V}$	Volumetric flow rate ($m^3\ h^{-1}$)
V_A	Molar volumes of asphaltenes in the asphaltene phase ($m^3\ mol^{-1}$)
V_d	Volume of deposit (m^3)
V_L	Molar volumes of asphaltenes in the liquid phase ($m^3\ mol^{-1}$)
V_t	Volume of channel (m^3)
v	Velocity ($m\ s^{-1}$)
w	Width of nozzle rim (m)
W	Power (W)
X_{A_i}	Fraction of non-associated molecules (–)
x	Horizontal coordinate (–)
x	Simulated value
$\tilde{x}$	Measured value
x_e	Distance expanded by element (m)
y	Youth variable (–)
y_0	Initial value of youth variable (–)
z	Axial coordinate (m)
z	Vertical coordinate (–)

GREEK

α	Local heat transfer coefficient ($W\ m^{-2}\ K^{-1}$)
α	Angle of nozzle entry (°)
α_{id}	Heat transfer coefficient for ideal cross–flow ($W\ m^{-2}\ K^{-1}$)
α_t^0	Tube–side heat transfer coefficient in clean conditions ($W\ m^{-2}\ K^{-1}$)
β	Linear expansion coefficient (K^{-1})
β	Constant in Ebert and Panchal model (–)
β	Thermocapillarity term (K^{-1})
γ	Suppression parameter ($m^4\ K\ N^{-1}\ J^{-1}$)
$\dot{\gamma}$	Local shear rate (s^{-1})
δ	Thickness of deposit (m)
δ_A	Average solubility parameters in the asphaltene phase ($Pa^{1/2}$)
δ_L	Average solubility parameters in the liquid phase ($Pa^{1/2}$)
ε	Temperature percentage residuals (–)
ζ	Data filtering procedure parameter (%)
η	Dimensionless function (–)
η_{mech}	Mechanical efficiency (–)
Θ	Non-dimensional parameter that depends on L_X (–)
θ	Characteristic formation time of polyaromatic structure in fouling layer (s)
θ	Fractional surface coverage (–)
θ_{tp}	Tube pitch angle (°)
λ	Thermal conductivity ($W\ m^{-1}\ K^{-1}$)
λ	Length of nozzle entry (m)
λ^0	Thermal conductivity for freshly deposit material ($W\ m^{-1}\ K^{-1}$)
λ^∞	The maximum value of thermal conductivity ($W\ m^{-1}\ K^{-1}$)
λ_f	Thermal conductivity of foulant layer ($W\ m^{-1}\ K^{-1}$)

μ	Dynamic viscosity (Pa s)
μ_0	Viscosity of fresh foulant material (Pa s)
ν	Kinematic viscosity (St)
ξ	Structure parameter (–)
ρ	Density (kg m^{-3})
ρ	Molar density in Section 5.2 (mol m^{-3})
σ	Interfacial tension between oil phase and asphaltene phase (N m^{-1})
σ	Measurement variance (K)
τ	Surface shear stress (Pa)
τ_{ya}	Surface shear stress along the arc of inner annulus (Pa)
ϕ_c	Characteristic value of chemical potential (kg m^2 s^{-2})
ϕ_{cr}	Fouling rate decrease by conversion of tarry layer to coke (m^2 K W^{-1} s^{-1})
ϕ_d	Fouling deposition term (m^2 K W^{-1} s^{-1})
ϕ_r	Fouling removal term (m^2 K W^{-1} s^{-1})
ϕ_{tr}	Fouling rate decrease by mass transfer of tars to the bulk fluid (m^2 K W^{-1} s^{-1})
φ	Heat balance closure parameter (%)
χ_i	Weight fraction for component i (–)
Ω	Model domain (–)

SUBSCRIPTS

0	Initial
b	Bulk
c	Cold
cr	Conversion to coke
d	Deposit
f	Fouling, film
h	Hot
i	Inner, i-th component, interface
in	Inlet
L	Fouling layer
n	Pass number
o	Outer
out	Outlet
p	Precursor
s	Shell-side
T	Tube-side
T	Total
tr	Tar layer
W	Wall

SUPERSCRIPTS

0	Initial state of deposit
∞	Final state of deposit
*	Asymptotic value
m	Reaction order for reactant one
n	Reaction order for reactant two
n	Fluid parameter for aging model
l	Reaction order for precursor

ABBREVIATIONS

2D	Two-dimensional
3D	Three-dimensional
AFM	Atomic force microscopy
ATR	Attenuated total reflection
CDU	Crude distillation unit
CFD	Computational fluid dynamics
CG	Coarse grained
C–H	Cahn–Hilliard
CII	Colloidal instability index
CIT	Coil Inlet Temperature
CLSM	Confocal laser scanning microscopy
COT	Coil Outlet Temperature
CV	Control valve
DFT	Density functional theory
DIM	Diffuse interface method
DNS	Direct numerical simulation
DPD	Dissipative particle dynamics
EOS	Equation of State
ETS	Emissions Trading Scheme
FDG	Fluid dynamic gauging
FTIR	Fourier transform infrared spectroscopy
HEN	Heat exchanger network
HEX	Heat exchanger
HiPOR	High Pressure Oil Rig
HTRI	Heat Transfer Research Incorporated
LES	Large eddy simulation
LLE	Liquid–liquid equilibria
LMTD	Logarithmic mean temperature difference
LVDT	Linear variable displacement transducer
N–S	Navier–Stokes
NRV	Nonreturn valve
PHT	Pre heat train

RANS Reynolds-averaged Navier–Stokes
SAFT Statistical associating fluid theory
SARA Saturate, Aromatic, Resin and Asphaltene
SEC Size-exclusion chromatography
SEM Scanning electron microscopy
TGA Thermogravimetric analysis
UV-F Ultraviolet fluorescence
VdW Van der Waals
VGO Virgin gas oil
VLE Vapor–liquid equilibria
VLLE Vapor–liquid–liquid equilibria
VOF Volume of fluid
XRD X-ray diffraction

CHAPTER ONE

Introduction

F. Coletti[1], H.M. Joshi[2], S. Macchietto[3], G.F. Hewitt[4]
[1]Chief Technology Officer, Hexxcell Ltd, UK
[2]Principal Heat Transfer Engineer, Shell, USA
[3]Professor of Process Systems Engineering, Imperial College London, UK
[4]Professor of Chemical Engineering, Imperial College London, UK

Stringent environmental policies and tightening refining margins in a highly competitive market are forcing oil companies worldwide to increase their focus on the efficient utilization of energy. Refinery efficiency is so critical that it has been reported to be capable of moving the share price of major oil and gas companies. In this perspective, the efficiency of heat exchange equipment and heat exchange networks is of paramount importance. Fouling, the deposition of unwanted material on heat transfer surfaces, is a major cause of inefficiencies with severe effects on refinery economics, operability, health and safety, and environmental impact.

Fouling is a long-standing problem in the process and energy industry. Taborek (1995) tracked the origin of the first industrial concern about fouling back in the 1880's USA power industry and the first mention of fouling in the open literature is in a paper by Orrok (1910). Fouling has been described both as "the major unresolved problem in heat transfer" (Taborek et al., 1972) and "a nearly universal problem in heat exchanger equipment design and operation" (Watkinson, 1988). Indeed fouling is ubiquitous in the oil industry.

In upstream operations, hydrate formation, asphaltene precipitation, and wax deposition from crude oil not only reduce the thermal efficiency of heat exchangers but also, more importantly, restrict flow, causing blockages that significantly impact operations. In some cases also oil pipelines are affected to the extent of becoming plugged. Wax deposition typically occurs when crude is cooled to a point at which the paraffins normally contained in solution start aggregating and eventually depositing on the surfaces.

In oil refineries, the feed/effluent heat exchangers in naphtha hydrotreaters, the slurry exchangers in fluid catalytic cracking units, and the furnace and exchangers in the visbreaker unit are all known to be affected by severe fouling problems. However, the largest share, about 50% (Van Nostrand et al., 1981), of the total fouling-related costs for the whole

Crude Oil Fouling
ISBN: 9780128012567
http://dx.doi.org/10.1016/B978-0-12-801256-7.00001-4

refinery originates in the preheat train (PHT) of the atmospheric distillation unit (here referred to as crude distillation unit or CDU).

The PHT is an extensive network of heat exchangers used to reduce energy requirements in the CDU, which is where primary fractionation of all the crude processed in the refinery is performed. If the PHT is not working efficiently, more fuel must be burnt at the downstream furnace that heats the crude to the required temperature for the distillation. The average energy involved with this process has been estimated to be over 192 TWh per year (6.94×10^{17} J per year) for refineries in the United States (DOE, 2006). This makes the CDU the largest energy utilizer in the refinery and one of the largest in the whole process industry. By comparison, the total primary energy consumed by Ireland in 2013 was 169 TWh.

Although there are only a few—largely outdated—studies that deal with the economic costs of fouling in oil refineries, there is little doubt that fouling has an enormous impact on the refinery's bottom line. Van Nostrand et al. (1981) estimated that process-side fouling cost the United States refineries US$ 1.36bn per year, and US$ 861 MM of that in the crude PHT alone. Adjusting for inflation to 2014 this is equivalent to US$ 3.6bn and US$ 2.26bn, respectively.

The environmental impact is even more staggering with fouling in refineries estimated to be responsible for 88 MM t of CO_2, or 2.5% of total worldwide anthropogenic emissions in 2009 (Müller-Steinhagen et al., 2009a).

How much of this inefficiency can be eliminated, and at what cost, is the key question. Fouling mitigation can provide increased capacity and reduced greenhouse gas emissions without significant capital expenditure. A study made in 2006 for the US Department of Energy (DOE) indicates that potential fuel savings up to 55% can be achieved in oil refineries by improving operating practices and capital equipment (DOE, 2006). Among the suggested improvements it was found that fouling mitigation in the PHT and fired heater in atmospheric distillation units could lead to a 15% fuel saving (1/3 from existing technology, 2/3 from technology in the R&D stage). As fuel consumption in the atmospheric furnace represents around 4% of the total refinery throughput (Yeap et al., 2004), a potential saving of 15% equates to a sizable 500,000 bbl per day potential savings worldwide, equivalent to the daily production of a large refinery, or (for oil at US$80 bbl^{-1}) a value of US$14.6bn per year. However, much still needs to be done to tap into these savings, starting from the way heat exchangers are designed, crudes are blended, and operations are managed. To improve these aspects, it is widely recognized that a more fundamental understanding of the underlying fouling phenomena is needed.

Over the years, several projects have been coordinated to study the complex and interacting phenomena involved in different types of fouling (Pilavachi and Isdale, 1993; Pritchard, 1988a). For crude oil fouling, many experimental and theoretical studies have been produced by the University of British Columbia and the Argonne National Laboratory (Watkinson, 1988; Panchal and Watkinson, 1994; Kuru et al., 1997), the University of Bath (Crittenden et al., 1987a,b), the HTRI's Crude Oil Fouling Task Force (Bennett et al., 2006, 2009), IHS ESDU's Oil Industry Fouling Working Party (ESDU, 2000), and others. More recently, fouling research has moved from the isolated activities of an individual research group, specialized in a specific technique, to large multidisciplinary projects that coordinate the activities of multiple researchers, expert in different areas, into a multipronged attack to the fouling problem. These projects, such as the Crude Oil Fouling project (CROF) (Macchietto et al., 2009) and the UNIHEAT project (Imperial College London and Boreskov Institute of Catalysis), typically involve both experimental and modeling work examining the fouling problem across all scales, from the molecular to the process unit to the overall heat exchanger network.

Although chemical and physical phenomena underlying the fouling process are extremely complex and several details remain unknown, the overall understanding of the problem has improved significantly in the past few decades. For example, it is now known that fouling in the hot end of the PHT is originated by a series of chemical reactions triggered by the high operating temperatures. There is also experimental evidence that critical velocities exist above which fouling does not occur due to deposit removal by shear forces from the fluid flow (Ebert and Panchal, 1995). However, all this knowledge has so far not been fully exploited and translated into tangible benefits for refinery operators. It would ultimately be highly desirable to improve traditional design methodologies to find fouling-resilient arrangements that allow undisrupted operations with minimal loss of efficiency.

Traditional heat exchanger design methodologies largely rely on highly empirical, fixed, TEMA "fouling factors" (TEMA, 1941). These are basically "safety factors" that lead the designer to specify an oversized heat transfer area that, in theory, should compensate for reduced thermal efficiency caused by fouling. As will be seen in more detail in Section 5.1, several authors have highlighted how this approach ultimately turns out to exacerbate fouling rather than mitigate it. The fundamental limitations of the fouling factor approach are that it does not capture in any detail actual fouling dynamics and its strong dependence on exchanger design and process conditions.

The ability to capture at the design stage the dependence of fouling on process conditions and time becomes therefore pivotal to achieve fouling-resilient designs and retrofits of heat exchangers and networks, and to better support operating decisions. Butterworth (2004), in his process heat transfer outlook for 2010, noted that:

> *The next major attack on fouling will be as a result of a better design process. For some key applications, such as crude-oil, preheat chain, we have equations which predict (though not very accurately) the way fouling develops and the effect of temperature and velocity on this. What we need therefore to do is to get these equations into our design software and to become skilled in using this to design for minimum fouling.*

This book illustrates a multipronged attack to the crude oil fouling problem covering several scales of investigation (from micro to refinery scale), from both an experimental and a modeling approach. The focus is not only in presenting fundamental science underlying the fouling mechanisms and the most advanced technologies available for experimental characterization and mathematical modeling, but also in providing practical suggestions to refinery operators. Industrial case studies are used to show how the techniques described can be used to effectively mitigate fouling.

The present chapter of the book introduces the topic of crude oil fouling and starts with a description of fouling in typical crude distillation units. The chapter continues with a discussion of the impact of fouling on the operation of a crude oil distillation unit, from both an operational and an economic point of view. Health and safety aspects are also discussed.

Chapter 2 reviews the underlying science of the fouling process; basic fouling mechanisms are described as are the routes for chemical reaction fouling. The sequence of events occurring in fouling are discussed and the variables affecting these events are reviewed.

To understand key variables affecting fouling, it is necessary to be able to generate deposits at controlled conditions and accurately measure primary variables (i.e. temperature, pressure, etc.). Experimental work on generation of fouling deposit is described in Chapter 3. The first part of the chapter describes deposit formation on a very small scale in microbomb reactors. The second part of the chapter describes a novel experimental technique for studying fouling in which fouling layers are created in a rotating cell device under controlled conditions of temperature, heat flux, and shear stress. Finally, the chapter discusses the construction of a rig for studying fouling in large-scale geometries simulating those found in real heat exchangers.

If an attempt is to be made to understand the mechanisms behind the fouling process, it is important to determine the structure, chemistry and physical characteristics of the deposits generated in the lab or collected from industrial equipment. Chapter 4 begins with a review of industrial practice and findings in this context. The second part of Chapter 4 discusses the determination of the chemical structure and molecular weight of deposits and the third part introduces a novel (*chemical imaging*) technique by which the location of various chemical species in a surface layer can be established. Though the most important characteristic of a fouling layer is its thermal resistance, the *thickness* of the layer is also important because the restriction in channel size generated by the presence of the fouling layer can lead to increased pressure drop. The final part of Chapter 4 gives a description of a dynamic foulant thickness gauge.

The ability to model the fouling processes is clearly important if the results of the lessons from the experimental analysis are to be brought to bear on real problems. Work in Chapter 5 describes a multiscale approach to modeling of crude oil fouling focused on improving understanding from the molecular level to industrial scale systems. The molecular and small-scale modeling work allows the determination of key parameters, such as diffusion coefficients and fluid physical properties, which can be used in detailed fluid-dynamic models to predict fouling deposition in simple flows. On a large scale, advanced systems modeling has resulted in an ability to predict the future course of fouling in a heat exchanger. The last section of this chapter shows how these models can be used to accurately assess operational costs due to fouling, assisting in heat exchanger design and devising improved operating strategies that minimize costs.

The final chapter (Chapter 6) summarizes the current state of knowledge on crude oil fouling and attempts to predict how the subject will develop in the future.

1.1 CRUDE DISTILLATION UNITS IN OIL REFINERIES

The CDU is the first of several units the crude oil encounters in any refinery. A typical CDU (Figure 1.1) consists of a preheat train, a desalter, a preflash drum, a furnace, and an atmospheric distillation column.

A typical CDU capacity, which determines the total capacity of the refinery, ranges between 100,000 and 200,000 barrels per day (bbl per day). Large refineries that can process more than 500,000 bbl per day may have more than one crude distillation unit.

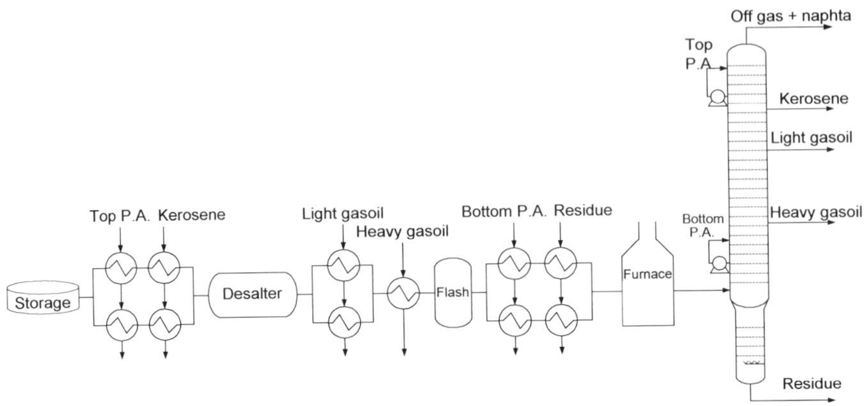

Figure 1.1 Schematic of a typical crude distillation unit.

In the atmospheric distillation column, crude is fractionated into its primary products. Hot distillates (kerosene, light and heavy gasoil) are withdrawn from different sections of the column while naphtha and liquefied petroleum gases (LPG) are removed from the column head. The number of trays in the distillation column typically varies between 30 and 50 trays, depending on the degree of fractionation required and the number of side streams. The final products, characterized by their boiling ranges, are reported in Table 1.1.

In United States refineries the average CDU's energy consumption has been estimated at 31.9 kWh (114.9 GJ) per barrel processed (DOE, 2006), roughly 98% of which is due to fuel consumption at the furnace and the remaining 2% due to electricity (including also the energy used for desalting). Although this is relatively low compared to other conversion processes downstream (Figure 1.2(a)), because it processes all incoming crude oil, with a United States average refinery consumption estimated in 190 TWh per

Table 1.1 Typical atmospheric distillation product boiling ranges, Jones (1995)

Stream	**Product name**	**Boiling range [°C]**
Off gas	LPG	Up to 30
Overhead distillate	Full-range naphtha	30–190
First side-stream	Kerosene	190–250
Second side-stream	Light gasoil	250–320
Third side-stream	Heavy gasoil	320–365
Residue	Fuel oil	365–500
Residue	Bitumen	>500

LPG, liquefied petroleum gases.

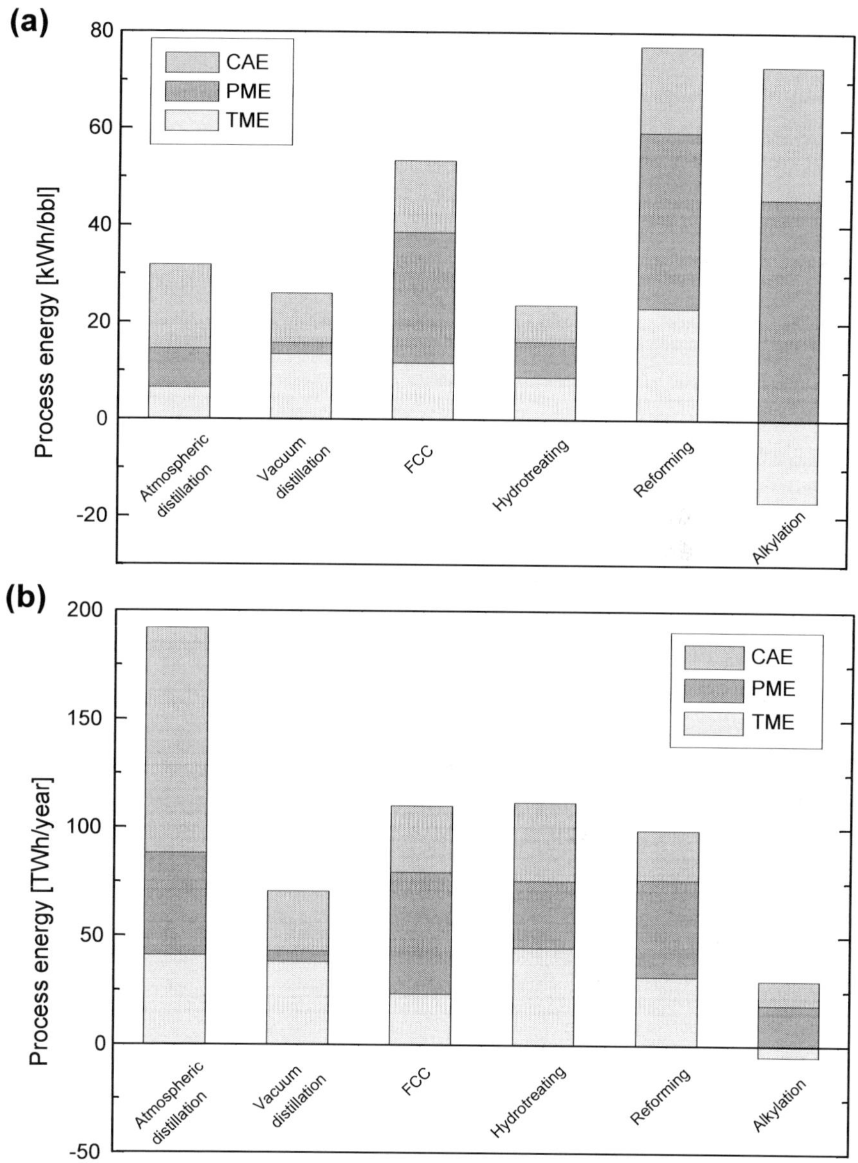

Figure 1.2 Average energy use in oil refineries (a) per barrel of feed processed basis and (b) production per year basis *(adapted from DOE, 2006)*. TME, theoretical minimum energy required; PME, practical minimum energy; CAE, current average energy.

year (6.86 × 10^{17}J per year), the CDU is the largest energy user in the whole refinery (Figure 1.2(b)).

Figure 1.2 considers the theoretical minimum energy required under ideal conditions (TME), the practical minimum energy required under nonideal conditions (PME), and the current average energy consumed under actual plant conditions (CAE). The difference between PME and CAE is a measure of potential gains that may be achieved, according to the DOE (2006) study, by improving current technologies, process design, or operating practices.

Given the large quantity of energy involved, the design of crude distillation systems—and the potential saving associated—has attracted the interest of both industrial and academic communities. In late 1970s, early researchers focused on improving the distillation column design (e.g., internal geometries, use of pump arounds, etc.). When in the early 1980s the use of Pinch Technology became widespread, the focus shifted to improving heat integration. However, it was only through the mathematical methodologies and tools developed in the 1990s that researchers were finally able to design, retrofit, and optimize systematically and simultaneously both a distillation column and its associated heat exchanger network as a complex and integrated system (Liebmann and Dhole, 1995; Liebmann et al., 1998; Suphanit, 1999; Bagajewicz and Soto, 2000; Ji and Bagajewicz, 2002a,b; Basak et al., 2002; Gadalla et al., 2003, 2005).

The next sections provide a brief overview of the key elements of a typical CDU with focus on their design and operations with respect to fouling.

1.1.1 The Preheat Train

To reduce the energy requirements at the furnace, the crude oil pumped from storage is heated in a series of heat exchangers, the PHT, against overhead, product, and pump-around streams from which nearly 60–70% of the heat necessary for primary fractionation is recovered (Panchal and Huangfu, 2000). Although a few examples exist of refineries using a compact plate and frame design (Andersson et al., 2009), a typical preheat train is made up of as many as 60 shell-and-tube heat exchangers (Panchal and Huangfu, 2000) with the crude usually processed on the tube-side.

In the PHT the crude reaches the highest temperature possible for single phase flow at which it enters the furnace (coil inlet temperature or CIT). Typically CIT ranges between 240 and 270 °C, at a coil inlet pressure between 12 and 15 bar, whereas the typical coil outlet temperature (COT) is around 365 °C.

Following the changes in crude temperature, the PHT is usually divided into sections referred to as cold (upstream of the desalter), intermediate (between the desalter and the preflash), and hot (downstream of the preflash).

1.1.2 Desalter

Crude oil contains varying amounts of inorganic salts such as NaCl, $CaCl_2$, and $MgCl_2$ (Bai and Wang, 2007). The presence of such salts not only plays an important part in the fouling process it also increases corrosion in the heat exchangers and poisons the catalysts in processing units downstream of the CDU. Liu et al. (2003) discussed the importance of reducing the concentration of salts in the crude stream. This is typically done through a desalting process by diluting the salt content with fresh water added before entering a desalter at a temperature between 90 and 120 °C with a recommended addition rate of 3.0–4.5% of crude flow (Choi, 2005).

Different types of industrial desalters exist. In the most common desalters—which operate at a pressure around 5–10 bar and temperature in the range of 100–150 °C—the gravity settling of salt-containing water droplets is enhanced by electrostatic precipitation. The water phase is sent to a sour water stripper and disposed whereas the oil phase is redirected to the intermediate section of the PHT.

It is important to note that desalter performance is strictly interlocked with thermal efficiency of the heat exchangers. Desalters usually operate within a restricted optimum temperature range thus improved heat exchangers operation (i.e., ensuring stable temperature outlets and less downtime for cleaning) affects in a positive way desalter control, which in turn is important for avoiding fouling.

1.1.3 Preflash Drum

The preflash separates lighter components of the oil before it enters the furnace, thus reducing its load. The use of the preflash in CDUs was originally proposed by Brugma (1941) and discussed in a number of publications (Feintuch et al., 1985; Yahyaabadi, 2005; Errico et al., 2009). Nowadays many refineries use it as a convenient way of debottlenecking the CDU and reducing the risk of a two-phase flow in the hot section of the train. While the gas phase bypasses the furnace and enters the distillation column directly, the liquid phase is further preheated in the hot end of the PHT before entering the furnace.

1.1.4 Furnace

The furnace is a fired heater where the crude is brought at the desired inlet temperature of the distillation column. The furnace is divided in two distinct sections: a convective and a radiant section. The crude, preheated in the PHT, enters in single-phase flow the convective section of the furnace where it is further preheated before entering the radiant section where boiling and evaporation occur.

Fouling in furnaces is also a major issue for refineries (Atkins, 1962; Takatsuka et al., 1989; Jegla et al., 2011; Morales-Fuentes et al., 2014). Unlike in heat exchangers, losses in heat transfer due to fouling are compensated by the large driving force produced by the radiant heat provided by the burners. However, the buildup of low thermal conductivity material produces an increase in the furnace tubes skin temperature, which limits the amount of heat that can be transferred.

1.2 IMPACT OF FOULING ON CRUDE DISTILLATION UNITS

As a result of the change in temperature that the crude oil undergoes when processed in the PHT of an oil refinery, fouling material progressively deposits on the heat exchanger surfaces. This process can occur in both the tube-side (Figure 1.3) and the shell-side (Figure 1.4) of the heat exchangers.

Fouling on the shell-side is very often associated with low velocity zones close to the baffles (Figure 1.4(a)), but it also affects significantly the outside of the tubes in the bundle (Figure 1.4(b)). As a result, the shell-side heat transfer coefficient can also be significantly affected.

Refinery operators are constantly faced with the problems caused by the buildup of fouling in the PHT, including:

1. Operating difficulties
2. Economic penalties
3. Increased environmental impact
4. Health and safety hazards.

Figure 1.5 shows a schematic of the cascade of problems generated by fouling in a typical oil refinery processing 100,000 bbl per day and associated estimates of costs. The following sections review in detail each of the aspects above.

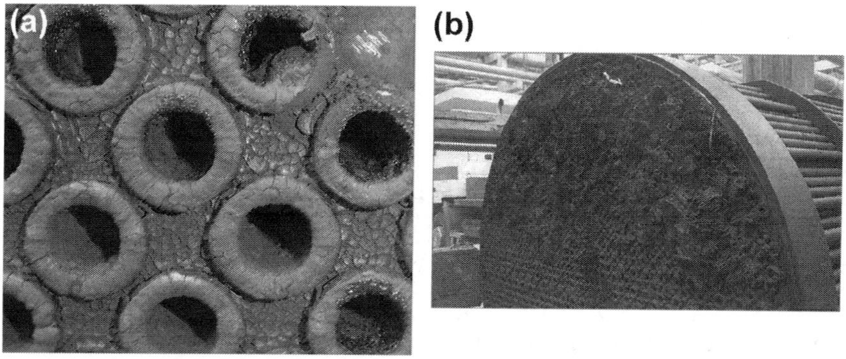

Figure 1.3 Photos of crude oil fouling on the tube-side of refinery heat exchangers: (a) detail of deposition at tube ends and (b) plugging of tubes.

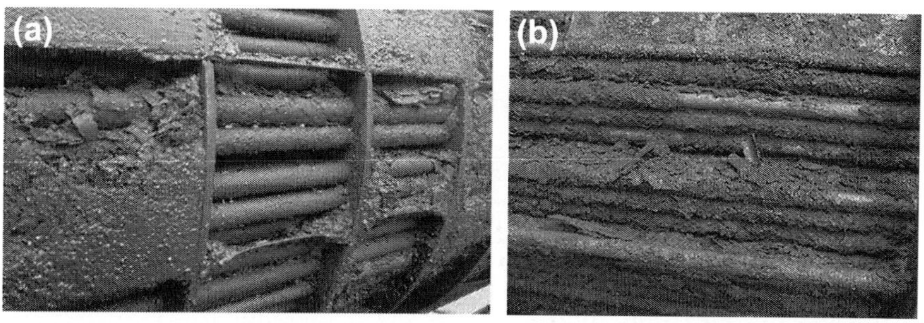

Figure 1.4 Photos of fouling on the shell-side of two refinery heat exchangers.

1.2.1 Operating Difficulties

Fouling material deposited on the heat transfer surfaces has a thermal conductivity—typically between 0.2 and 1 W m^{-1} K^{-1} (Watkinson, 1988)—up to two orders of magnitude lower than that of the tube metal wall (ca. 38 W m^{-1} K^{-1}, depending on the metal used and temperature). As a consequence, heat transfer is impaired and the energy recovered reduced. This progressive decay in performance results in a typical loss of 8–11 °C per year in the CIT. To keep a steady temperature at the column inlet, the decline in CIT must be countered by burning additional fuel in the furnace downstream of the PHT. Not only is this an energy cost, but it also has an impact on the environment due to increased release of greenhouse gas emissions.

Moreover, when the furnace is operating at its firing limit, additional thermal inefficiency in the PHT cannot be countered any further by

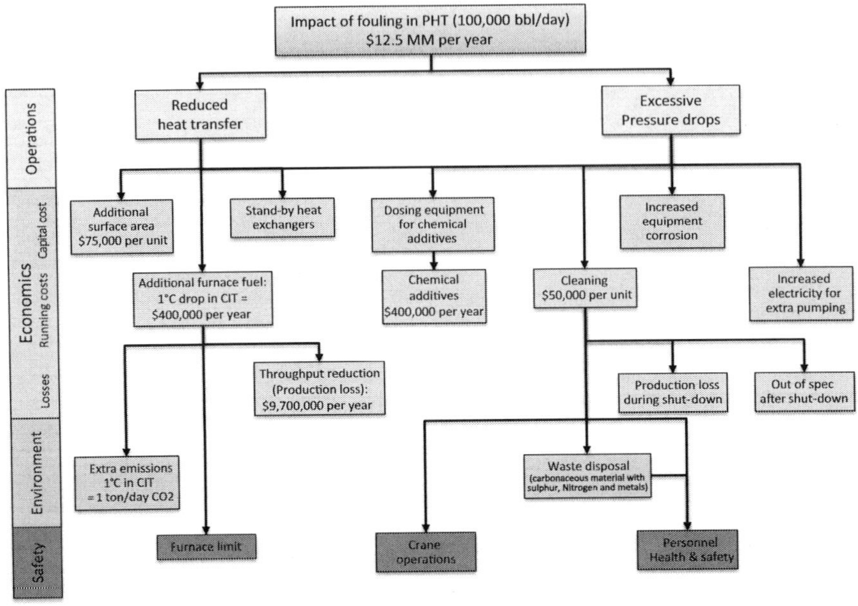

Figure 1.5 Impact of fouling on PHT and related costs for a typical oil refinery processing 100,000 bbl per day of crude oil. The figures, reported in US$, are adjusted for inflation to 2014. *Data from various sources, reported in Table 1.2.*

increasing the heat duty at the furnace. As a consequence, refineries are forced to reduce the throughput which, as will be shown later (Section 1.2.2.2), is the major cause of economic losses due to fouling.

There is little doubt that fouling affects greatly the CDU's energy efficiency. Estimations of the extra energy requirements caused by fouling deposition reported in the literature range from no less than 10% (Szklo and Schaeffer, 2007) to over 30% (Van Nostrand et al., 1981). Based on their experience Sikos and Klemes (2010) recently evaluated that refinery energy consumption was 10–20% higher because of fouling in CDUs. Thermal efficiency of the PHT is therefore a major concern and plants usually have targets for CIT. Refinery operators adopt different strategies to achieve such targets. For example, Zabiri et al., (2006) indicated that heat exchangers are cleaned when the overall heat transfer coefficient drops to a value below 30% of the clean value. Depending on a number of factors (e.g., refinery configuration, cost of fuel margins, etc.), some operators may decide to shutdown for cleaning when smaller drops of efficiency is reached. Often at 80–85% efficiency (taking into account the heat recovery in the entire PHT) the economic incentive is large enough to justify cleaning. Other operators monitor the fouling resistance as a means to decide when the right time for cleaning is. In this case when an excessive value of R_f is reached in one or more heat exchangers, these are taken out for cleaning.

Whereas it is more common for a PHT to be thermally limited, in some occasions, the hydraulic implications of fouling may be dominating. Usually refineries have a discrete hydraulic flexibility but problems can arise if the progressive reduction of the cross-sectional area available to the crude flow causes an excessive increase in pressure drops. If the throughput is to be kept constant, more pumping power must be spent until pump capacity is reached. In some extreme (but not rare) cases, plugging of tubes (Figure 1.3(b)) can occur, forcing a shutdown of the unit for cleaning.

When economic penalties become excessive, cleaning actions, which disrupt normal refining operations, are required. Refinery personnel give particular importance to the planning of cleaning actions. Dismantling units for cleaning is in fact a hazardous and costly operation (see Section 1.2.4) for which planning is paramount. Several authors dealt with different aspects of the cleaning scheduling of heat exchanger networks in general (Markowski and Urbaniec, 2005; Sanaye and Niroomand, 2007) and PHTs in particular (Wilson and Polley, 2001; Smaïli et al., 2001; Lavaja and Bagajewicz, 2004, 2005a,b; Rodriguez and Smith, 2007; Ishiyama et al., 2009a).

1.2.2 Economic Penalties: The Cost of Fouling

Estimating fouling costs is a complex activity, especially when antifoulants are used (Pritchard, 1988b), as estimates depend on many variable such as the current price of the crude barrel, market demand, the geographical region of the refinery, and the cost of applicable environmental policies.

As noted at the beginning of this chapter, there are only a few studies in the literature that report figures for fouling-related costs in oil refineries. To date, the most comprehensive one is the study by Van Nostrand et al. (1981) who collected information for crude distillation, hydrotreating, reforming, and visbreaking units and broke them down in three categories: extra energy consumption, throughput loss, and maintenance costs. The figures reported are staggering: process-side fouling in United States refineries was estimated US$ 1.36bn per year, US$ 861 MM of which is attributed to the crude PHT alone. Just adjusting for inflation[1] to 2014 this is equivalent to US$ 3.6bn and US$ 2.26bn, respectively. From some even older data for the United Kingdom (Thackery, 1979) it is possible to estimate the overall cost of

[1] Calculations performed using the Consumer Price Indexes by the US Bureau of Labor Statistics.

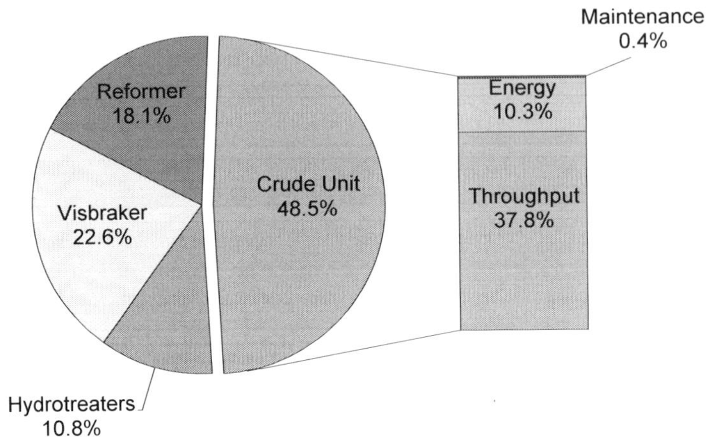

Figure 1.6 Hypothetical typical refinery fouling-related expenses. *Data from Van Nostrand et al. (1981).*

PHT fouling between £5 and 7 MM per year, which, in 2014 US$, would be equivalent to ca. 35–50 MM. For a single refinery processing 100,000 bbl per day, Van Nostrand et al. (1981) estimated fouling-related costs at US$ 25.7 MM per year (2014 basis) with almost 50% of it attributed to fouling in the PHT. The major loss was due to the reduction in throughput (ca. 40%), followed by the energy costs (ca. 10%), and only a small fraction (<1%) is due to maintenance (Figure 1.6).

Note that the estimations by Van Nostrand et al. were made at a time (1981) when environmental policies were not as strict as at present (e.g., the Emission Trading Scheme), the crude barrel was significantly cheaper (ca. US$30), and worldwide refinery throughput was considerably lower. Simply adjusting for inflation 30-year-old figures are not sufficient to capture fluctuations in market conditions (i.e., price of oil, demand) and improvement in technology and fouling management methodologies.

A recent life cycle cost analysis over 25 years in the hot end of the PHT of a 50,000 bbl day^{-1} refinery showed a US$25MM negative net present value (NPV). The analysis accounted for the cost of periodic cleanings and tube bundle replacements. It is estimated that if mitigation strategies are implemented and fouling is reduced by about 70%, the NPV can be improved to –US$4MM.

Table 1.2 summarizes the fouling-related costs in CDUs reported in the literature or collected as personal communications.

Table 1.2 Summary of typical economic losses associated to fouling in CDUs. Adapted from Coletti (2010).

Loss type and fouling location	Refinery size (bbl per day)	Original figure (US$)	Source	Adjusted to 2014 US$
Whole refinery	100,000	9.87 MM per year	Van Nostrand et al. (1981)	25.7 MM per year
PHT	100,000	4.78 MM per year	Van Nostrand et al. (1981)	12.5 MM per year
One branch of the PHT	360,000	1.8 MM per year	Liporace and de Oliveira (2007)	2 MM per year
Throughput reduction	160,000	1.5 MM in 3 months	ESDU (2000)	2 MM in 3 months
	100,000	3.73 MM per year	Van Nostrand et al. (1981)	9.7 MM per year
	n.a.	2 bbl^{-1}(Texas)	Polley et al. (2009a)	2.22 bbl^{-1}
	n.a.	10 bbl^{-1}(India)	Polley et al. (2009a)	11 bbl^{-1}
Extra fuel at furnace	50,000	–	This work[b]	1.1 MM yr^{-1}
	100,000	1.02 MM per year[a]	Van Nostrand et al. (1981)	2.7 MM per year
		400,000 per year	Panchal and Huangfu (2000)	550,000 per year
Extra CO_2 due to fouling	100,000	750 tCO_2 per year (=€22,500 per year)[c]	Yeap (2003)	30,400 per year
Drop of 1 °C in CIT	100,000	1 tCO_2 per day = 100,000 per year	Baudelet and Krueger (1998)	145,000 per year
Production loss due to HEX out of service	215,000	£250,000 per year	Personal comm. (2009)	400,000 per year
	100,000	20,000 per day = 80,000 per clean	Baudelet and Krueger (1998)	30,000 per day = 120,000 per clean
Cleaning of a unit	n.a.	30–50,000 per unit	Personal comm. (2009)	30–50,000 per unit
Antifoulants	100,000	155,000 per year	Van Nostrand et al. (1981)	400,000 per year

Note: Inflation calculated with the average Consumer Price Index for a given calendar year (data from US Bureau of Labor Statistics).
[a]Fuel cost assumed: US$9.55 MWh^{-1}.
[b]Fuel cost assumed: US$22 MWh^{-1}.
[c]Carbon tax rate assumed: €30 t^{-1}.

In the following sections a breakdown analysis highlights the individual contribution to the losses due to fouling in the CDU.

1.2.2.1 Fuel Costs

Van Nostrand et al. (1981) reported a typical decay rate in the coil inlet temperature due to fouling in the PHT of 4–10 °F per month (ca. 2–5 °C per month). In more recent discussions, experts from major oil companies were much more optimistic estimating this figure and reported a decay rate of 8–11 °C per year in heavy fouling trains.

The cost associated with the extra fuel burnt to counter the drop in CIT is very significant for the economics of the refinery. A 1 °C drop in CIT has been reported to cost to a United Kingdom refinery (215,000 bbl per day capacity) ca. £250,000 per year (ca. US$400,000 per year). Assuming an average 5 °C drop over a year, this would cost the refinery over US$2 MM. Other costs figures were reported in literature by Panchal and Huangfu (2000) who estimated for a smaller refinery, processing 100,000 bbl per day, an economic loss of about US$550,000 per year. Recently, Liporace and de Oliveira (2007) reported US$1.8 MM per year extra costs in fuel alone due to fouling in one branch of the PHT of a Brazilian refinery (360,000 bbl per day), which amounts to over US$2 MM in 2014.

A recent evaluation of operating data from ca. 15 heat exchangers in the hot end of the PHT in a 50,000 bbl day^{-1} CDU showed that the total heat duty loss due to the two worst fouling heat exchangers was 5.98 MW averaged over a 1 year period. This translates into an energy cost of US$1.1 MM per year, at US$22 MWh^{-1}.

1.2.2.2 Pumping power

If the CDU is not hydraulically limited (i.e., case 1 described in the section below), the increase in pressure drop across the heat exchangers in the PHT can be compensated by increasing the pumping power required to overcome them. In this case, the control valve placed between the last exchanger of the preheat train and the furnace drop, can be adjusted to compensate for the pressure drops increase in the heat exchanger network. This of course comes at a price, paid in terms of the extra electricity spent at the pump. However, as shown in a later chapter (Section 5.4), pumping power costs are typically dwarfed by the costs related to thermal inefficiencies. As a result, unless fouling causes blockage of the entire exchanger, the increased pumping power associated with it is rarely a concern for the refinery.

1.2.2.3 Reduction in Throughput

Refinery operating strategy is normally that of maximum throughput. The reduction in thermal and hydraulic efficiency caused by fouling is paid not only at the furnace or at the pump as extra energy but, more importantly, as loss in production. Production loss due to fouling can occur for two reasons:

1. Hydraulic limitations of the PHT. The volumetric flowrate, provided by the pump is given by:

$$\dot{V} = \frac{W\eta_{mech}}{\Delta P} \tag{1.1}$$

   where η_{mech} is the mechanical efficiency of the pump, W the pumping power and ΔP the differential pressure. As discussed in Section. As discussed in Section 1.2.2.2, the pumping power, required to maintain a constant flowrate as fouling builds up in the PHT, needs to increase over time. However, when the maximum capacity of the pump is reached, the system is not able to compensate for the extra pressure drops and the throughput is reduced.

2. Thermal limitations of the CDU. Any furnace used in refinery applications has a maximum heat duty achievable (often referred to as furnace firing limit), which is typically constrained by the maximum skin temperature possible in the tubes and by the temperature of the flue gases in the chimney:

$$Q_{furnace}^{max} = \dot{M}c_p(\mathrm{COT} - \mathrm{CIT}) \tag{1.2}$$

   where $\dot{M}$ is the crude mass flow rate in the furnace and c_p its specific heat capacity. Given that the COT must be kept constant while the CIT decreases because of fouling, the heat duty increases (i.e., extra fuel must be burnt). With a large decline in CIT due to fouling, the furnace hits its firing limit. At this point, the throughput must be reduced causing loss of production, which is responsible for the major share of the total costs related to fouling in the CDU.

Thackery (1979) suggested that loss in production is the most difficult cost to estimate because it depends on the current level of market products demand. Regional differences also affect the cost of loss in production: Polley et al. (2009a) reported that at the beginning of 2009, this cost was US\$2 bbl^{-1} in some United States refineries, over US\$5 bbl^{-1} in Asian refineries, and over US\$10 bbl^{-1} in one Indian refinery. The importance of including throughput loss in scheduling and operability considerations has been highlighted by several authors (Lavaja and Bagajewicz, 2005a; Polley et al., 2009b). Overall, Van Nostrand et al. (1981) estimated the throughput loss in US$9.4 MM per year (2014 dollars) for the 100,000 bbl per day refinery they considered. Coletti and Macchietto (2010) used an advanced mathematical model of refinery preheat trains to quantify and predict the impact of fouling and broke this down into the different contributions given by fuel, pumping, and environmental and production loss costs. They have shown how the reduction in throughput is, in fact, the most significant cost for the refinery related to fouling.

1.2.2.4 Maintenance (cleaning)

If the throughput is reduced because of the fouling-related increase in pressure drops, or if the costs incurred for extra fuel burnt at the furnace to compensate the energy inefficiency caused by fouling are deemed too high by the operator, selected units are usually taken off-line and cleaned. Worrell and Galitsky (2005) reported the results of an audit of the Martinez refinery (California) where it was found that regular cleaning of heat exchangers and maintenance of insulation would result in estimated annual savings of over US$14 MM at a total expenditure of US$9.85 MM with a payback period of around 8 months (DOE, 2002). In this context, monitoring of fouling (see Section 5.1) is a key activity for identifying units that need maintenance.

Cleaning is usually performed via dismantling a chosen unit and using water jets—typically at 1000 bar although in some special cases 2000 bar is required—to mechanically remove deposits from the heat transfer surfaces. The cost to clean a single unit via this methodology is very much location dependent, but could be between US$10,000 and US$50,000. To this figure, the cost due to production loss and increased fuel consumption while the heat exchanger is out of service for cleaning must be added. This has been estimated in US$30,000 per day in a 100,000 bbl per day refinery (Baudelet and Krueger, 1998). Considering that to clean each tube bundle requires up to a week before it is in service again, the total cost due to maintenance of a single heat exchanger can add up to US$150,000–200,000.

An alternative to the expensive (in terms of time, safety, and money) mechanical cleaning is chemical cleaning, performed by flowing a chemical solvent through the exchanger without the need for disassembly. The chemical action of the solvent dissolves some or all of the constituents of the solid foulant deposit. However, Joshi and Brons (2003) reported that attempts to chemically clean heat exchange equipment have been successful only in a limited number of cases, and even then, with benefits sustained for only a relatively short period compared to mechanical cleaning.

The limited success mentioned above is a result of several factors; the most important of those being the flow rate of the circulating solvent, the time between cleanings, and the chemical used. First, high volume circulation is required for a successful cleaning, and the small piping connections that are typically used to provide the circulation are not adequate. Second, as a general rule of thumb, chemical cleaning for heavily fouling heat exchangers should be carried out at intervals of 12 months or shorter. Attempts to clean after 18 or 24 months of operation have most often proved unsuccessful possibly because of the hardening of the fouling deposits over time (i.e., the ageing process described in Section 2.3.5) makes the chemicals less effective. Lastly, if the cleaning chemical is selected without knowing (or just guessing) the composition of the foulant deposit, it is often the case that unsuitable chemicals are used.

1.2.2.5 Antifoulants

In the 100,000 bbl per day refinery monitored by Van Nostrand et al. (1981) the decline rate of the PHT heat duty was 1.5 MW per month when no chemicals were used whereas the decay rate was reduced to 0.56 MW per month when antifoulants were added to the crude stream. They estimated the cost of chemicals at around US$392,300 per year (2014 basis), which was paid back by the 65% reduction in fouling and associated decrease in the number of cleanings required per year.

Despite some success stories such as the one above and the optimism of Canapary (1961) who stated that "a complete antifouling program will solve process side fouling in oil refineries", a chemical solution capable of eliminating fouling seems to be still lacking. The complexity of the crude compositions, their variability of the blends processed over time, and the fact that fouling mechanisms are not yet fully understood (see Chapter 2) force suppliers to provide ad hoc solutions for each refinery. As a result, plant experience indicates that very often chemical treatment does not work for extended periods of time and in many cases is just ineffective.

1.2.2.6 Capital Costs

Heat exchanger design methodologies deal with fouling by simply compensating for the reduced heat transfer efficiency with the addition of larger heat transfer surfaces (see Section 5.2). However, not only is this practice far from being optimal it also increases the total capital expenditure. Moreover, examining 600 downstream process projects, Jones (1995) showed how project completion time is also affected by overdesign. In his study, project completion time for heat exchangers designed for capacity can be reduced by 17% because of the smaller equipment being built, transported, and installed.

The cost of a PHT heat exchanger, typically sized between 300 and 900 m^2, can be roughly estimated at US$74,000–166,000 using a cost per unit area in the range of US$188–249 m^{-2} (ESDU, 1994)[2]. Considering that most heat exchangers are 30–50% oversized at the design stage due to the fouling factors design approach,[3] the capital costs associated with the extra area can be evaluated in the range of US$22,000–83,000 per unit. As already noted, a typical PHT can use 16 to 60 heat exchangers; therefore, the total extra capital cost ranges from US$354,000 to US$5 MM. If special material of construction is used to avoid corrosion, these figures may grow substantially.

An extra capital expense is incurred if antifoulants are used. In fact, the costs associated to antifoulants are not limited to the chemicals themselves but also include the capital cost of dosing equipment necessary to inject them at the correct point in the crude stream.

1.2.3 Environmental Impact

The environmental impact of fouling in refineries has been estimated as being responsible for 88M t of CO_2, accounting for 2.5% of all worldwide anthropogenic emissions in 2009 (Müller-Steinhagen et al., 2009a). Cap-and-trade regulations such as the Emissions Trading Scheme (ETS) enforced in Europe give a tag value to each ton of CO_2 emitted by the refinery, adding to the total cost of fouling. According to ETS regulations, each installation is allocated allowances for emission of carbon dioxide. If the refinery exceeds the allocated allowances, it must buy extra ones in the market. The reverse is also true: if the allocated allowance is not reached, the excess allowance can be sold. Table 1.2 reports calculations made with a carbon tax rate at €30 t^{-1} which is the peak value reached in April 2006. Phase two of ETS started in January 2008 with a price fixed at €25 t^{-1} which dropped to €10 t^{-1}.

[2] Original figures, £60–80 m^{-2}, have been adjusted for inflation and converted at 2014 exchange rates without including changes in capital cost for raw materials, etc.

[3] As a general rule, Müller-Steinhagen (2000) suggests that an excess area of 30–40% may correspond to 25% additional capital costs without including increased shipping and installation expenses due to the larger dimensions.

Ecological aspects involve also the disposal of the carbonaceous deposits formed on the heat transfer surfaces which can be substantial. To illustrate the quantity of deposits that must be removed from the PHT of a refinery, Murphy and Campbell (1992) considered a refinery processing 27,000 t per day (ca. 200,000 bbl per day) of crude that contains 0.05% of what they call "sediments" and 10 ppm of salts. The potential deposits each year are 5000 t of sediments and 100 t of salts. More optimistically, Watkinson (2005) assumed that only one part per million by weight of the oil would deposit producing 15 t per year of solid carbonaceous material in a 300,000 bbl per day refinery. These deposits can contain sulfur, nitrogen, and metals and must be disposed in an adequate way.

1.2.4 Health and Safety Hazards

Aspects often overlooked in the literature but very much present in the minds of refinery operators are the health and safety hazards involved in dismantling for cleaning a pressurized vessel processing hydrocarbons. During the cleaning process, refinery procedures require a safety observer in 100% attendance to supervise the following risky activities:

1. Opening of the units. Although units are flushed with steam, flammable atmospheres and pyrophoric material are often present.
2. Extraction of the bundle with cranes (typically units are 6.1 m long and up to 1.5 m diameter) in confined spaces. Electrical hazards are also involved.
3. Transportation on trucks to storage.
4. Cleaning with 1000 bar water jets (although ultrahigh pressure pumps can provide up to 2800 bar for heavily fouled tube-side surfaces) with associated noise and toxic material being thrown in the atmosphere.
5. Delivery and reinsertion of the bundle with crane and trucks movement.

Safety is the primary concern for the oil industry. It is therefore not surprising that safety considerations, let alone the large costs involved, make crude oil fouling one of the biggest and more important challenges the industry is facing today.

1.3 CONCLUDING REMARKS

Given the impact of crude oil fouling on refinery operations, economics, safety, and environmental impacts shown above, the relevance and importance of tackling this problem are evident. However, as it has been anticipated in the Introduction of this chapter and as will be seen in more detail in the following ones, crude oil fouling is an extremely difficult problem to solve as it involves complex chemical and physical phenomena interacting at different scales of investigation. The scientific community, both in industry and in academia, agrees that only via a better understating of the fundamental causes and mechanisms of fouling will it be possible to mitigate—if not completely solve—this challenging and long-standing problem, at least as old as the refining process.

The following chapter will provide the reader with the necessary background on the science of the fouling process, which will be instrumental when looking at the experimental and modeling techniques used in crude oil fouling research and applications.

CHAPTER TWO

Basic Science of the Fouling Process

F. Coletti[1], B.D. Crittenden[2], S. Macchietto[3]
[1]Chief Technology Officer, Hexxcell Ltd, UK
[2]Professor of Chemical Engineering, University of Bath, UK
[3]Professor of Process Systems Engineering, Imperial College London, UK

The first attempt to rationalize and categorize fouling was made by Epstein (1983). He summarized the level of understanding in different aspects of the fouling process in what is now known as the 5 × 5 fouling matrix (Figure 2.1(a)). The five columns of the matrix list the different fouling mechanisms and the five rows indicate the subprocesses involved with it. The 5 × 5 fouling matrix has been updated over time to reflect the current level of understanding by Bohnet (1987) (Figure 2.1(b)) and, more recently, by experts in various aspects of the field (Figure 2.1(c)) participating in the "Heat Exchanger Fouling and Cleaning VIII" conference held in Austria in June 2009 (Müller-Steinhagen et al., 2009b). Over the years, many gaps in research seem to be filling. In particular transport, which was already identified as the

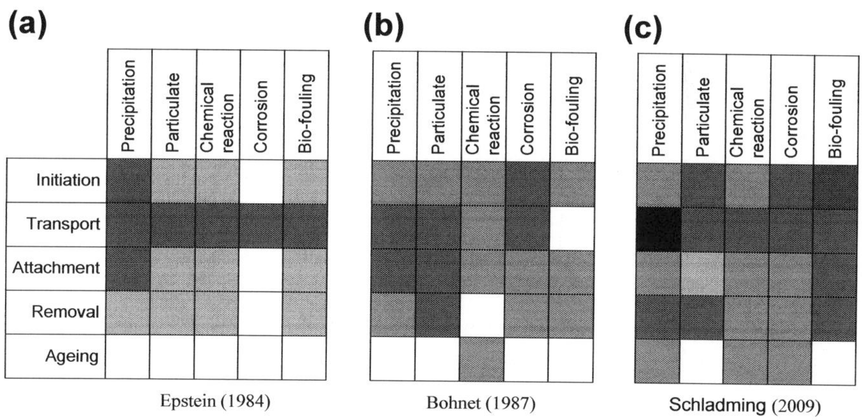

Figure 2.1 Evolution over time of the 5 × 5 fouling matrix summarizing the current level of understanding of fouling mechanisms and processes: (a) original 5 × 5 fouling matrix Epstein (1983); (b) interpretation by Bohnet (1987) and (c) by researchers attending the 2009 Heat Exchanger Fouling and Cleaning conference in Austria (Müller-Steinhagen et al. 2009b). Light to dark shading indicates increase in level of understanding (Coletti, 2010).

Crude Oil Fouling
ISBN: 9780128012567
http://dx.doi.org/10.1016/B978-0-12-801256-7.00002-6

most studied subprocess in the original Epstein matrix, has gained a remarkable level of attention across all fouling mechanisms types. At the other end of the scale, ageing was identified as the topic having received the least attention.

There are several reasons why the fouling problem has not been solved so far. There is disagreement between the researchers on the mechanisms involved in producing the deposits, on what the foulant precursors are, on the location of the reactions taking place (whether these are in the bulk of the fluid or at the tube surface) and, quite disappointingly, even the chemical structure of the species that constitute the fouling deposits is disputed. However, there is common ground on the fact that mitigation of fouling in industrial heat exchangers can only be achieved if a better understanding of the underlying science is achieved.

A number of review articles exist on fouling of heat transfer surfaces (De Deus, 1980; O'Callaghan, 1980; Collier, 1980; Epstein, 1981b; Somerscales, 1981; Bohnet, 1987; Bott, 1988b; Watkinson, 1988; Somerscales, 1990; Knudsen, 2002) and more specifically on crude oil fouling formation (Crittenden (1987a,b); Watkinson (1988, 1992); Crittenden et al. (1997); Panchal (2001); Watkinson and Wilson (1997).

This chapter aims at providing the reader with the basic concepts in fouling science, starting from the possible fouling mechanisms (Section 2.1)—with particular focus on chemical reaction fouling (Section 2.2)—and the subprocesses involved in fouling (Section 2.3). Then, a description of the variables affecting fouling and how these interact with each other is provided (Section 2.4) before a final conclusions (Section 2.5).

2.1 FOULING MECHANISMS

The columns in the Epstein matrix (Figure 2.1) represent the five major categories into which fouling has been divided according to the mechanism that produces it:

1. Crystallization fouling — precipitation and deposition of dissolved salts that, under process conditions, become supersaturated (Hasson, 1981; Bott, 1988a; Yiantsios et al., 1995).
2. Particulate fouling — deposition of suspended particles on heat transfer surfaces (Gudmundsson, 1981; Epstein, 1988; Karabelas et al., 1997).
3. Chemical reaction fouling — deposition resulting from one or more chemical reactions between reactants contained in a flowing fluid creating accumulation on the heat transfer surface (Watkinson, 1988; Crittenden et al., 1997; Watkinson and Wilson, 1997).

4. Corrosion fouling — deposition produced by a chemical reaction that involves a reactant and the metal surface. The increased roughness of the surface may also promote fouling due to other mechanisms (Lister, 1981; Somerscales, 1997, 1999).
5. Biological fouling — formation of organic films consisting of microorganisms that promote attachment of macroorganisms such as mussels, algae, etc. (Kent, 1988; Melo, 1997).

Crude oil fouling in refinery preheat trains can happen via any of the above mechanisms, except biofouling. Crystallization may occur when salts are deposited from the aqueous fraction of crude oil upstream of the desalter. For example, Joshi (2001) showed the case of one heat exchanger processing crude at about 88 °C where salt deposition was the suspected cause of fouling. Particulate fouling may occur when dirt, clay, or rust suspended in the oil are transported to the heat transfer surface where they may become attached. Chemical reactions between the different species contained in the crude may be triggered and follow a variety of reaction pathways depending on crude composition and process conditions:

- Sulfur compounds in the oil can react with metal surfaces leading to corrosion fouling.
- Oxygen present in concentrations as low as a few ppm (Watkinson, 2005) in storage tanks might form insoluble gums primarily by the autoxidation mechanism.
- Polymerization of monomeric compounds can happen when cracked feedstock have been added to the crude.
- Components in the oil might undergo thermal decomposition (cracking) at high temperatures to ultimately form coke.

Another important mechanism in crude oil fouling is asphaltene precipitation. In this case, changes in composition, temperature, and pressure might cause asphaltenes or resins present in the oil to precipitate out. Although this is a physical phenomenon, it is often a series of chemical reactions that lead to the formation of the precursors responsible for asphaltene deposition.

Establishing which is the overarching mechanism is paramount in order to take remedial action at both design and operation stages. However, whereas for some fluids that undergo chemical reaction fouling (e.g., milk fouling, Beuf et al., 2003; Bansal and Chen, 2006) the species involved, reaction pathways, and underlying kinetics are well defined, crude oil has a complex and diverse composition that makes it extremely challenging to determine exact information for all these aspects. In fact, even if the species involved can be identified via chemical and physical characterization, it is extremely difficult to determine their exact role in the deposition process.

Very often, analysis of the deposits only reveals the product of several reactions which prevents the possibility of establishing a link between precursors and deposits, thereby making it virtually impossible to identify the underlying fouling mechanisms. Moreover, to complicate even further the task of identifying the root causes of fouling, it is most likely a combination of mechanisms (mixed fouling) that is actually responsible for deposition (Crittenden et al., 1992). So far, it has only been possible to interpret experimental and field data to find indications of potential causes of fouling without finding conclusive theories on the actual deposition mechanisms. As a result, only conjectures on possible reaction mechanisms exist for crude oil and hydrocarbon fouling (Panchal and Watkinson, 1994; Watkinson and Wilson, 1997).

Industrial experience (see Section 3.1) shows that two mechanisms for organic fouling are more likely to occur in refinery preheat trains:

1. A combination of corrosion and chemical reaction fouling via the autoxidation mechanism.
2. Precipitation of asphaltenes that could adsorb or otherwise adhere to the heat transfer surface.

Although it is possible that operating conditions might be consistent with the latter mechanism, there is also evidence to suggest that the first mechanism is present. In this case, iron sulfide corrosion may have an important role to play in the induction period (see Section 2.3.1) and even in later steps of crude oil fouling. Hydrogen sulfide formed by the breakage of C–S bonds is believed to react with iron containing surfaces to form iron sulfide and this type of corrosion is known to promote fouling by increasing the surface area and decreasing the interfacial tension for stronger adhesion to insoluble asphaltenes (ESDU, 2000). Wang and Watkinson (2011) suggested that iron sulfide particles could be formed in suspension by the reaction of H_2S or organic sulfides with iron oxides formed elsewhere in the process leading to particulate fouling. Their alternate view was that iron sulfide first forms on the heat transfer surface itself. Iron sulfide is black and, without chemical analysis, could be mistaken for carbonaceous material. Whichever mechanism is operative, Joshi et al. (2009) reported that fouling deposits from crude oil exchangers often contain a substantial (35%+) amount of iron sulfide. They also reported that the rate of fouling can be reduced by changing the tube metallurgy. These observations prompt the suggestion that corrosion leads to a roughening of the heat transfer surface and that this roughening is then responsible for promoting the accumulation of fouling precursors. In confirmation, a number of studies have shown that significant amounts of both iron and sulfur are present in crude oil deposits (Wang and Watkinson, 2011; Young et al., 2011).

Details of the possible routes of crude oil fouling formation are presented below.

2.2 ROUTES TO CRUDE OIL FOULING FORMATION

2.2.1 Corrosion

The general electrochemical mechanism of corrosion involves the removal of a metal ion from an anionic site leaving behind excess electrons on a metal. For example, with iron:

$$Fe_{metal} \rightarrow Fe^{2+}_{soln} + 2e_{metal}$$

Electrons are simultaneously consumed in a balancing reaction at adjacent cathodic sites. For conventional oxidative corrosion the balancing reaction is usually the dissolution of dissolved oxygen and thence the formation of metal oxides or hydroxides:

$$O_2 + 2H_2O + 4e \rightarrow 4OH^-$$

For heat exchangers left outside in damp environments a certain amount of corrosion would occur (Figure 2.2) on metal surfaces that almost certainly would provide sites for other types of fouling to occur once installed in the process.

Wang and Watkinson (2011) studied the mixed organic/inorganic nature of deposits, noting that the iron and sulfur contents in crude oil fouling deposits from three studies ranged from 21% to 49% and 12–22%, respectively, with Fe/S ratios ranging from 0.5 to 1.4. Without doubt, iron sulfide corrosion was occurring. It is considered that the formation of iron sulfide could occur in parallel or in series with the formation of organic deposits.

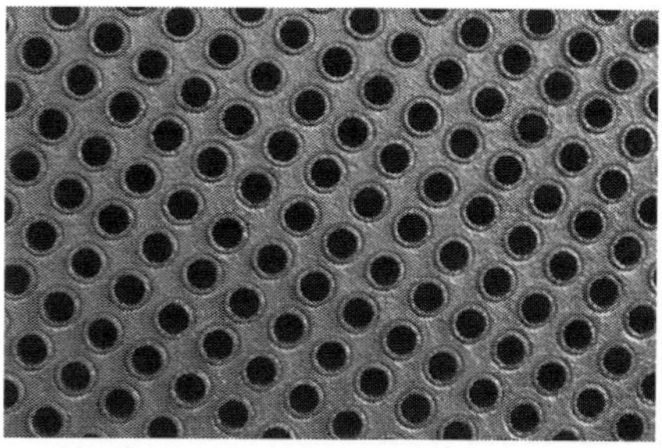

Figure 2.2 Refinery shell and tube heat exchanger tube sheet left outside after cleaning. (*copyright B. D. Crittenden*).

Sulfur is present in all crude oils at various percentage levels in the form of mercaptans, sulfides, disulfides, etc. Iron is present at ppm levels in crude oils but Wang and Watkinson (2011) suggested that corrosion upstream can provide additional dissolved iron sources such as napthenates. In the absence of oxygen, the formation of iron sulfide on a metallic heat transfer surface can proceed via (1) corrosion of iron by organic acids upstream, (2) thermal decomposition of acid salts, and (3) reaction of iron oxide with organic sulfur or hydrogen sulfide:

$$2RCH_2COOH + Fe \rightarrow (RCH_2COO)_2Fe + H_2$$

$$2RCH_2COOH + Fe^{2+} \rightarrow (RCH_2COO)_2Fe + 2H^+$$

$$(RCH_2COO)_2Fe \rightarrow RCH_2COCH_2R + FeO + CO_2$$

$$FeO + R'{-}S{-}R'' \rightarrow FeS + R'{-}O{-}R''$$

$$FeO + H_2S \rightarrow FeS + H_2O$$

The last four reactions (where R'–S–R'' represents an organic sulfide) can create FeS in the fluid bulk such that it can deposit elsewhere in a heat exchanger.

2.2.2 Autoxidation

Autoxidation is a free radical chain reaction mechanism that can operate in oils at moderately low temperatures (Crittenden, 1988a). Like all oxidation reactions, autoxidation is exothermic. Oxygen is required and is likely to be present dissolved at low concentrations in crude oils in storage. The first step is the abstraction of hydrogen from an organic molecule R–H by a free radical $X^{\bullet}$.

$$R{-}H + X^{\bullet} \rightarrow R^{\bullet} + XH$$

The chain reaction involves the reaction of molecular oxygen with the substrate radical $R^{\bullet}$ and further hydrogen abstraction creating peroxy radicals and hydroperoxide molecules:

$$R^{\bullet} + O_2 \rightarrow ROO^{\bullet}$$

$$ROO^{\bullet} + RH \rightarrow ROOH + R^{\bullet}$$

Once the autoxidation process has been initiated, homolysis of the weak O–O bond in the hydroperoxide leads to the formation of more free radicals that can take part in hydrogen abstraction. The reaction thereby breeds free radicals.

$$ROOH \rightarrow RO^{\bullet} + OH^{\bullet}$$

With a good supply of oxygen the concentration of substrate molecules would be low. Hence, of the following three possible free radical termination reactions, the last one is probably the most important:

$$R^{\bullet} + R^{\bullet} \rightarrow R{-}R$$

$$R^{\bullet} + ROO^{\bullet} \rightarrow ROOR$$

$$ROO^{\bullet} + ROO^{\bullet} \rightarrow ROOR + O_2$$

Trace amounts of soluble metal salts, particularly of cobalt, manganese, iron, copper, chromium, lead, and nickel, can catalyze the rate of hydroperoxide homolysis and may also take part in initiation reactions:

$$ROOH + CO^{2+} \rightarrow RO^{\bullet} + OH^{-} + CO^{3+}$$

$$ROOH + CO^{3+} \rightarrow ROO^{\bullet} + H^{+} + CO^{2+}$$

$$RH + CO^{3+} \rightarrow R^{\bullet} + H^{+} + CO^{2+}$$

Oxidation of aromatics and primary C–H bonds in alkanes is relatively difficult but becomes easier with secondary and tertiary C–H bonds. Thus, the presence of aromatics or naphthenic species can inhibit the deposition rate from straight chain alkanes whereas the rate is considerably higher if there is a C=C bond or an aromatic nucleus α to the group undergoing oxidation (Crittenden, 1988b). Hence the rate of autoxidation can be high for cracked feedstocks containing olefinic species.

As will be seen in Section 4.1, this an important route generally assumed to be responsible for chronic fouling affecting the long-term performance of refinery preheat trains.

2.2.3 Polymerization

Polymerization mechanisms involve the conversion of individual compounds (monomers) into a large multiple of themselves. Copolymerization can occur when two or more such monomers are involved in the creation of a high molar mass compound. Polymerization can be expected if there are olefinic compounds in the feedstock. Hence, polymerization as a fouling mechanism is probably limited to crude oils to which cracked feedstocks have been added.

Polymers can be created by a wide variety of processing techniques that include bulk, solution, suspension, emulsion, and precipitation methods (Crittenden, 1988a). The general mechanism of polymerization is broadly similar to that of autoxidation and involves radical formation and chain initiation, reaction chain propagation, and chain termination. The rate of each step and the length of the polymer chains depend on a number of factors including the concentration and activity of the reactants. Similar to autoxidation, species such as oxygen, halides, sulfides, nitrogen compounds, certain metals, and metallic compounds are all known to be capable of initiating polymerization. Under certain conditions, mercaptans and sulfides can take part in corrosion reactions with the heat exchanger surface to produce free radicals that could initiate polymerization reactions, for example,

$$\mathrm{Fe + R{-}SH \rightarrow FeSH + R^{\bullet}}$$

If the polymer being formed becomes insoluble in its solution then it will precipitate out. This would occur wherever processing conditions were favorable, for example, in the bulk fluid, in the heat transfer film, or at the heat transfer surface.

2.2.4 Thermal Cracking

Thermal cracking of hydrocarbons tends to be restricted to temperatures in excess of 650 °K and is therefore more likely to arise in the crude distillation furnace rather than in the preheat exchangers. Even so, cracking and its associated mechanism of coking could occur in deposits that spend long durations on hot heat transfer surfaces in which temperatures increase as fouling proceeds.

In this mechanism, higher molar mass hydrocarbons are broken down into lighter alkanes and alkenes, the latter being able to take part in both autoxidation and polymerization mechanisms. Although a vast number of individual reactions can occur in the thermal cracking of complex hydrocarbon mixtures, the overall mechanism comprises the steps of initiation, hydrogen abstraction, radical decomposition, radical addition, and termination. Initiation involves a single molecule breaking into two free radicals at a C–C bond:

$$\mathrm{R{-}R' \rightarrow R^{\bullet} + R'^{\bullet}}$$

Only a small fraction of the feed might participate in the initiation step. Each free radical may then abstract hydrogen from a hydrocarbon molecule turning it into a free radical:

$$\mathrm{R^{\bullet} + R''R''' \rightarrow RH + R''R'''^{\bullet}}$$

The more complex free radical can then decompose into a hydrogen free radical and an alkene:

$$R''R'''^{\bullet} \rightarrow R''{=}R''' + H^{\bullet}$$

Alkenes can react with free radicals to form larger free radicals, which can cause aromatic products to be formed when heavier feedstocks are involved:

$$R''R'''^{\bullet} + R''''{=}R''''' \rightarrow R''R'''R''''R'''''^{\bullet}$$

Two common types of termination reaction are possible, namely recombination in which two free radicals combine to create a larger molecule and recombination in which one free radical transfers a hydrogen atom to the other one, yielding an alkene and an alkane. For example,

$$R^{\bullet} + R''R'''^{\bullet} \rightarrow RR''R'''$$

$$R''R'''^{\bullet} + R''R'''^{\bullet} \rightarrow R''{=}R''' + R''R'''$$

Thermal cracking is endothermic (i.e., ΔH is positive) but is favored thermodynamically (i.e., the free energy change ΔG is negative) because although the bond dissociation energy for a C–C bond is quite high the large and positive entropy change ΔS that results from the breakage of large molecules into several smaller ones means that the $T\Delta S$ term in the Gibbs free energy equation is larger than the ΔH term:

$$\Delta G = \Delta H - T\Delta S \tag{2.1}$$

This explains why thermal cracking occurs at higher temperatures. At high temperature, coke can be formed via the secondary or synthesis reactions of the products of the primary or cracking and dehydrogenation reactions. Figure 2.3 (Crittenden et al., 1988a, adapted from Fitzer et al. (1971)) summarizes the primary and secondary reactions. In the context of crude oil fouling, coke formation is noted again under ageing in Section 2.3.5.

2.2.5 Asphaltene Precipitation

Several authors (Wiehe, 2001a; Mason and Lin, 2003; Stark and Asomaning, 2003; Saleh et al., 2005a) indicated that asphaltene flocculation and deposition caused by asphaltene/oil incompatibility are responsible for crude oil fouling, especially at high temperatures (Eaton and Lux, 1984; Lambourn and Durrieu, 1986; Dickakian, 1989).

Asphaltenes are defined as the *n*-alkane-insoluble/toluene-soluble fraction of the crude (Watkinson, 2008) and form one of the four key classes

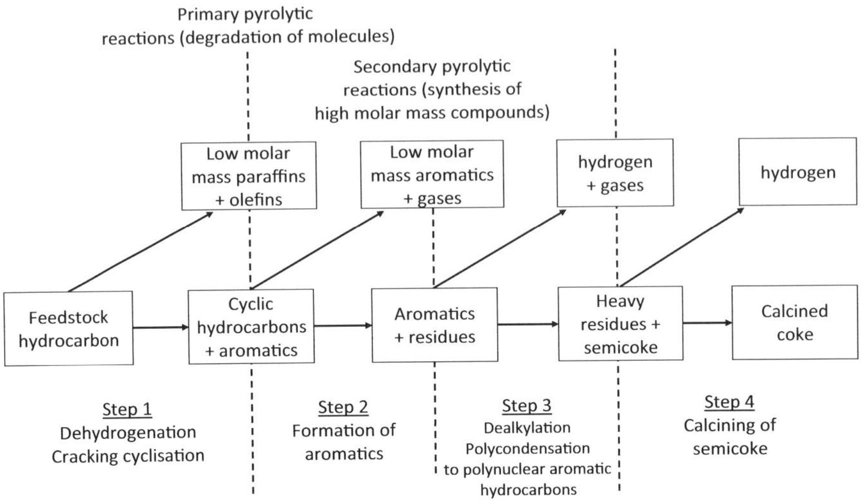

Figure 2.3 General schematic of hydrocarbon pyrolysis (cracking and coking). *Crittenden et al. (1988a), adapted from Fitzer et al. (1971).*

of organic compounds that characterize the polarizability and polarity of crude oils according to the SARA (Saturate, Aromatic, Resin, & Asphaltene) analysis. Although asphaltenes have been studied in a number of works (Dickakian and Seay, 1988; Buenrostro-Gonzalez et al., 2004; Aguilera-Mercado et al., 2006), there is ongoing debate among the scientific community on their exact structure (Durand et al., 2010) and on whether an archipelago (Karimi et al., 2011; Murgich et al., 1999; Sheremata et al., 2004), continental or an island model (Groenzin and Mullins, 1999; Mullins, 2010, 2011; Sabbah et al., 2012) is more appropriate to describe them. More details on this debate and aggregation mechanisms are reported in Section 5.2.1.

Different ratios of SARA components produce different effects on the stability of asphaltenes in the crude. This plays a key role when two or more crudes are blended together. For example, both the resin and the asphaltene fractions of crude oils contain polar constituents with the former acting as a dispersant for the latter. If incompatible oils are mixed together, asphaltenes may flocculate and come out of solution resulting in severe fouling events. For this reason, major research efforts have been made to establish the effect of crude compatibility on asphaltene stability (Wiehe and Kennedy, 2000; Gabrienko et al., 2014).

Various ways of characterizing compatibility have been proposed. Asomaning and Watkinson (1997) quantified compatibility via the colloidal instability index (C.I.I.):

$$\text{C.I.I.} = \frac{\text{Saturates} + \text{Asphaltenes}}{\text{Aromatics} + \text{Resins}} \tag{2.2}$$

When C.I.I. is less than unity, the combined fractions of resins and aromatics are considered sufficiently high to keep the asphaltene fraction in solution or dispersion. On the other hand, when C.I.I. is greater than one, asphaltenes are expected to precipitate out. For such unstable oils, it has been found that precipitated asphaltenes have compositions that are similar to those of fouling deposits (Watkinson, 2005) and it has been assumed (Watkinson, 2005) that asphaltene fouling is simply a case of particulate deposition complicated by precipitation and agglomeration processes. Although Equation (2.2) provides a useful indication of stability, it has been argued that its use alone is not sufficient to predict asphaltene precipitation over a wide range of compositions. Thus Al-Atar and Watkinson (2001) included also the resin/asphaltene ratio as a parameter influencing fouling.

In view of the complexity and potential importance of asphaltenes in crude oil fouling, attempts have also been made to understand the underlying fundamental mechanisms involved in the phase boundaries of asphaltene–crude oil chemical systems (reviewed in Section 5.2).

Although oil blend compatibility plays an important role in fouling in PHTs, this is not the only cause of asphaltene deposition. Wiehe (2001b) showed that asphaltenes do not require being insoluble to produce fouling, pointing out that crudes exist that are self-incompatible (i.e., they do not need to be blended to produce fouling) and that even compatible oils can undergo asphaltene fouling. Complex chemical reactions may be triggered by the process conditions inside the heat exchangers leading to asphaltene deposition from crude oils. Eaton and Lux (1984) proposed the following series of degradation reactions:

$$\begin{matrix}\text{Saturated}\\\text{hydrocarbons}\end{matrix} \xrightarrow{\text{Inorganic acids}} \begin{matrix}\text{Unsaturated}\\\text{hydrocarbons}\end{matrix} \xrightarrow{O_2} \begin{matrix}\text{Organic}\\\text{acids}\end{matrix} \xrightarrow{\text{Metals}} \begin{matrix}\text{Resins \&}\\\text{asphalthenes}\end{matrix} \xrightarrow{\text{Wall}} \begin{matrix}\text{Coke-like}\\\text{deposits}\end{matrix}$$

Besides the limitations of analytical techniques (e.g., instrument resolution, limited solubility of the samples), the number and complexity of the reaction and physical processes involved make it very difficult to confirm the above mechanisms and even harder to measure the individual reactions rates and their respective kinetic parameters.

2.2.6 Generalised mechanism

Previous sections have highlighted that several uncertainties are involved with the identification of the mechanism overarching the fouling process.

However, from all the experimental evidence reported in the literature, there seems to be agreement on a generalized reaction fouling mechanism which comprises the following three steps:

$$\text{Reactant } A \xrightarrow{r_1} \text{Precursor } B \xrightarrow{r_2} \text{Foulant } C \xrightarrow{r_3} \text{Aged deposit } D$$

In the first step, soluble precursors are formed via a reaction (or series of reactions) among different species in the crude. In some cases, the formation of the precursors can happen before the crude enters the heat exchanger and Panchal and Watkinson (1994) found indications that these can have a more significant effect than the precursors generated in the exchanger itself. In the second step, the precursors react giving insoluble foulant species that will then deposit on the heat transfer surface. A third step involves chemical and physical changes of the deposits caused by the high wall temperatures. This last process goes under the name of ageing (see Section 2.3.5) and, as will be shown in later chapters, it plays an important role in the overall heat exchanger performance.

According to Panchal and Watkinson (1994), reactions r_1 and r_2 can take place in the bulk, in the boundary layer, or at the wall. Figure 2.4 summarizes the three possible scenarios:

- Case 1: Precursor generation in the bulk. In Case 1a, reaction r_2 occurs on the wall surface whereas in Case 1b, reaction r_2 occurs in the bulk itself.
- Case 2: Precursor generation in the boundary layer. In Case 2a, reaction r_2 occurs on the wall surface whereas in Case 2b, reaction r_2 occurs in the boundary layer itself.
- Case 3: Precursor generation on the wall surface.

Panchal and Watkinson (1994) developed a mathematical model for fouling in a single tube based on the reaction scheme shown in above. They compared model simulations in the different cases with experimental data finding a relatively good match to Case 1a and Case 2, suggesting that fouling may occur via reaction in the bulk or in the boundary layer. However, admittedly, they neglected reaction r_3 (i.e., ageing) which may affect the accuracy of their final conclusions. Indeed, the question of where the reactions take place is still open. Ebert and Panchal (1995) and Srinivasan and Watkinson (2005), for example, suggested that the fouling reaction takes place in the bulk film and not at the wall surface, whereas Crittenden et al. (1992) argued that most of the deposits is generated in the highest temperature region, i.e., at (or near) the surface.

A detailed description of the events in fouling formation is given in the following section.

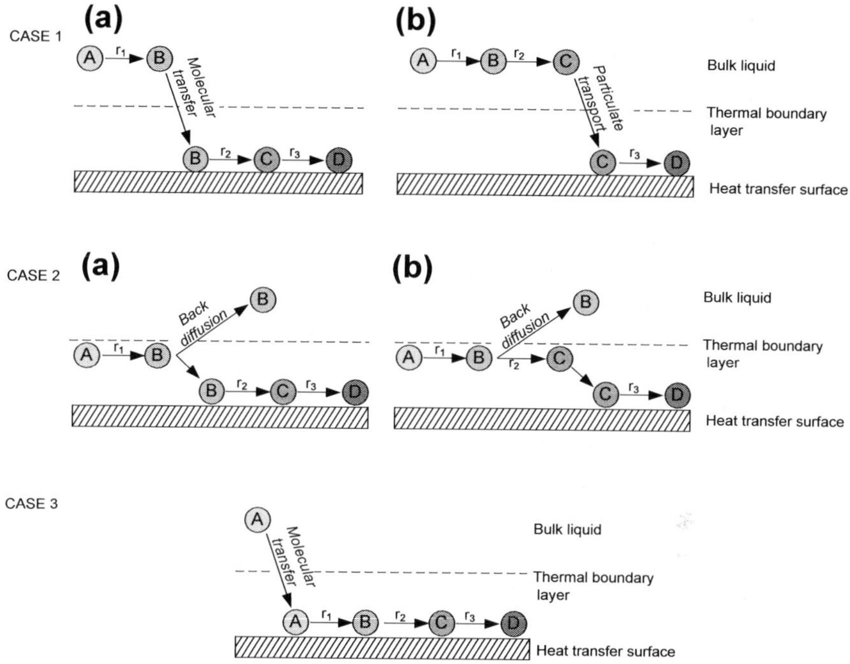

Figure 2.4 Deposition mechanism where precursor generation is in the bulk (Case 1), in the boundary layer (Case 2), or at the wall surface (Case 3). *Adapted from Panchal and Watkinson (1994).*

2.3 EVENTS IN CRUDE OIL FOULING

Fouling is a dynamic process. Figure 2.5 shows idealized fouling curves where the fouling resistance is plotted against time, starting from a clean surface. A lag time, t_l, is often seen before the layer starts growing on the initially clean heat transfer surface. Depending on process conditions, fluid composition, and fouling mechanism, the fouling resistance evolves in different ways. In some cases, fouling just keeps growing at a linear rate, in some others, the thickness of the layer keeps increasing but the rate of deposition decreases over time (falling rate). If the conditions are such that the forces at which the foulants are deposited on the surface are balanced by the forces that either suppress deposition or remove deposited species, the observed behavior is defined as asymptotic.

Epstein (1983) considered that each of five generic types of fouling (crystallization, particulate, chemical reaction, corrosion, and biological) comprise five distinguishable steps (reported in the rows of his 5 × 5 matrix in Figure 2.1):

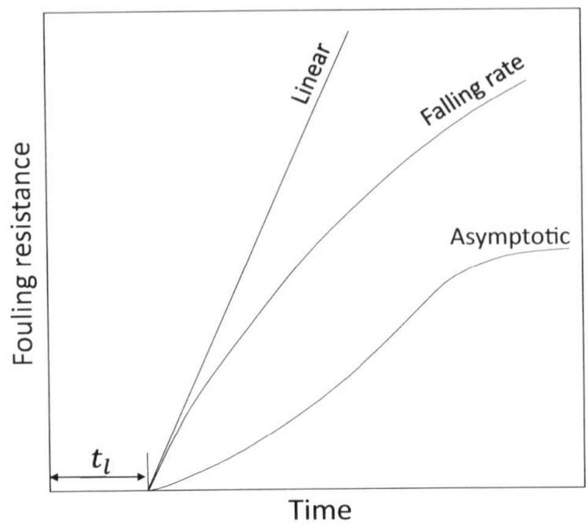

Figure 2.5 Idealized fouling curves.

- Initiation (Yang et al., 2009a)
- Transport to surface (Panchal and Watkinson, 1994)
- Attachment (Visser, 1988a,b)
- Removal (Yiantsios and Karabelas, 1994; Bohnet et al., 1997)
- Ageing (Wilson et al., 2009; Coletti et al., 2010).

Although there must be some sequential element to any overall fouling process, in reality many of these steps could, and probably would, operate in parallel. When uncertain reaction mechanisms are taken into account, it is clear that crude oil fouling becomes a very complex physico-chemical process.

When material of increasing molar mass and structural complexity exceeds its local solubility in crude oil then it will form a solid. This could happen at the heat transfer surface or in a reaction zone within the fluid and close to the heat transfer surface where process conditions are favorable enough. Reactants (or fouling precursors) must be transported to the reaction zone by convection mechanisms and reaction products could be convected away perhaps to be deposited, or to take part in other reactions, elsewhere within the heat exchanger. The possibility also exists for the action of fluid shear to remove solid deposits in a wholesale fashion from the heat exchanger surface for them to be deposited elsewhere. Clearly, initiation must precede ageing but, once started, the steps of initiation, transport, attachment, removal, and ageing could in fact all proceed in parallel. For crude oil fouling, transport, attachment, and removal steps have been commonly deemed to occur in parallel. In contrast, the steps of initiation and ageing remain poorly understood. Although chemical reaction does not feature as a distinguishable step in Epstein's analysis, reactions are most likely to occur in the initiation, attachment, and ageing steps for crude oil fouling. It is unlikely that chemical reactions are involved in transport and removal.

Long before Epstein distinguished steps in fouling processes, Nelson (1934a) proposed a simple mechanism to account for fouling of oil refinery heat exchangers in which none of the steps were considered to be discrete and sequential. It was considered that the rate of fouling would be directly dependent on the volume of fluid in a heat transfer film which would be at a temperature higher than that of the bulk fluid. Hence, the volume of hydrocarbon likely to react would be reduced at higher velocity. In turn, this would mirror the common observation that fouling in oil refinery heat exchangers could be mitigated by using higher velocities. Furthermore, a higher velocity would not only reduce the residence time of the thinner film again leading to less fouling but also if there were a deposit removal mechanism taking place, then the rate of removal almost certainly would be increased by a higher velocity because the shear stress on the heat transfer surface would be higher. Nelson's idea clearly integrated the transport, attachment, and removal steps. Nevertheless, Nelson's simple concept dealt neither with initiation nor with ageing.

A brief discussion for each of these subprocesses is provided in the following sections.

2.3.1 Initiation

When a new or clean heat transfer surface is commissioned there might be little change in the overall heat transfer coefficient, U, for some time, creating the impression, almost certainly a false one, that nothing is happening. Lasting anything up to several days, the length of the initiation period is related to the nature of the surface (see Section 2.4.4) as well as to process conditions such as surface temperature and fluid velocity. In some cases initiation might actually increase rather than decrease the overall heat transfer coefficient, U due to the creation of roughness on the surface (Crittenden and Alderman, 1988). Such roughness creation might come from deposition of chemical elements within the crude oil and/or from corrosion-type reactions on the surface. In reality, very little is known about the initiation, or induction, of crude oil fouling and indeed there is often difficulty in even interpreting when the initiation step has started and ended. Saleh et al. (2003) arbitrarily used the time at which the change in reciprocal overall heat transfer coefficient ($1/U$) was 5% or more of the total change in $1/U$ to be the end of the induction period.

Yang et al. (2012) adopted a more generic and arguably a more scientific approach in which it was assumed that active foulant species stick to the surface and gradually cover it, the rate of change in surface coverage $d\theta/dt$ being proportional to the fractional free surface area $(1-\theta)$. It was further assumed that the foulant already on the surface could act as a seed, thereby attracting more foulant such that the growth rate was first order in θ with a rate constant equal to k_1. In addition, by adopting a removal mechanism akin to that in adsorption phenomena, the removal rate of the coverage was assumed to be proportional to θ with a rate constant equal to k_2. A combination of these three assumptions led to the following relationship:

$$\frac{d\theta}{dt} = k_1\theta(1-\theta) - k_2\theta \tag{2.3}$$

According to Equation 2.3, the fouling layer grows on this covered surface at a rate equal to $\theta dR_f/dt$, where R_f is the fouling resistance and dR_f/dt can be any established expression that accounts for the normal period of fouling of the heat exchanger. It was found that the surface coverage in the induction period θ was related to the rate constants and time t by the following equation:

$$\theta = \frac{k_1 - k_2}{k_1}\,\frac{1}{1 + ce^{-(k_1-k_2)t}} \tag{2.4}$$

Here, c is a constant of integration. Equation (2.4) produces a smooth sigmoidal curve ranging from $\theta = 0$ at the beginning of the induction period to $\theta = 1$ at its end. A time $t_{0.5}$ is defined to be the time when θ is equal to half its maximum value.

$$t_{0.5} = \frac{\ln c}{k_1 - k_2} \tag{2.5}$$

Yang et al. (2012) used data obtained during induction periods for quite different types of fouling including crude oil, whey protein, and calcium salt scaling. The induction period model supported experimental observations in which shorter induction periods are found with higher surface temperatures. The activation energy E for the induction period of the crude oil study was found to be 78.2 kJ mol^{-1}; this value is important in determining what effect the surface temperature might have on the length of the induction period. Yang et al. (2012) were able to show also that the length of the induction period was much reduced when the heat transfer surface was not clean initially. They claimed that fouling threshold phenomena could be captured by their interpretation of the induction period. Thus, if k_2 were to remain close to the value of k_1 then the surface coverage θ would remain low and the induction period would, accordingly, be very long. Hence, a boundary could be drawn between a fouling zone in which $k_1 > k_2$ and a nonfouling zone in which $k_1 < k_2$. What this theoretical interpretation of the induction period does not do is identify what physical and/or chemical processes are occurring at the heat transfer surface in the very early stages of crude oil fouling.

2.3.2 Transport

In all reaction mechanisms, the fouling precursors (foulants) must be transported by convective mechanisms from the bulk of the fluid toward the zone in which they are converted into deposits by attachment or adhesion to the heat transfer surface. In a similar manner, the foulant or deposit, if mobile enough, could be transported back toward the fluid bulk, perhaps to take part in further reactions elsewhere in the equipment. The transport of foulants and their precursors toward the heat transfer surface is traditionally based on diffusion and the film theory of convective mass transfer and the mechanisms would be different for suspended particulates and dissolved precursors. The transport of precursors is governed by the difference between their bulk concentration c_b and their concentration at the heat transfer surface c_s. The lumped parameter k_m is the mass transfer coefficient. Strictly, this equation should describe the mass (or molar) flux of foulant precursors:

$$\frac{dm}{dt} = k_m(c_b - c_s) \tag{2.6}$$

The mass transfer coefficient k_m can be obtained from any appropriate correlation that relates the Sherwood number to the Reynolds and Schmidt numbers. If the deposit properties (density and thermal conductivity) were to remain independent of time then the rate of fouling would be proportional to the rate of mass deposition and hence the rate of fouling would be given by:

$$\frac{dR_f}{dt} = k'(c_b - c_s) \tag{2.7}$$

Here, k' would be a coefficient that incorporates the mass transfer coefficient, k_m as well as the density and thermal conductivity of the deposit.

For particles already present or formed in the fluid bulk, fouling can be considered to comprise two distinct steps, namely transport to the wall region followed by adhesion or attachment to the wall (Watkinson, 2005).

If the transport step controls the overall rate of deposition then the fouling rate should only be a weak function of surface and bulk temperature. The fouling rate should increase with increasing velocity, linearly in turbulent flow and more weakly in laminar flow (Watkinson, 2005). Epstein (1981) has summarized the mechanisms that could be involved in the transport of particles toward a surface (particulate fouling). Settling is unlikely to be relevant to crude oil fouling and so the principal reason as to whether a particle can get to a surface and remain there is related to the surface forces such as the London–van der Waals forces, which are always attractive. Electrical double-layer forces are attractive if the particle and surface have zeta potentials of opposite sign. A hydrostatic viscous interaction force may also be operative in which the fluid friction on the particle increases significantly as it moves normal to the plane of the surface. When interaction barriers are additive then there is a large energy barrier which must be overcome, which is equivalent to the activation energy of a chemical reaction. It is for this reason that the deposition flux is commonly assumed to comprise two resistive components:

$$\dot{m}_d = \frac{c_b}{(1/k_m) + (1/k_r)} \tag{2.8}$$

Here k_r is a rate constant applied to the energy barrier and commonly assumed to apply to a first-order process. The rate constant is expected to be a strong function of temperature; the relationship is commonly expressed in the form of an Arrhenius-type equation:

$$k_r = k_{r\infty}\, e^{-E/RT_w} \tag{2.9}$$

Where $k_{r\,\infty}$ is the pre-exponential factor, T_w is the wall temperature and R the universal gas constant.

The mass transfer coefficient again can be obtained from correlations involving the Reynolds and Schmidt numbers. The particle diffusivity, which is required to calculate the Schmidt number, can be obtained from the Stokes–Einstein equation (Epstein, 1981). Clearly, the interactions change when deposits have already been established on the heat transfer surface. Although electrophoresis has been suggested as an important mechanism for particulate deposition (Epstein, 1981), it is unlikely to be operative in crude oil systems unless there is a significant aqueous fraction. Thermophoresis, a mechanism that can exert a force on a submicrometer particle due to a temperature change across it, is also unlikely to be effective in crude oil fouling. Because it is close to a hot surface when crude oil is being heated, the effect would be to repel the particle.

2.3.3 Attachment

Once transported, foulants either attach or adhere to the heat transfer surface or they leave the surface such that they may be deposited elsewhere. Factors affecting adhesion include surface energies, surface temperature, and shear forces acting at the surface, as well as the nature and compositions of previously deposited layers.

Little is known about the fundamental details of the adhesion of deposits from crude oils on heat transfer surfaces. Simple theories have been proposed to describe this phenomenon (Bennett, 2012) but they are difficult to validate given the challenges associated with collecting the relevant experimental data. While it is the complexity of crude oils which makes this an uncertain subject, some lessons can be drawn from fundamental studies of the other major type of reaction fouling, namely that from food processing. Here, the basic fundamentals of adhesion and the associated subject of removal have been provided by Visser (1988a,b).

Should the adhesion step control the overall fouling rate if it is a particulate-based mechanism then the overall fouling rate should increase strongly with surface temperature because the adhesion step would be sensitive to temperature. The overall fouling rate would not be strongly dependent on bulk temperature but should decrease with increase in velocity because less time would be available for adhesion to occur (Watkinson, 2005).

2.3.4 Removal

Removal mechanisms are poorly understood in comparison with deposition. Models of crude oil fouling commonly assume that the overall or net rate of deposition is the difference between a rate of deposition and a rate of removal. However, such a distinctive approach is almost certainly not the way crude oil fouling proceeds. Rather, it is almost certainly the case that many mechanisms including those responsible for deposition and those responsible for removal will operate together in a continuous manner. Epstein (1981) prefers to use the term "reentrainment" rather than "removal." One reason for this is that on a clean heat transfer surface there cannot, in principle, be any removal since there would be no deposit to remove. Nonetheless, once deposits have started to form on the heat transfer surface the possibility then exists for them to be removed by the surface shear stress in turbulent flow or by diffusion should they be soluble in the chemical bulk. The latter mechanism becomes a possibility when deposition is the result of chemical reactions because the concentration of reaction products in the bulk fluid would be low.

Epstein (1981) summarized the reentrainment mechanisms that previous researchers had used in mathematical models of fouling processes. They include spalling, erosion, bond fracture, and dissolution. Each, in principle, could be considered applicable to crude oil fouling and may, of course, operate in parallel. Generally, models of removal have the rate of mass removal being directly proportional to the mass of deposit on the surface.

2.3.5 Ageing

The exposure of the fouling layer to high wall temperatures over extended periods can trigger chemical and physical transformations which alter the structure and properties of the deposits. Nelson (1934a,b) described initial deposition of crude oil fouling as involving a gel which changes its structure over time to a harder material similar to coke. These transformations not only alter the rheology of the deposit layer (Sileri et al., 2009) but also its thermal conductivity, and thus:

- The overall thermo-hydraulic behavior of the exchanger.
- The nature of the material recovered for analysis with analytical techniques. This makes it difficult to identify the material originally deposited (Wilson et al., 2009).
- The ease of deposit removal (Wilson, 2005). Ageing therefore has important implications for the development of cleaning strategies.

Within a fouling layer, the material is subject to a range of temperature histories so that ageing is non-uniform in space and highly time dependent. Timescales that enable appreciation of the effects of ageing are rather long and laboratory experiments rarely report deposit ageing (Wilson et al., 2009). Unlike the induction period, which happens at much shorter timescales, ageing plays a more important role in industrial equipment than in laboratory experiments. However, despite the importance of these effects, relatively little attention has been paid in the literature to these phenomena, particularly in the area of chemical reaction fouling (Figure 2.1).

Nelson (1934a) was perhaps the first to report on ageing through a semi-quantitative model of crude oil furnaces. Atkins (1962) subsequently was the first to propose that a hydrocarbon fouling layer (for a fired heater) might comprise two layers: a porous, tarry layer and a hard crust layer. From this concept, Crittenden and Kolaczkowski (1979b) developed a mathematical model of fouling for two such layers whence the subject of ageing has been largely ignored, perhaps because it is so hard to study experimentally.

More recently, Ishiyama et al. (2010) developed a lumped parameter, first-order kinetics ageing model that is based on a series of layers laid down at fixed time steps to form a series of annuli with varying history. Once deposited, each layer is covered by the next layer with deposition occurring only at the deposit–oil interface. Although it was considered that the thickness of each layer would not change with time, it was assumed that a change in thermal conductivity of a layer would constitute the ageing process. Although the approach adopted is still idealistic, Ishiyama et al. were able to make exploratory studies of the outcomes from various scenarios including deposition with no ageing and deposition with ageing for operation with either constant wall temperature or constant heat flux. The model was then extended by Coletti et al. (2010) to account for the effect of ageing in distributed systems as summarized in Section 5.1. No experimental studies are known to validate these various ageing models.

Ageing in crude oil deposits almost certainly follows a coking mechanism (Section 2.2.4). According to Fitzer et al. (1971), the secondary reactions in a cracking and coking mechanism (Figure 2.3) involve cyclization of hydrocarbon chains, thereby forming aromatics and subsequently condensation of the aromatics to form polynuclear aromatic structures of high molar mass. Although the formation of aromatics tends to occur preferentially at temperatures above 950 °K, aromatic structures will already be present in a fouling deposit that has been created through an asphaltene deposition mechanism. Once present in a deposit, chemical condensation of aromatics can take place in the liquid phase at temperatures in the range 650–800 °K, the lower end of which will be experienced in deposits formed in the crude oil preheat exchanger train. At temperatures below 650 °K, polynuclear aromatics can cross-link to form higher molar mass compounds via both Diels–Alder and free radical pyrocondensation mechanisms (Fitzer et al., 1971).

2.4 VARIABLES AFFECTING FOULING

Laboratory experiments have been performed in microbomb reactors (MBRs) (Section 3.1), stirred cells (Section 3.2), and recirculating closed loop apparatuses in laboratory equipment (Section 3.3) or in field units (Kuru et al., 1997) to assess the effects that different variables have on fouling. The key variables identified with these experiments are:

1. Crude oil composition and inorganic contaminants
2. Bulk and wall temperatures
3. Velocity (shear stress)
4. Surface conditions (e.g., roughness and roughness dynamics).

Notably, pressure is missing from the list above. Although it can have a significant effect in asphaltene precipitation during recovery of heavy crude oils (Asomaning et al., 2000), it is not known whether pressure could have an influence under typical refinery operating conditions (i.e., below 30–35 bar).

Through the experimental results available in literature, it can be established that mass transport in crude oil fouling increases weakly with both wall and bulk temperature and increases nearly linearly with velocity. Attachment, on the other hand, has been shown to increase strongly with wall temperature but only weakly with bulk temperature and decreases with increased velocity (Watkinson, 2008). The fact that the fouling rate decreases with increasing velocity suggests that crude fouling is controlled by adhesion or reaction.

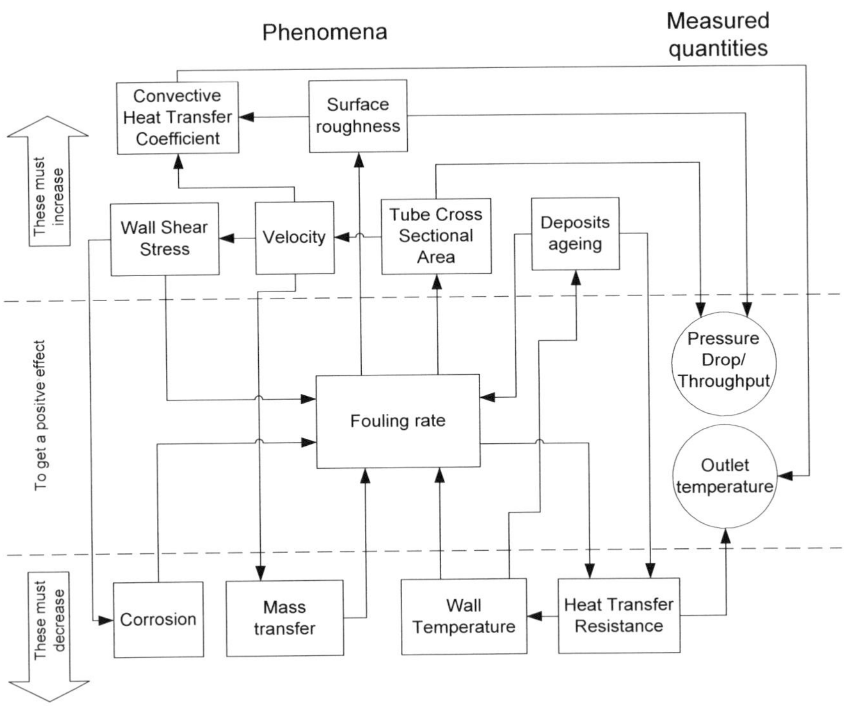

Figure 2.6 Schematic of phenomena occurring in a tubular heat exchanger undergoing crude oil fouling. Boxes on the top of the figure show the phenomena whose increase reduces fouling rates and have a positive impact on the measured quantities: pressure drop and coil inlet temperature. Boxes at the bottom of the figure show phenomena that should decrease in order to mitigate fouling. Arrows indicate relations between phenomena (square boxes) and measured variables (circles). *Reprinted with permission from Coletti and Macchietto. A dynamic, distributed model of shell-and-tube heat exchangers undergoing crude oil fouling, 2011. Ind. Eng. Chem. Res. 50(8): 4515-4533. Copyright (2011) American Chemical Society.*

Figure 2.6 summarizes the interactions of the above variables involved in the tube-side crude oil fouling process and their combined effect on the two main measured quantities: pressure drop and outlet temperature.

The overall effect can be significant: deposition of foulants increases the resistance to heat transfer, but it also affects the fluid dynamics of the system by reducing the cross-sectional area which in turn increases the velocity, hence the convective heat transfer coefficient. The variables affecting fouling will be explored more in detail in the next sections.

2.4.1 Crude Oil Composition and Inorganic Contaminants

Crittenden (1988b) reviewed the extensive literature on how the presence of trace species can affect the deterioration of feedstocks when autoxidation is the dominant mechanism.

2.4.1.1 Oxygen

Deterioration and gum formation are not just dependent on dissolved oxygen concentration but also on the presence of chemically bound oxygen, nitrogen, and sulfur, as well as the presence of trace metal contaminants. Rigorous exclusion of dissolved oxygen (or air) can substantially reduce—or eliminate—deposition from hydrocarbons in the liquid phase, although the extent depends on the type and concentration of sulfur species present. As might be expected, the addition of organic peroxides to deoxygenated feedstocks has been found to be highly deleterious.

2.4.1.2 Sulfur

The importance and role of sulfur in crude oil fouling have been highlighted in Section 2.2.1 when discussing fouling mechanisms.

The effect of sulfur is complex (Crittenden, 1988b). Thiols, sulfides, disulfides, and some condensed thiophenes which can break down to form free radicals can be problematic but diphenyl sulfide and dibenzothiophene are not. Free sulfur, disulfides, polysulfides, and thiophenol promote sludge formation in storage tanks whereas thiophenes, aliphatic mercaptans, and aliphatic sulfides show little effect. The organo-sulfur structure is important as well with the thermal decomposition of alkanes being promoted by sulfur atoms in branched structured compounds. The presence of small amounts of nitrogen-containing compounds such as pyridines and pyrroles can be deleterious to hydrocarbon stability (Crittenden, 1988b).

2.4.1.3 Metals

Metal chlorides of iron, calcium, and magnesium hydrolyze to produce acids that promote hydrocarbon fouling substantially while the nature of metal surfaces themselves can strongly influence deposition rates with copper and vanadium being highly deleterious (Crittenden, 1988b). When present in the form of acetyl-acetonates, low levels of metals can dramatically increase deposition rates.

2.4.1.4 Asphaltenes

As reported in Section 2.2.5, Asphaltenes are defined as the *n*-alkane-insoluble/toluene-soluble fraction of the crude (Watkinson, 2008) and their role in crude oil fouling has been studied by several authors.

Using their patented batch stirred cell (Section 3.2), Eaton and Lux (1984) found that the addition of asphaltenes with molar mass in the range 1000 to 300,000 indeed promoted crude oil fouling whereas the addition of resins (molar mass <5000) had less serious effects. Both the concentration and the solubility of asphaltenes were found to be important factors. On the other hand, Venditti et al. (2009a) reported on preliminary experiments using a MBR, described in Section 3.1, that showed the role of asphaltenes to be somewhat ambiguous in deposit formation from the bulk at temperatures in the range 280–390 °C. There was evidence for the formation of larger molecules when the oil was heated even when no solids were observed at lower temperatures and, accordingly, there was evidence of chemical transformations. The formation of deposits in the MBR was favored by high temperatures and long residence times but, irrespective of the temperature, no deposits were found at low residence times (1 h). There was also evidence that carbonaceous deposits originated not only from the asphaltene fraction of crude oil but also from the crude after it had been deasphalted. Moreover, it was found that, under the same operating conditions, the deasphalted crude under testing could produce a larger amount of deposits than the original crude oil (Venditti et al., 2009a). Discussions with refiners reveals that even crudes with just traces of asphaltenes do produce fouling in the hot end of the PHT. This is confirmed by industrial observations and experimental work by Saleh et al. (2005b) who studied the fouling behavior of an Australian crude oil with low sulfur, ash, and asphaltene levels.

It should be noted that crude oil composition and its contaminants are not variables that can be generally manipulated whereas the process conditions of temperature and, in particular, velocity can be adjusted by the designer in order to mitigate fouling.

2.4.2 Temperature

As a general rule, high temperatures are usually associated with the promotion of chemical reaction fouling and corrosion and lead to shorter induction times. Furthermore, temperature has a direct effect on the ageing rate of the foulant layer. The deposit may be hardened and become difficult to remove or, alternatively, become weaker and tend to spall under the influence of temperature and time.

There has been some debate (Polley et al., 2002a) in the literature over whether the rate of the fouling reaction is a function of surface (wall) temperature, T_w, or rather the film temperature, T_f, between the bulk temperature, T_b, and the wall (Ebert and Panchal, 1995):

$$T_f = 0.55(T_b + T_w) \tag{2.10}$$

The temperature chosen to describe fouling rate dependence is associated with the zone where the reaction is assumed to take place (i.e., Case 1, 2, or 3 in Section 2.2.5).

2.4.3 Velocity and Shear Stress

Of all the variables, velocity is the one on which the designer has the greatest control (Bott, 1995b). An increase in velocity has a double-pronged effect on deposition of increasing the wall shear stress and increasing the convective heat transfer coefficient. The first is perhaps the more obvious effect on fouling: increased shear forces at the deposit/fluid interface impede fouling deposition and may even result in foulant removal. The role of shear stress in fouling behavior is highlighted by a plot by Joshi et al. (2009) that relates fouling rates to shear stress for several heat exchangers in six different PHTs, covering a range of tube-side velocities from 0.9–2.7 m s^{-1}. The majority of the exchangers considered had a design shear stress range of 5–8 Pa. The graph shows that the higher the shear stress, the lower the fouling, as expected, but it is surprising that temperature appears to have no effect on fouling rates, thereby apparently attributing the dominant role to shear stress alone.

The second is a less obvious—although important—effect of velocity on fouling. An increase in velocity, assuming that everything else remains constant, produces an increase in the convective heat transfer coefficient and thereby a reduction in total thermal resistance. As a result, the wall temperature is reduced, generating a beneficial effect on fouling rate.[1] This is the principle exploited by fouling mitigation devices based on a fixed wire

[1] See numerical example in Bott (1995b) in which a twofold increase in velocity leads to 25 °C lower wall temperature.

matrix inserted on the tube-side of industrial units (Ritchie et al., 2009). On the other hand, if the deposition involves mass transfer of insoluble species (e.g., Case 1a and 2a in Figure 2.4), higher velocities will increase the diffusion toward the surface in the presence of a concentration gradient, thereby enhancing deposition.

Controversial effects of velocity on initial fouling rate have been reported by Crittenden et al. (1988a, 2009) in both model chemical and crude oil systems. At low surface temperatures the initial fouling rate decreased with increasing flow rate. At high surface temperatures the opposite dependency was observed. They concluded that a combination of high surface temperatures and low flow rates can lead to mass transfer of deposit precursors to the surface becoming the rate-determining step. This clashes with other observations, reported at the beginning of Section 2.4, that suggested adhesion or reaction is the rate-determining step. Industrial observations are that if the velocity is reduced, even for short periods, the result can be an increased deposition rate and, as a consequence, highly fouled surfaces. Once this has happened, even after the heat exchanger is brought back to normal operation, the deposits may have hardened their structure (aged) and not respond to increased velocities by detaching. This confirms the importance of maintaining the highest possible velocities allowed by other constraints given by an arbitrary earlier choice of allowable pressure drop (Butterworth, 2004) or by corrosion and vibration problems that may occur at large velocities (ESDU, 1989).

2.4.4 Surface Conditions and Roughness Dynamics

The importance of surface conditions has been addressed in several articles (Crittenden and Kolaczkowski, 1979a; Zhao and Müller-Steinhagen, 2001; Förster and Bohnet, 2001; Santos et al., 2004; Kukulka et al., 2010; Coletti et al., 2010). If the surface is corroded, it not only provides resistance to heat transfer but it also creates sites that encourage deposition. Additionally, corrosion products become particulate fouling for downstream sections of the heat exchanger.

Another important aspect of the surface is its roughness as it affects the convective heat transfer coefficient (Walker and Bott, 1973). The disruption of the viscous sublayer caused by a rough surface generates an increase in the turbulence level compared to that of smooth surfaces (Figure 2.7) which in turn produces higher heat transfer coefficients (Yaglom and Kader, 1974). A constant value of equivalent sand roughness is typically used, calculated from shorthand correlations. However, not only the initial

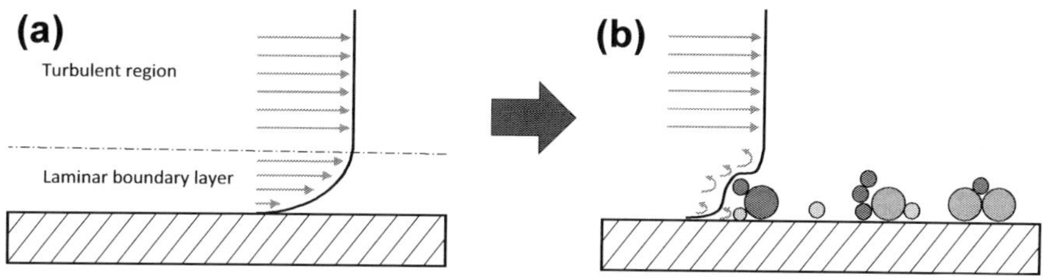

Figure 2.7 Effect of the growth of the deposits on a thermal surface. In clean conditions (a) the surface is relatively smooth and the boundary layer is well defined. Once deposition begins the deposits enhance surface roughness (b) and the laminar boundary layer is disrupted with related enhancement of the convective heat transfer coefficient.

status of the clean surface matters, but also its changes over time. The accumulation of foulant material on heat transfer surfaces often results in a progressive increase in roughness, with associated effects on heat transfer coefficient and pressure drop. Although in laminar flow the surface roughness has little effect on the friction factor and heat transfer coefficient (Shah and Sekulic, 2003), it plays an important role in the turbulent regime typical of industrial applications.

The effect of roughness on the heat transfer coefficient is largely acknowledged in literature, but the effect of fouling on surface roughness dynamics has received relatively little attention (Ceylan and Kelbaliyev, 2003). Enhancement of heat transfer due to an initial increase in surface roughness generated in fouling tests has been reported and analyzed for particulate fouling by Crittenden and Alderman (1988), for crystallization fouling by Albert et al. (2009), and for corrosion fouling by Panchal (1988). For chemical reaction fouling, most of the reported cases are qualitative. Bott (1990) noted that, due to the complex interaction of variables, no systematic study on the effects of roughness on fouling had been undertaken. However, some quantitative results have been presented. Cousineau et al. (1988) reported the increase in heat transfer coefficient for a Bayer process in which sodium aluminosilicate scales deposit with kinetics of the second order. The apparent negative fouling resistance reported in the initial period (up to 40 h) of testing a crude oil by Knudsen et al. (1997) could be attributed to the progressive increase in surface roughness. Wilson and Watkinson (1996) reported both thermal and hydraulic roughness effects (from analysis of pressure drop measurements) in studies of autoxidation reaction fouling, while Asomaning et al. (2000) observed these in testing crude oils.

It should be pointed out that the dynamics of surface roughness are expected to apply mainly to laboratory testing, where more precise measurements of heat transfer are made, over short timescales. Industrial applications feature larger timescales, so that short-term variations in surface roughness are not noticed. Nonetheless, being able to capture this phenomenon is essential for the correct interpretation of accelerated fouling tests performed in pilot plants.

Quantitative modeling of roughness dynamics is in its infancy: for example, in their analysis of crude oil heat exchangers subject to fouling, Yeap et al. (2005) changed the surface roughness of clean tubes to that reported for bitumen by Kern (1988) to represent the foulant. An approach to modeling the increase in roughness due to fouling on tube surfaces was developed by Yiantsios and Karabelas (1994). They proposed a population balance model based on a population of roughness elements to describe deposition and removal in crystallization fouling. However, their focus was on the removal process and they did not discuss roughness in terms of hydrodynamics, pressure drop, and more importantly, impact on heat transfer. The first model for roughness dynamics in chemical reaction fouling was proposed by Coletti et al. (2010), which captured the effect of increased surface roughness due to fouling deposition on heat transfer coefficient and pressure drop. This model will be reviewed in Section 5.1.

2.5 CONCLUSIONS

In this Chapter, the complexity of the crude oil fouling problem has been described. In particular, it has been shown that despite the large research activity on the subject, much remains to be understood starting from the nature and composition of the crude oil itself and the chemical reactions that produce both precursor and fouling species. The chemical and physical phenomena that take place and interact with one another are extremely difficult to identify and isolate. Laboratory measurements and field experience have provided some useful indications on how key variables such as temperature and velocity affect fouling rates within the heat exchanger. However, further experimental work in obtaining fouling deposits under controlled conditions as well the development of analytical techniques that allow a better characterization of both crude oil and its deposits is needed. These will be described in Chapter 3 and Chapter 4, respectively.

CHAPTER THREE

Experimental Generation of Fouling Deposits

B.D. Crittenden[1], G.F. Hewitt[2], M. Millan-Agorio[3], K. Rostani[4], S. Venditti[5], M. Yang[6]

[1]Professor of Chemical Engineering, University of Bath, UK
[2]Professor of Chemical Engineering, Imperial College London, UK
[3]Reader in Chemical Engineering, Imperial College London, UK
[4]CO2 Management R&D, PETRONAS Research Sdn Bhd, Malaysia
[5]R&D Engineer, Resource Centre for Environmental Technologies, Luxembourg
[6]Research Associate, University of Bath, UK

Research on crude oil fouling using actual plant data is not only slow to progress but can also be affected by a variety of logistical and operational requirements that do not lend themselves very well to fundamental scientific studies (Crittenden et al., 1992). In addition, the use of plant measurements can create difficulties in the interpretation of the data (Takemoto et al., 1999). Although many problems relating to plant studies can be eliminated by using well-designed laboratory experiments, such methods in turn suffer from the disadvantage that the crude oil is not exposed to the time–temperature–flow history experienced within industrial equipment. Accordingly, concerns arise that crucial fouling precursors could become depleted in a laboratory apparatus in which crude oils are used continuously or repetitively, either in batch systems or in recirculating flow loops. On the other hand, laboratory-scale studies can be carried out with precise control over operating conditions and they can provide excellent access to the heat transfer surface, its deposits, and the fluids from which the fouling originates. Laboratory-scale methods can also facilitate study of the effects of the heat transfer surface, that is, its composition and morphology, as well as the use of devices that might be used to simultaneously improve heat transfer and reduce fouling (Crittenden et al., 1993).

Several well-designed experimental devices have been created to study crude oil fouling in the laboratory (Eaton and Lux, 1984; Chenoweth, 1988b):

1. *Hot wires*. A preheated oil stream contacts a hot metal wire, the temperature of the wire representing the surface temperature of the heat exchanger of interest. The temperature of the wire increases as fouling progresses. The flow field is quite different from that of an industrial heat exchanger and

Crude Oil Fouling
ISBN: 9780128012567
http://dx.doi.org/10.1016/B978-0-12-801256-7.00003-8

hence the surface shear stress will also be quite different. Accordingly, although the apparatus is simple, the effect of velocity on deposit growth is poorly defined and difficult to interpret (Stephenson et al., 2011).

2. *Microbomb reactors.* Microbomb reactors are small-scale closed systems, typically holding up a few milliliters of crude oil, which allow the study of the effect of temperature and oil composition in the absence of any flow-related effects. Microbomb systems have been developed at Imperial College London (Venditti, 2010; Rostani, 2014) and are described in Section 3.1.
3. *Batch stirred cells.* Also known as an autoclave, the batch cell is essentially a pressure vessel that contains a magnetically coupled stirrer and a test surface heated by an embedded cartridge heater. The stirrer can be a rotating cylinder that creates a well-defined flow field. The temperature and velocity fields can be well defined and modeled by computational fluid dynamics (CFD). Furthermore, the volume of crude oil required for fouling tests can be small, of the order of 1 L. Other advantages of this type of apparatus are that the fouling deposits can be inspected by opening the cell at the end of a fouling test and that the test surface can be changed quite easily. Examples of this type of apparatus can be seen in Eaton's patent (Eaton, 1983) and in Young et al. (2011). It is relatively easy to create bulk and surface temperatures and surface shear stresses that are representative of industrial heat exchangers but, of course, the flow field is not representative of industrial practice and nor is the temperature–time history of the crude oil. The stirred batch cell developed at Bath University is described in Section 3.2.
4. *Flow loops.* A quantity of oil is pumped through a heated test section which can be of tubular or annular cross-section. Tubular flow loops with two parallel test sections have been used at the University of Bath (Crittenden et al., 1993, 2009; Takemoto et al., 1999). Flow loop systems comprising either or both annular and tubular test sections have been selected by Imperial College London (Macchietto et al., 2011) and Heat Transfer Research Inc. (Bennett et al., 2009; Smith, 2013). Although these systems are able to closely replicate real plant conditions, they require relatively large quantities of oil and the flow loop itself can comprise complex and expensive equipment including pumps, flow meters, flow controllers, and heating sources (either Joule heating or imbedded heating elements). The HIPOR apparatus developed at Imperial College London is described in Section 3.3.

The following sections describe three different types of experimental apparatuses developed for the generation of fouling deposits under controlled conditions, namely microbomb reactors, stirred cells, and flow loops.

3.1 SMALL SCALE, ACCELERATED CONDITIONS: MICROBOMB FOULING TESTS

M. Millan, S. Venditti, K. Rostani

One of the challenges posed by experimental fouling research is to produce relevant data using small scales, in terms of both volume of sample and time, in comparison with the times and volumes of crude at which fouling appears in refineries. Therefore, small-scale experiments often resort to accelerating the fouling process by employing higher temperatures than those used in the field. As will be discussed below, another characteristic of laboratory-scale experiments is that the amount of deposits produced may be a significant fraction of the original crude. The aim of experiments at laboratory scale is not necessarily to reproduce the actual conditions in a heat exchanger or other refinery unit but to identify and isolate some of the variables that play a role in deposition as well as enabling a procedure for comparing the behavior of different crude oils under well-defined conditions, identical for all crudes.

This section focuses on laboratory-scale fouling equipment used at Imperial College London in the study of the chemistry of oil deposition in the absence of any flow-related effects. By contrast with other fouling test rigs described in later sections, the microbomb is not a flow system and therefore it enables the study of the effect of the chemistry and operating conditions on deposition in isolation from flow-related shear stress.

Two types of batch microbomb test apparatus were employed in these studies: a N_2-pressurized design, where N_2 as an inert gas is used to reduce or suppress sample volatility at a free crude oil surface, and a closed liquid system design, where the use of gas is not necessary and the crude oil pressure is controlled by means of a valve downstream of the test section.

The systems employed are batch without recirculation and typically operate with sample sizes between 1 and 10 g of sample. The N_2-pressurized microbomb is described in Section 3.1.1 and the closed liquid system fouling test microbomb is described in Section 3.1.2. The former configuration was used in the study of deposition from some crude oils of interest and the latter on fouling due to incompatibility of crude oil blends.

3.1.1 The N_2-Pressurized Microbomb Fouling Apparatus and Its Applications

3.1.1.1 Equipment Description and Experimental Procedure

The N_2-pressurized microbomb is shown in Figure 3.1. It was originally developed as a high-pressure high-temperature reactor for hydrocracking

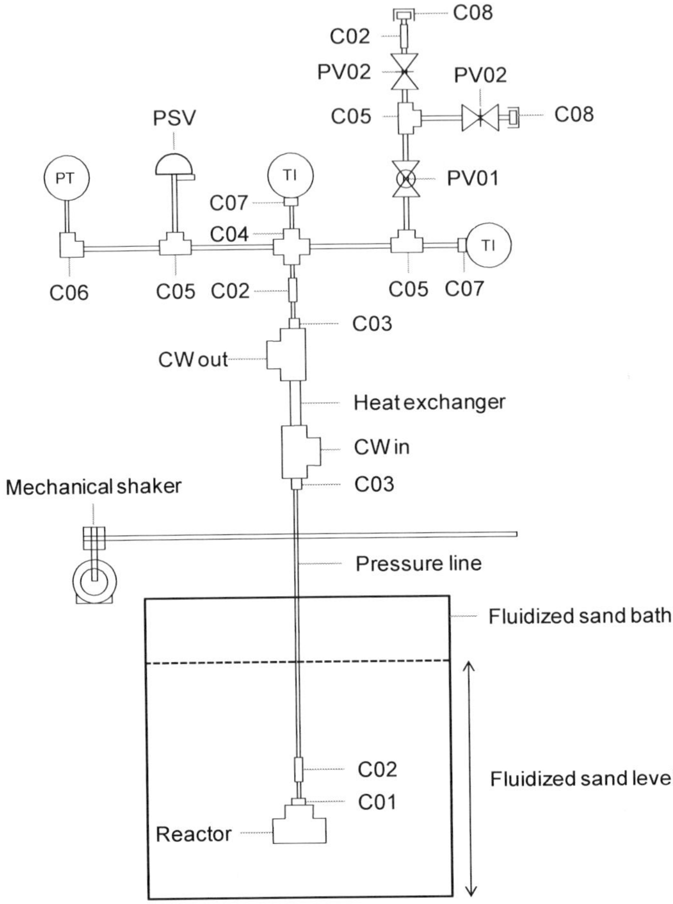

Figure 3.1 N_2-pressurized microbomb diagram. Here PT is pressure transducer, TI is temperature indicator, PSV is relief valve, PV01 is ball valve, PV02 is needle valve, C01 is ½ in to ¼ in reducing port connector, C02 is ¼ in straight union, C03 is ¼ in to ½ in reducer, C04 is ¼ in union cross, C05 is ¼ in union tee, C06 is ¼ in union elbow, C07 is ¼ in to thermocouple reducer, and C08 is ¼ in cap.

(Zhang, 1996) and since then it has been used, with some modifications to its configuration, in multiple applications such as catalyst screening (Puron, 2013), heavy oil upgrading in sub- and supercritical water (Pinilla, 2013), in addition to fouling experiments. The apparatus is made of 316 austenitic stainless-steel Swagelok components and has been described in more detail in the literature (Zhang, 1996). In fouling applications, the crude oil sample is loaded into a ½ in bored-through union tee with plugs in the two horizontal ends and the third attached via a reducing port connector to a ¼ in o.d. pressurization line. A pressure transducer and a pressure relief valve sit at the top of the pressurization line. A K-type thermocouple runs through the ¼ in line and measures the temperature in the tee. A heat exchange section is built in the ¼ in line by means of an external ½ in concentric tube connected to two ½ in tees, each with a closed end and a connection for water inlet/outlet in the other, which is used as coolant. This heat exchanger is intended to condense any vapors leaving the high-temperature section and prevent the pressure transducer from being exposed to temperatures above 70 °C, which may damage it. A second K-type thermocouple monitors the temperature at the top of the rig as a safety precaution in case overheating takes place. This assembly is held on top of a fluidized sand bath such that the ½ in tee is immersed in the bath. The fluidized sand bath provides good heat exchange and brings the microbomb quickly up to temperature, typically in around 3 min. The sand bath temperature is controlled to a setup point and related to the temperature in the tee through a calibration. Typically, the sand bath temperature is 15 °C higher than the crude oil temperature.

The whole assembly can be shaken to provide good mixing in the tee and enhance heat transfer between the tee and the sand bath. Shaking rates applied are around 50 cpm. The microbomb and heating system are located in a purpose-built steel cabinet for safe operation at high pressures. In fouling studies, this microbomb was pressurized with N_2 to suppress volatilization of the crude light ends. The system can withstand operating temperatures of up to 500 °C and 200 bar of gas pressure.

The experimental procedure is summarized as follows. Each experiment employs around 3 mL of crude oil. The crude is loaded in the ½ in stainless-steel tee, which is then connected to the pressurization line. Pressurization with N_2 follows and next the tee is immersed in the sand bath already set at the desired temperature. Cooling water and the shaker are then switched on. The exposure time at temperature is measured from the time the system reaches the operating temperature. After the targeted time at temperature has elapsed, the microbomb is cooled to room temperature and the products are separated. The deposits are recovered with the help of a spatula and washed with a solvent ($CHCl_3$) to remove any crude oil adsorbed on them. Thus, following each run, it is possible to determine the resulting amounts of liquid and solid, and perform further analysis in these samples. The next section will present the trends on the formation of solids obtained with this equipment.

3.1.1.2 Fouling Studies Using the N_2-Pressurized Microbomb Apparatus

This section aims to illustrate how the laboratory-scale experimental setup described in the previous section can be employed to obtain data on changes in the oil, leading to the formation of deposits and coke, and to measure rates at which these processes occur. The characteristics of the crudes used in the study are presented in Table 3.1. Most of this work was performed with a 24 °API crude oil (named Crude B).

Pressure in the range 20–80 bar was observed to have an effect on the amount of deposits obtained (Venditti, 2010) which was thought to be related to the suppression of volatilization from the crude due to the free surface between the oil and the inert gas. In these experiments, the N_2 pressure applied was 80 bar in order to minimize volatilization of the crude. Experiments were carried out at temperatures between 200 and 390 °C for holding times at temperature of 4, 5, 6.5, and 24 h to determine the effect of these conditions on deposition. A comparison between deposition from Crude B and the other two crudes was carried out in experiments at 320 °C. The heptane-soluble fraction of Crude B was also used as feed in some experiment to determine its fouling propensity. The results of these studies are presented below.

Formation of deposits as a function of temperature and time at temperature is presented in Table 3.2 for Crude B. No deposits were observed to form from the crude exposed for 24 h at temperatures of 280 °C and below. However, a significant amount of deposit appeared when the temperature was increased to 320 °C for the same time at that temperature (24 h). These results match the temperature threshold at which thermal C–C bond breaking becomes significant. When the temperature was subsequently increased to 390 °C, the formation of deposits becomes much faster. These conditions approach those found either near the wall of a heat exchanger at the hot end of a preheat train, in the furnaces of atmospheric and vacuum distillation units or further downstream of the refining process.

Table 3.1 Properties of Crudes A, B, D, and Oriente used in studies presented in Chapter 3.1 and in Chapter 4.1

Crude	API gravity (°API)	Asphaltene content (wt%)
A	28	6.7
B	24	5.4
D	46	0.06
Oriente	23	7.8

Table 3.2 Deposits formed (as wt% of the original Crude B) as a function of temperature and holding time in experiments carried out in the N_2-pressurized microbomb apparatus

Temperature (°C)	Holding time (h)	Weight of deposits (%)
200	24	–
280	24	–
320	24	1
390	4	2
	5	6
	6.5	12
	24	19

Experiments in the N_2-pressurized microbomb apparatus can also be designed to measure fouling rates under different conditions of temperature, residence time, and crude oil or crude oil blend composition, among other variables. As an example, average fouling rates for the period between consecutive measurements at 390 °C in Table 3.2 are plotted in Figure 3.2.

It can be observed that the average rate of fouling for the initial 4 h was relatively small but it increased by nearly one order of magnitude between 4 and 5 h and was still at high levels between 5 and 6.5 h. The existence of an induction period (section 2.3.1) has been observed in the literature (Asomaning, 2000). It is likely that the initial formation of deposits triggers an autocatalytic effect providing sites for further deposit formation and growth. There is also a contribution from processes taking place in the liquid phase that tend to increase the concentration of fouling precursors, as will be discussed below. Table 3.3 shows how at 390 °C the amount of asphaltenes in the liquid (heptane insolubles) decreased with time while the deposits increased. The combined amount of asphaltenes and deposits also increased steadily with time.

The amounts of asphaltenes and deposits obtained at various temperatures after 6.5 h at temperature are shown in Figure 3.3. Although there were no deposits at the two lower temperatures, there was a slight increase in the amount of asphaltenes in the liquid. As these asphaltenes further react at higher temperatures (390 °C) to form deposits, their content in the liquid dropped. However, the combined amount of asphaltenes (heptane insolubles in the liquid) and deposits increased.

The decrease in the rate of deposition between 56 and 24 h can be related to the decrease in concentration of foulants in the crude oil as

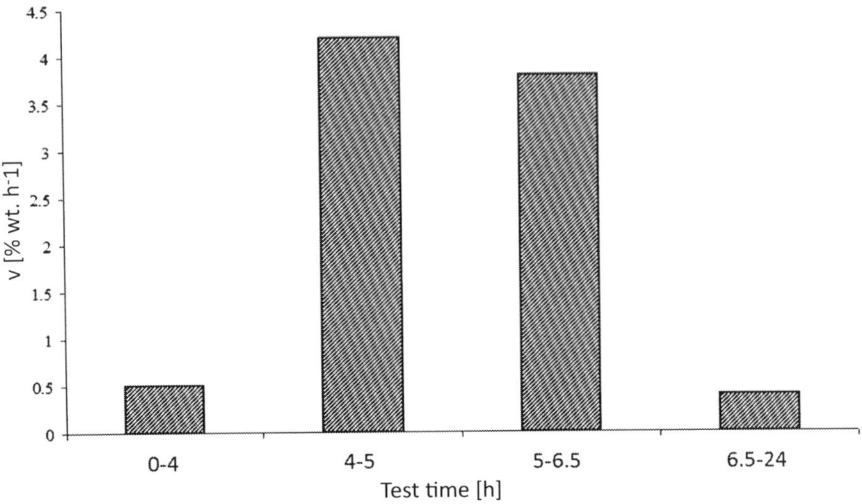

Figure 3.2 Average fouling rates (as wt% of the crude per hour) as a function of residence time in the N_2-pressurized microbomb for experiments carried out with Crude B at 390 °C.

Table 3.3 Heptane insoluble present in the liquid phase (HIL) and deposits (both as wt% of the original Crude B) for runs at 390 °C in the N_2-pressurized microbomb

Holding time (h)	HIL (wt%)	Deposits (wt%)	HIL + deposits (wt%)
4	7.7	2	9.7
5	5.5	6	11.5
6.5	2.2	12	14.2
24	0.6	19	19.6

they are removed from the liquid as deposits. The analysis (not shown) of the remaining liquid following the formation of deposits indicated a clear shift toward lighter products with smaller molecular weight distributions. As this decrease is a consequence of the depletion of fouling precursors in the crude, it would not be observable in continuous operation, for example, in the refinery itself, as the fraction of crude that forms deposits is very small and there is a continuous supply of fresh feed.

A comparison between the amounts of deposits obtained from different crudes is presented in Table 3.4. These experiments were carried out at 320 °C and 24 h. It is clear that deposition values are strongly dependent on the oil employed. Crude D, which is the lighter crude of these three, did not produce any deposits in this time scale, while fouling was observed in experiments with the other two crudes.

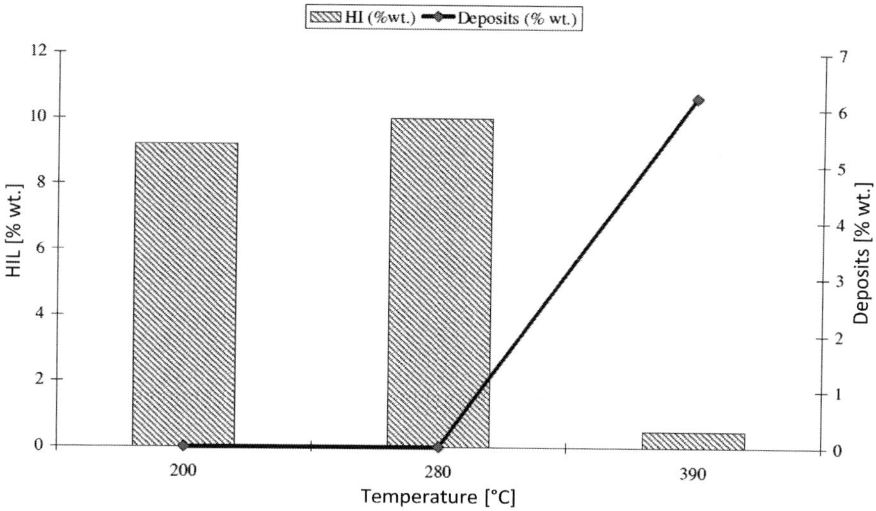

Figure 3.3 Comparison between deposits and heptane insoluble (HI) both as wt% of the original crude recovered from 6.5-h-long experiments in the N_2-pressurized microbomb.

Table 3.4 Deposits formed (as wt% of the original Crude B) as a function of the crude oil used in experiments carried out in the N_2-pressurized microbomb at 320 °C and 24 h

Crude oil	Deposit (wt%)
Crude D	0
Crude B	1
Oriente	2

It is usually assumed that the asphaltenes are the fraction that contributes the most to formation of deposits. The contribution from the deasphalted (heptane-soluble fraction) to deposition was further investigated in experiments conducted with the heptane-soluble fraction from Crude B as feed. Experiments were carried out with this fraction at temperatures between 280 and 390 °C.

As Table 3.5 shows, the heptane solubles did not form deposits at 280 °C. However, an interesting trend is observed at 320 °C and above as the heptane-soluble fraction was observed to form more deposits than the whole Crude B under the same conditions. This finding suggests that the combined presence of asphaltenes and heptane solubles in the feed slows down deposition, which could be linked to stabilization of fouling

Table 3.5 Deposits formed (as wt% of the original feed) as a function of temperature and holding time in experiments carried out in the N_2-pressurized microbomb apparatus with Crude B and its heptane-soluble fraction as feed

Sample	Holding time (h)	Temperature (°C)	Weight of deposits (%)
Crude B	24	390	19
		320	1
		280	0
Crude B heptane solubles	6.5	390	3.4
	24	390	26
		320	10
		280	0

Table 3.6 Heptane-insoluble present in the liquid phase (HIL) and deposits (both as wt% of the original Crude B) for runs at 390 °C in the N_2-pressurized microbomb with Crude B heptane-soluble fraction as feed

Holding time (h)	HIL (wt%)	Deposits (wt%)	HIL + deposits (wt%)
6.5	1.9	3.4	5.3
24	1.6	26	27.6

precursors in the crude. Moreover, analysis of the liquid (Table 3.6) showed that the heptane-soluble fraction formed asphaltenes (heptane insolubles), which further reacted and deposited.

3.1.2 The Closed Liquid System Microbomb and Its Applications in Fouling Studies

A second fouling apparatus was designed and commissioned with the objective of eliminating the need for an inert gas while maintaining the main characteristics of these experiments, namely the use of relatively small amounts of sample and the ability to expose the crude to temperatures in excess of 400 °C in the absence of flow. The applicability of this closed liquid system microbomb in fouling research will be demonstrated in work with crude oil blends as a quick test for establishing crude oil compatibility and incompatibility regions.

3.1.2.1 Equipment Description and Experimental Procedure

The apparatus is shown in Figure 3.4. To minimize or eliminate the liquid-free surface, the test section of the microbomb is entirely filled with sample, and the oil pressure is controlled by a downstream needle valve and monitored by a pressure transducer, instead of being determined by the pressure

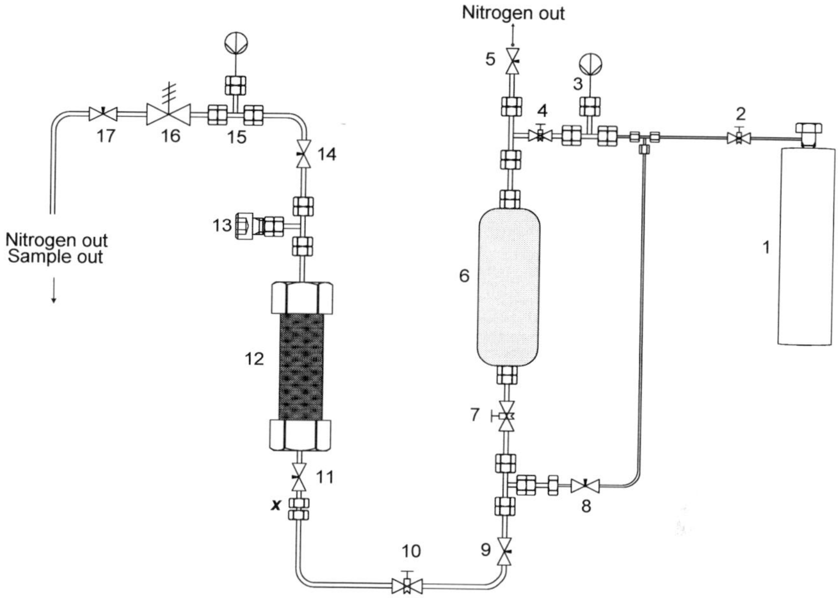

Figure 3.4 Schematic diagram of the closed liquid system microbomb. [1] Nitrogen gas cylinder; [2] ball valve; [3] pressure transducer; [4] ball valve; [5] needle valve; [6] sample cylinder; [7] ball valve; [8] needle valve; [9] needle valve; [10] ball valve; [11] needle valve; [12] test section; [13] thermocouple assembly; [14] needle valve; [15] pressure transducer; [16] pressure relief valve; [17] needle valve.

of an external inert gas. This section presents an overview of a closed liquid system microbomb (Rostani, 2014), and a case study of its application into studying the fouling potential of crude oil blends is presented in Section 3.1.2.2.

The test section is made of a ½ in outer diameter stainless-steel tube with a thickness of 0.048 in. The length of the test section tube is 1050 mm, allowing this microbomb to accommodate 8 cm^3 of sample. The test section can withstand pressures up to 197 bar at a temperature of 538 °C, a range much higher than that needed in fouling studies. Both ends of the test section are connected to ¼ in outer diameter stainless-steel tube, and a K-type thermocouple is positioned inside the reactor from the top end. Another K-type thermocouple is attached to the outer wall of the test section, which measures the wall temperature of the tube. The temperature is controlled by a standard heating tape controller. The heating tape covers the whole part of the tube including the upper part of the thermocouple assembly and the connecting tube leading to the needle valve attached at the bottom of the test section. Products are collected at the end of each experiment and analyzed following a similar procedure to that described in Section 3.1.1.1.

3.1.2.2 Crude Incompatibility Studies Using the Closed Liquid System Microbomb

The fouling test rig described in the previous section was employed in the determination of fouling from blends of Crudes A and D, whose properties are displayed in Table 3.1. The experimental results were compared to the predictions for these two crudes from the oil compatibility model developed by Wiehe et al. (2000). Next, the compatibility and incompatibility regions between these two crudes according to the model are calculated. The experimental results will be presented later in this section.

Establishing the compatible region of the blend requires the experimental determination of two parameters, the insolubility number, I_N, which is related to the minimum volume of toluene that avoids asphaltene flocculation in an infinitely diluted oil solution in a heptane–toluene mixture, and the heptane dilution number, H_D, which is the minimum volume of heptane that causes asphaltene precipitation in the crude oil in the absence of toluene (Wiehe, 2000). These incompatibility parameters are determined through solubility tests as follows. The crude was mixed with a test liquid consisting of heptane–toluene mixtures with different compositions. For a given oil to test liquid ratio, the minimum amount of heptane in the test liquid that led to deposition was determined. Precipitation of asphaltenes was observed by optical microscopy, where the presence of asphaltene agglomerates could be visualized. This onset of asphaltene precipitation as a function of the amount of heptane in the test liquid was determined for 20 and 50 vol% of oil in the oil and test liquid solution. H_D was measured directly through heptane dilution tests that quantify the maximum amount of heptane, V_H, which can be added to the crude oil without precipitating asphaltenes. This procedure added a third point (H_D, at the x-axis intercept) to each of the curves shown in Figure 3.5. I_N for each crude was obtained by extrapolation of the relevant line as the y-intersect.

The solubility parameters determined above were used to calculate the solubility blending number, S_{BN}, for each crude according to the equation:

$$S_{BN} = I_N\left[1 + \frac{100}{H_D}\right] = I_N\left[1 + \frac{V_H}{V_{oil}}\right]_{V_T=0}$$

The summary of results for Crude A, B, and D is presented in Table 3.7.

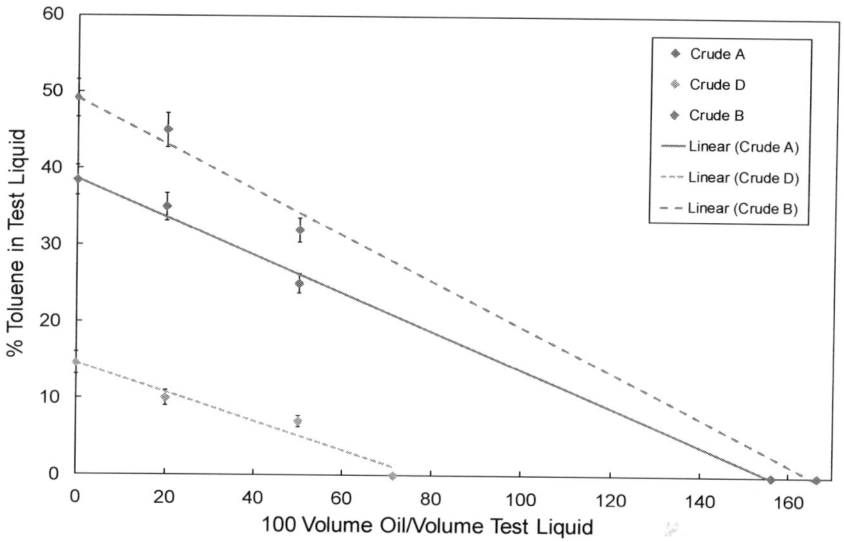

Figure 3.5 Determination of I_N (insolubility number) at the *y*-intercept and H_D (heptane dilution value) at the *x*-intercept for each crude (A, B, and D).

Table 3.7 Heptane dilution value (H_D), insolubility number (I_N), and solubility blending number (S_{BN}) for Crudes A, B, and D

Crude oil	H_D	I_N	S_{BN}
A	156	38	62
B	167	49	78
D	71	15	36

The solubility blending number of the mixture, $S_{BN,mix}$, can be calculated from the volumetric average of the oil mixture:

$$S_{BN,mix} = \frac{V_1 S_{BN1} + V_2 S_{BN2}}{V_1 + V_2}$$

where V_1 and V_2 are the volumes of crude oil 1 and 2, respectively, and S_{BN1} and S_{BN2} their solubility blending numbers. A crude oil blend is predicted as incompatible by the model if the $S_{BN,mix}$ is smaller than the maximum I_N between the two crudes:

$$S_{BN,mix} < I_{N,max}$$

The incompatibility regions between Crude B and Crude D were the subject of a previous study in the literature in which the predictions of the model were successfully correlated to experimentally determined regions of incompatibility measured by ellipsometry (Álvarez, 2012) for these two crude oils. The present work aims to determine experimentally the incompatibility region between Crude A and Crude D using the microbomb described in Section 3.1.2.1.

As the insolubility number of Crude A (38) is greater than the solubility blending number of Crude D (36), the existence of an incompatibility region is clear as the blend becomes richer in Crude D. If Crude D is progressively added to Crude A, the mixture remains compatible up to the point where $S_{BN,mix}$ equals the I_N value for Crude A. Figure 3.6 shows the compatibility and incompatibility regions between Crude A and Crude D. The incompatibility region occurs at Crude D fractions higher than 86 vol%. Asphaltene precipitation is favored if the composition of the blend falls within this region. This precipitation is likely to trigger the formation of deposits when the blend is heated.

Once the compatibility behavior predicted by Wiehe's model was determined, experiments were carried out in the closed liquid system microbomb to establish if the formation of deposits correlated with these regions. The ratio between Crude A and Crude D in the blend was changed and the amount of deposits was quantified in each case. The Crude A to Crude D

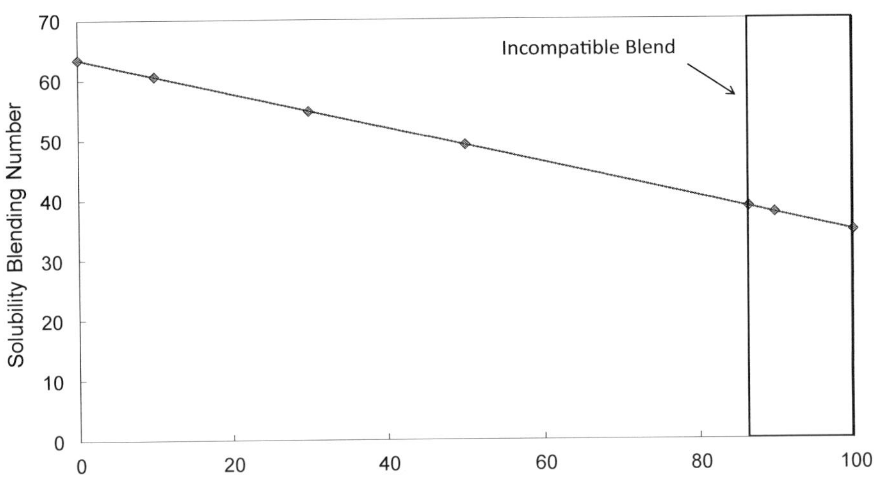

Figure 3.6 Representation of blending behavior of Crude A and Crude D: solubility blending number of the mixture as a function of volume of Crude D in the blend. The area representing the incompatibility region of the blend is highlighted.

ratios employed were 50:50, 40:60, 30:70, 20:80, 10:90, and 5:95. The former four fall within the region where the blend is predicted to be compatible, whereas the blends containing 90% and 95% Crude D are incompatible. A 30 mL total sample was prepared for each blend. Before the sample was fed into the test section, the mixture was stirred for 20 min to ensure good mixing of both crude oils. All experiments were performed with a wall temperature of 350 °C and a holding time of 48 h at an operating pressure of 10 bar. During blend preparation, Crude D was added to Crude A to avoid momentarily forming blends with less than 85% Crude D, which would fall within the incompatible region and may distort the results.

The amount of deposits produced for each oil ratio is presented in Figure 3.7. Values shown are averages of two runs. It is observed that the amounts of deposits recovered for oil ratios of 50:50 to 30:70 were similar to each other and relatively low, as expected for blends whose composition is within the compatible region. These amounts are relatively lower than the amount produced for neat Crude A (not shown) whereas no deposits were formed for neat Crude D. An increase in the amount of deposits is observed as the blend composition approached that of the incompatible blends. The amount of deposits had a further significant increase once the ratios were within the incompatible blending region, 90% and 95% Crude D. The higher deposit formation was observed for the 10:90 blending ratio. The value decreases slightly at the 5:95 ratio, probably due to the decreasing

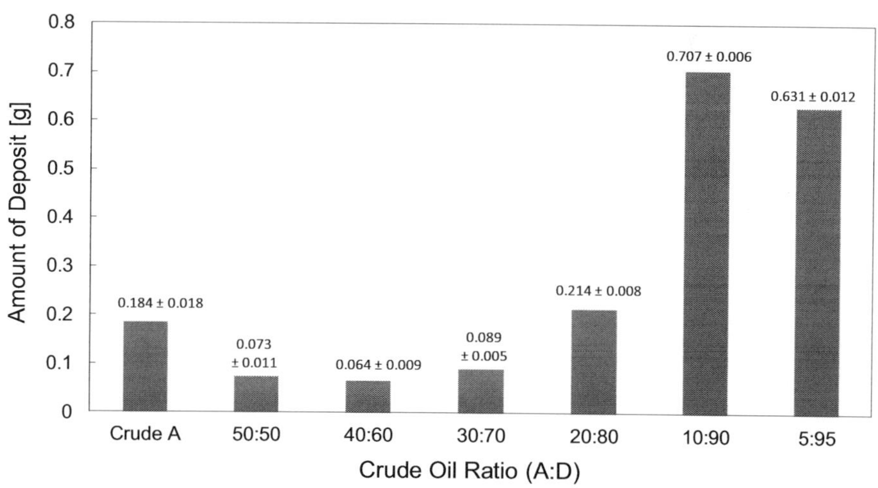

Figure 3.7 Amount of deposit produced for each Crude A:Crude D ratio in experiments carried out in the closed liquid system microbomb at 350 °C and 48 h.

amount of Crude A asphaltenes in the solution, which reduced the concentration of fouling precursors. These experiments show a good correlation between that predicted by Wehie's compatibility model and that measured in a fouling test in a microbomb.

3.1.3 Summary on the Use of Small-Scale Experiments

The use laboratory-scale fouling testing configurations has been shown as a valuable tool for studying the fouling propensity of several crude oils and their mixtures. Although these laboratory-scale experiments do not attempt to recreate refinery operating conditions, they are useful tests that may provide a benchmark for comparison between crudes, to establish the chemical behavior of a given crude as a function of its temperature–time history and to measure incompatibility regions between two or more crudes. The next section will describe a compact stirred batch system used to produce fouling deposits with flow conditions.

3.2 BATCH SYSTEM: STIRRED CELL AT THE UNIVERSITY OF BATH

M. Yang, B.D. Crittenden

The design of the batch stirred cell system at the University of Bath (Figure 3.8) closely follows that designed and operated by Eaton and Lux (1984). The cell itself comprises a pressure vessel made in-house from a block of 304 stainless steel and it is fitted with a top flange. The maximum operational pressure and temperature are 30 bar and 450 °C, respectively. Pressure is maintained by means of a nitrogen blanket. Air can be admitted to sparge the crude oil in advance of a fouling run if so desired. The base of the stainless-steel vessel houses an upward pointing test probe that is internally heated using a cartridge heater. The heat flux to the crude oil is fixed by controlling the power to the cartridge heater (Figure 3.9). About 1 L of the chosen crude oil is agitated by a downward facing cylindrical stirrer that is mounted coaxially with the test probe and driven by an electric motor via a magnetic drive. The stirrer speed can be adjusted by means of a variable frequency drive.

Initial heating to the vessel and its contents is provided by external band heaters. Rejection of the energy at the same rate as it is inputted via the cartridge and band heaters is facilitated by an internal cooling coil which uses a nonfouling fluid (Paratherm®). Heat is rejected from the Paratherm® to the atmosphere by means of an air-cooled heat exchanger. The batch stirred

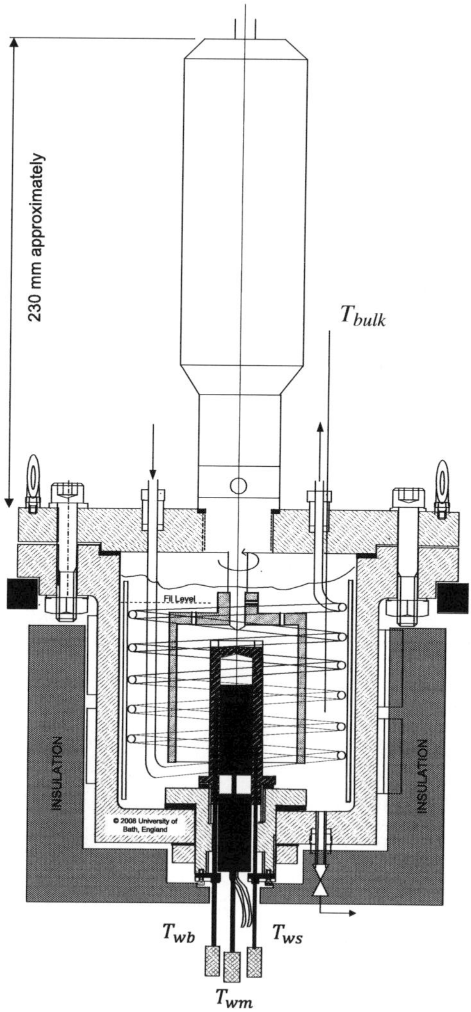

Figure 3.8 University of Bath batch stirred cell. *Reproduced with permission from Young et al., Characterization of crude oils and their fouling deposits using a batch stirred cell system. Heat Transfer Eng. 2011, 32 (3–4), 216–227. Copyright (2011) Taylor & Francis.*

vessel is fitted with a pressure relief valve and there is a single thermocouple to measure the crude oil bulk temperature which is controlled by regulating the flow rate of the fluid in the cooling coil. Control is affected by means of a pump driven by a variable frequency driver which is controlled by a PID controller. A constant voltage power source is used to supply electrical power to both the cartridge heater and the stirrer motor, thereby helping to minimize fluctuations in power input caused by random variations in the mains power voltage.

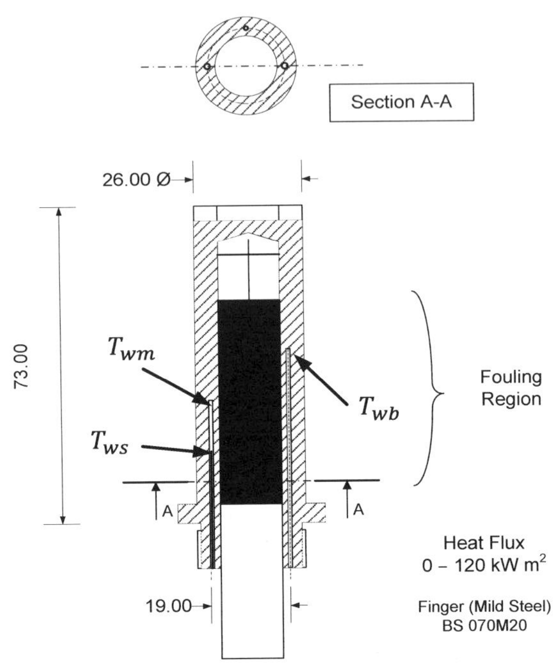

Figure 3.9 The heated probe. *Reproduced with permission from Young et al., Characterization of crude oils and their fouling deposits using a batch stirred cell system. Heat Transfer Eng. 2011, 32 (3–4), 216–227. Copyright (2011) Taylor & Francis.*

The standard test probe that fits tightly over, and is heated by, the cartridge heater is manufactured from low carbon steel with a material specification similar to that of a standard A179 heat exchanger tube. However, any other type of metal can be used in order to study the effect of metallurgy. A coherent laser light scanning technique (Scantron Industrial Products, Taunton, England) is used to obtain the initial surface roughness e^0. Furthermore, the probe can be modified with a wire nest attachment, as seen in Figure 3.10, in order to make a crude representation of a wire tube insert such as hiTRAN®. The probe can also be designed as a package consisting of a fixed probe body with an exchangeable sleeve, as sketched in Figure 3.10. This design may significantly minimize the cost in manufacturing of a number of probes of different materials and different surface characteristics, and provides a simple platform for applying coatings to the surface. In addition, this design offers flexibility in changing the material of the sleeves.

Three small diameter type J thermocouples (supplied by TC Thermocouples Ltd, UK) are located at various heights within the walls of the test probe. These thermocouples, labeled T_{wb}, T_{wm}, and T_{ws} in Figures 3.8 and 3.9, are

Figure 3.10 Probe modifications: probe with wire attachment (a); probe sleeve (b). *Reproduced with permission from Yang et al.,* Applied Thermal Engineering, *2013, 54, 516–520. Copyright (2013) Elsevier.*

used to follow the progress of the fouling at different axial locations. The stirrer speed is monitored using a Hall-effect sensor and data are acquired and interpreted using Labview software.

The vessel is filled initially with the crude oil sample to be tested. The oil is then usually sparged with air at 1 bar while being stirred at the desired speed for 30 min following which the vessel is purged and blanketed with nitrogen at 20 bar. The vessel and its contents are then brought up to the bulk operating temperature by means of the external band heaters; the heating rate is normally 15 °C min^{-1}. When the oil is at 200 °C the stirrer is started and heating with the band heaters is continued until the desired bulk operating temperature (typically around 260 °C) is reached. Power is then applied to the cartridge heater either at the desired heat flux or to create the desired surface temperature, as appropriate. Also at this point the cooling circuit is started to develop a steady-state heat transfer rate in the apparatus. Control of steady-state heat transfer conditions is achieved by a bespoke combination of band heating and cooling coil operation under traditional PID control.

A single fouling test commonly lasts for less than 10 h, at the end of which the power to the cartridge heater is turned off, the stirrer speed is reduced, and the vessel is allowed to cool down by maintaining flow through the cooling coil. This arrangement should prevent the fouling deposit from being removed during shutdown so that the fouling run could be resumed the next day if required. When a linear fouling rate has been observed for a sufficiently long period then the fouling run is terminated. The oil is drained and the test probe and cartridge heater assembly are removed for inspection and deposit characterization. If a previously fouled probe is to be used for a subsequent run it is thoroughly cleaned by brushing with ethanol, kerosene, *para*-xylene, and finally with acetone before it is reused.

Experimental errors are due to a number of variations and uncertainties in the operational conditions including the stirrer speed and heater power input, as well as errors in thermocouple readings (in the range of 1–2 °C). In practice, however, interest centers on the accuracy of the initial fouling rate and not on the accuracy of an absolute fouling resistance at any particular time. Fouling rates are always found to be linear after any induction period (Young et al., 2011). Accordingly, any local variations or scatter of experimental results are smoothed out by linear regression of data over a period of several hours. Table 3.8 lists typical operating conditions.

A particular advantage of the batch stirred cell is that fouling rates are very easy to determine under the closely controlled operating conditions. The fouling resistance R_f is simply calculated from Equation (3.1) (Crittenden et al., 1987a; Bennett et al., 2009):

$$R_f = \left(\frac{T_s - T_b}{q''}\right)_t - \left(\frac{T_s - T_b}{q''}\right)_{t=0} \tag{3.1}$$

In this equation, q'' is the heat flux and T_s and T_b are the local surface and bulk temperatures, respectively, at time t and time $t = 0$, as appropriate. The advantage of using the batch stirred cell now becomes evident since the use of Equation (3.1) avoids the need to calculate film heat transfer coefficients. Since the apparatus is operated at constant q and constant T_b then an additional advantage arises because Equation (3.1) becomes simplified as follows:

$$R_f = \left(\frac{T_{st} - T_{s0}}{q''}\right)_t \tag{3.2}$$

Here, T_{st} and T_{s0} are the surface temperatures at time t and zero, respectively. Of course, in this simplification, the assumption is made that the film

Table 3.8 Typical operating conditions

Operational parameter	Units	Range
Bulk temperature	[°C]	240–280
Average heat flux	[kW m^{-2}]	85–122
Surface temperature	[°C]	345–420
Stirring speed	[rpm]	100–400
Pressure	[Bar]	24–28

heat transfer coefficient does not alter as fouling proceeds. Surface temperature measurements are made using thermocouples T_{wb}, T_{wm}, and T_{ws} located within the probe wall (Figure 3.9). CFD is used to convert these temperatures to those at the actual interface between the metal heat transfer surface and the crude oil (Young et al., 2011). Provided that the assumption about the film heat transfer coefficient not altering as fouling proceeds is valid, then the initial interfacial temperature will remain constant throughout a fouling run (Young et al., 2011). This argument effectively assumes that the roughness of the deposit does not alter during an experiment and is the same as the roughness of the clean heat transfer surface. In practice, the fouling deposit thickness is usually less than 20 μm which supports the view that the roughness change and its influence would indeed be minimal. Hence, from Equation (3.2), the instantaneous rate of fouling is now proportional only to the instantaneous rate at which the local surface temperature changes with time. Hence, any errors in the fouling rate are dependent only on the errors in the heat flux and in the rate of change of surface temperature with time.

In summary, the advantages of the batch stirred cell system design are:

1. Only a small volume of crude oil is required for fouling tests
2. It is quick and easy to change crude oil samples
3. It is quick and easy to alter operational conditions
4. It is quick and easy to modify the fouling surface
5. There are well-defined temperature and velocity fields which can be modeled with the help of CFD simulation
6. It is quick and easy to disassemble the cell in order to gain access to the fouling deposits

3.2.1 CFD Simulation

Both the velocity and the temperature fields in the stirred cell can be described quantitatively using CFD. CFD can also provide the surface shear stress distribution. This is important because surface shear stress, which is impossible to measure experimentally within the batch stirred cell, is at least as important as surface temperature in understanding how process parameters affect local fouling rates. COMSOL Multiphysics software is used at the University of Bath to determine such distributions of temperature, flow, and shear stress (Yang et al., 2009) and examples of velocity and temperature fields for crude oil are shown in Figures 3.11 and 3.12, respectively.

Validation of the CFD simulation results has been provided in part by the fair agreement between the experimental measurements and the simulation

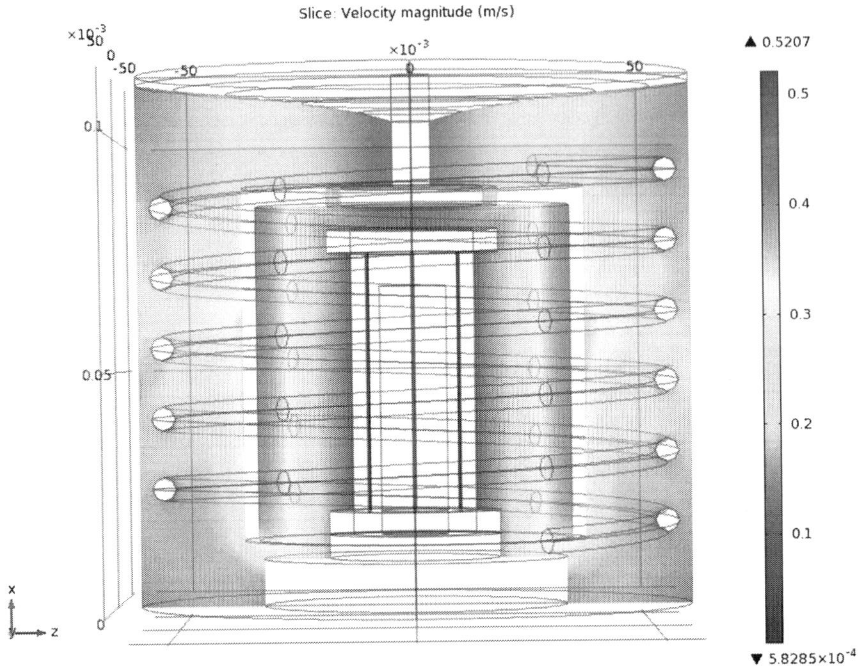

Figure 3.11 Velocity field. Bulk temperature 258 °C; stirring speed 200 rpm, heat flux 79 kW m^{-2}. *Reproduced with permission from Yang et al.,* Applied Thermal Engineering, *2013, 54, 516–520. Copyright (2013) Elsevier.*

predictions of temperature (Yang et al., 2009). Nonetheless, it is not possible to validate the accuracy of the CFD simulation of the shear stress due to lack of a viable technique to measure the local surface shear stress experimentally.

3.2.2 Equivalent Reynolds Number and Velocity

A common criticism leveled at the batch stirred cell approach is that the flow fields do not represent those found in industrial heat exchange equipment. To counter this criticism, the concept of equivalent Reynolds number was developed in order for fouling rate data generated using the batch stirred cell to be applied to other heat exchanger geometries, whether these be shell-and-tube or plate configurations, etc. (Yang et al., 2009b; Young et al., 2011). The equivalent Reynolds number is defined to be the Reynolds number in a standard round tube that gives the same shear stress as that which occurs over the stirred cell test probe surface. The concept was developed further for use in situations in which round tubes are fitted with heat

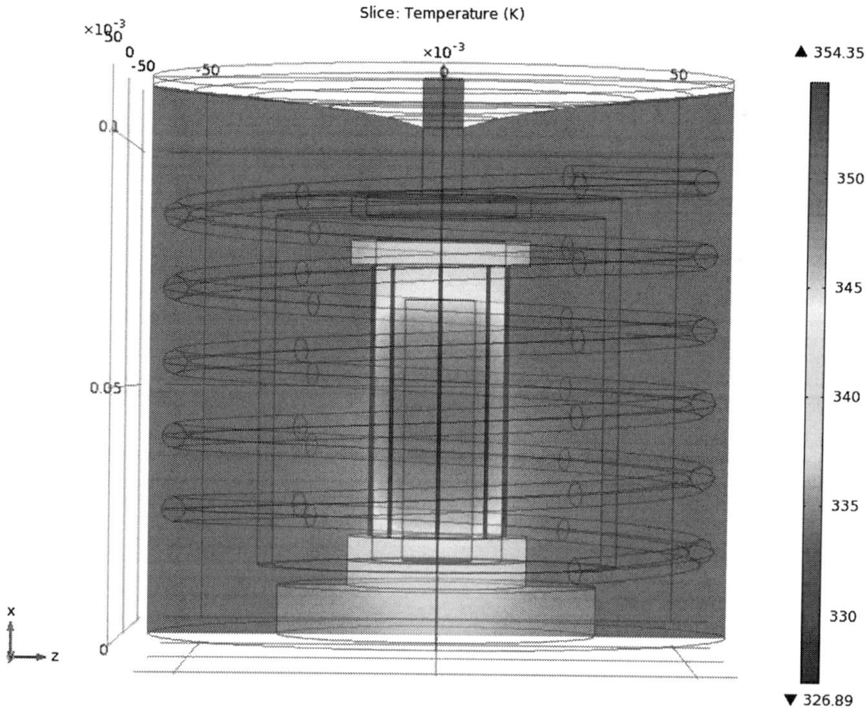

Figure 3.12 Temperature field. Bulk temperature 258 °C; stirring speed 200 rpm, heat flux 79 kW m^{-2}. *Reproduced with permission from Yang et al.,* Applied Thermal Engineering, *2013, 54, 516–520. Copyright (2013) Elsevier.*

transfer inserts such as hiTRAN® (Yang and Crittenden, 2012). Shear stress data at a given linear velocity, or at a given stirrer speed, are generated by CFD simulation and a correlation is established to relate the operational parameters to an equivalent Reynolds number (or velocity). For the fluid flow in the batch stirred cell, example average surface shear stresses around plain and wired probes (Figure 3.10) are compared in Figure 3.13 from which the equivalent Reynolds numbers for this type of application can be generated.

3.2.3 Fouling Rates

Early data on crude oil fouling research using a batch stirred cell apparatus were given by Eaton and Lux (1984). Two experimental methods were described. In the first, the net weight of deposit during a given time was used as a measure of the fouling severity. Although this provided a simple and a direct measurement of fouling, very little information was obtained about the in situ fouling process. In the second method, the fouling factor

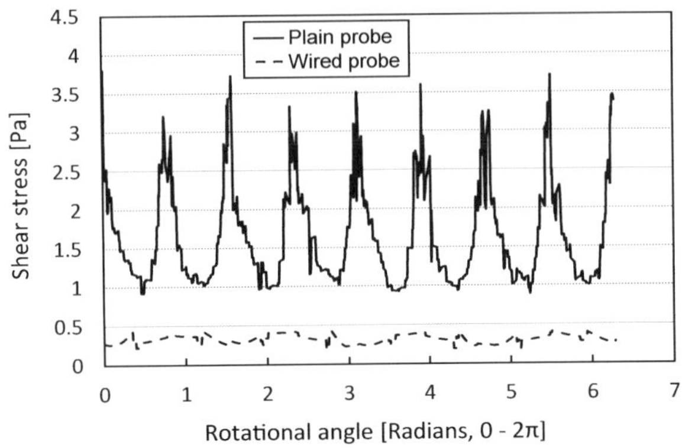

Figure 3.13 Shear stress distribution around the probe surface at the mid vertical position with respect to the rotational angle from zero to 2π. Bulk temperature is 258 °C; stirring speed 200 rpm; heat flux 79 kW m^{-2}. The continuous line shows the shear stress around the plain probe at $z = 0.04$ m and the dashed line shows the shear stress around the wired probe at the same location ($z = 0.04$ m). *Reproduced with permission from Yang et al.,* Applied Thermal Engineering, 2013, 54, 516–520. Copyright (2013) Elsevier.

or severity was measured continuously with time through in situ measurement of the change in heat transfer, and hence in surface temperature, caused by the fouling deposit on the probe surface. This provided an indirect measurement of the deposit formation with time. Most results were obtained using the second method and allowed the effects of resins, asphaltenes, temperature, and other parameters on crude oil fouling to be investigated. The research work on crude oil fouling using the batch stirred cell at the University of Bath closely follows the second method used by Eaton and Lux (1984).

As an example, three crude oil samples (Table 3.9) have been tested and their fouling rates compared. Crudes A and B are untreated raw blends of sour crudes blended in the supplier's laboratory whereas Crude DS is a desalted blend. Analyses shown in Table 3.9 were provided by LGC Ltd (Teddington, UK). Given the advantages of the short experiments, small sample volumes, and easy alterations to operational conditions, the batch stirred cell has not only been used to determine the effects of principal operational parameters such as surface temperature, bulk temperature, and surface shear stress on fouling rates, but also to determine some deposit and fluid characteristics. The cell has also been used to obtain data for investigation of fouling induction periods (Yang et al., 2012). A typical fouling curve is shown in Young et al. (2011).

Table 3.9 Properties of crude oil blends tested

	Crude A	Crude B	Crude DS
API	27.5	24.43	30.92
Viscosity (cst) @ 80 °C	15	8.8	2.9
Viscosity (cst) @ 260 °C	1.74	0.745	0.58
Total sulfur (%wt)	2.82	3.65	2.1
Iron (ppm)	4	4	1
Nickel (ppm)	42	25	18
Vanadium (ppm)	226	52	55
Saturates (%)	28.36	21.68	29.2
Aromatics (%)	56.87	65.63	63.65
Resins (%)	6.8	5.68	4.28
Asphaltenes (%; IP143)	8	7	2.87
CII	0.56	0.4	0.48

An example of how the surface temperature increases as fouling proceeds is shown in Young et al. (2011). It was concluded that the fouling rate is increased with an increase in surface temperature and decreased with an increase in stirrer speed (and hence in surface shear stress).

Induction periods were always observed when the fouling test was started using a new or a thoroughly cleaned probe surface. For a given surface, the induction period has been found to be a strong function of surface temperature. This is shown, for example, in Figure 3.14 for Crude B (Table 3.9).

Further analysis of the effects of operational conditions resulted in the creation of a generic model of fouling induction periods (Yang et al., 2012). This model, reported in more detail in Section 2.3.1 of this book, quantitatively describes the influence of the surface temperature on the length of the induction period for crude oil (among other fluids). The model also describes in a semi-quantitative manner the influence of velocity on the induction period.

3.2.4 Deposit Analysis

Figure 3.15 shows a fouled test probe from which samples of deposit were easily removed for analysis. Example analyses shown in Table 3.10 reveal that deposits contain significant quantities of both sulfur and iron, suggesting the formation of iron sulfide during the fouling process. This is in line with results from Wang and Watkinson (2011). The atomic H:C ratios were almost identical for the two oils at 1.37 for Crude A and 1.36 for Crude B. These values are within the range of 1.0–1.5 for asphaltenes as reported in the literature (Speight, 1980).

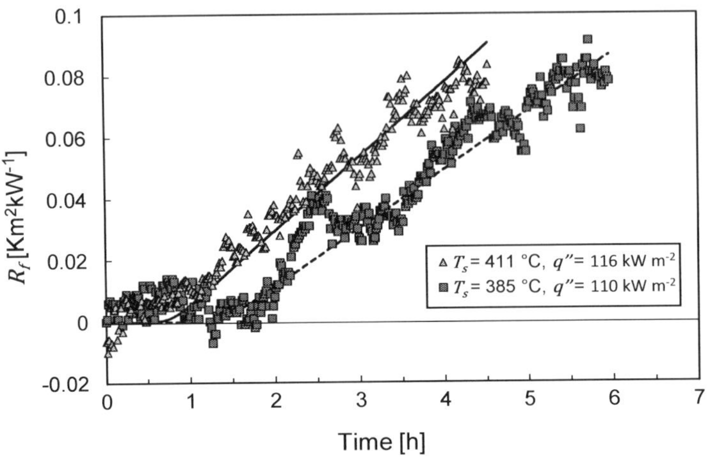

Figure 3.14 Fouling curves and the effect of temperature on induction period for Crude B. Squares: surface temperature 385 °C, average heat flux 110 kW m^{-2}; triangles: surface temperature 411 °C, average heat flux 116 kW m^{-2}; lines: model fits. *Reproduced with permission from Yang et al., Modelling fouling induction periods. Int. J. Thermal Sciences, 2012, 51: 175–183. Copyright (2012) Elsevier.*

Figure 3.15 Fouling deposit on a heated probe. *Reproduced with permission from Young et al., Characterization of crude oils and their fouling deposits using a batch stirred cell system. Heat Transfer Eng. 2011, 32 (3–4), 216–227. Copyright (2011) Taylor & Francis.*

Table 3.10 Elemental analyses of deposits (conducted by Medac Ltd, Egham, UK)

Element	C	H	N	S	Fe
Crude A	36.12	4.11	<0.1	16.61	13.71
Crude B	53.98	6.11	0.48	13.82	7.53

3.2.5 Surface Profiling

Figure 3.15 shows that the deposit is formed somewhat unevenly on the probe surface. Young et al. (2011) show typical results for the actual deposit thickness profile over the length of the probe found by using the Proscan laser microscan technique. The scan shows that fouling is more severe in the middle of the probe, but that there is less deposition toward both the top and the bottom of the probe's heated section.

To interpret this behavior, the deposit thickness profile has been compared with the CFD-predicted temperature profile as shown in Figure 3.16 in which it can be seen that the fouling layer thickness profile over the probe surface is strikingly similar in form to the temperature profile, with both their maxima occurring near the middle of the heated probe surface. The CFD simulations confirm that the shear stress over the probe surface where the fouling zone is located is relatively constant and hence the fouling profile is solely determined by the surface temperature. Accordingly, the similarity of the temperature and the fouling layer thickness profiles can be used to explain qualitatively the fouling profile observed in the experiment.

Since the fouling rate is constant (that is, the increase in fouling resistance with time is linear) and since all points on the surface have had their deposits grown over the same period of time, then the fouling rate at each position on the surface can be expressed in terms of the fouling layer thickness at that position. A correlation between the fouling deposit thickness (and hence the fouling rate) with the surface temperature at each position can now be obtained, leading to an Arrhenius plot, as shown in Figure 3.17. The Arrhenius

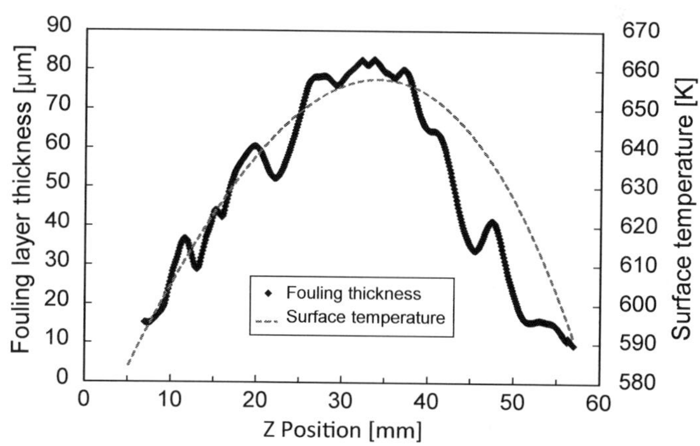

Figure 3.16 Fouling layer thickness by Proscan (symbols) and simulated surface temperature profiles by Comsol simulation (dashed line) (Yang et al., 2009b.).

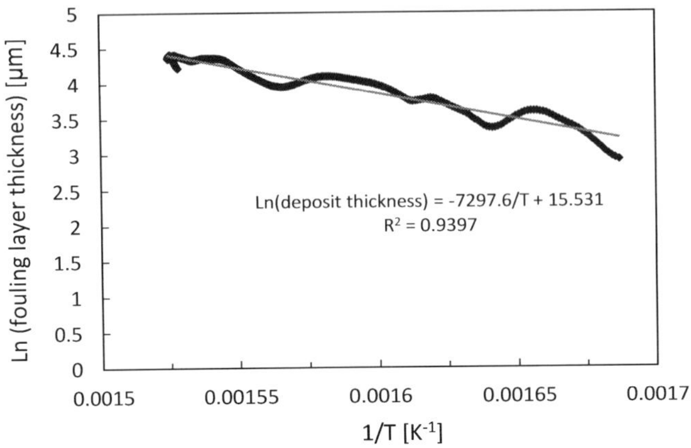

Figure 3.17 Arrhenius plot for Petronas B crude oil. Blue: log of measured fouling layer thickness in micrometers against 1/T in K; red line: linear plot. (For interpretation of the color in this figure legend, the reader is referred to the online version of this book) (Yang et al., 2009b).

plot obtained in this way is in a continuous form demonstrating a further particular advantage of using the batch stirred cell in conjunction with both Comsol CFD simulation for surface temperature and Proscan technique for deposit thickness measurement (Yang et al., 2009). Ordinarily, in all other types of fouling apparatus, including plant equipment, Arrhenius plots can only be constructed following a series of individual fouling experiments. The batch stirred cell, nevertheless, allows this to be done in a single experiment.

3.2.6 Effect of Shear Stress

Partial removal of a fouling deposit by temporarily increasing the surface shear stress by making a step change in the stirrer speed is shown in Young et al. (2011). In this experiment, the stirrer speed was first kept constant at 200 rpm so that the fouling rate remained constant over a period of around 3 h. The stirrer speed was then increased from 200 to 550 rpm such that the surface shear stress was increased to 2.5 Pa for a short period of time. The stirrer speed was then reduced back to 200 rpm such that the surface shear stress was reduced back to its original value of 0.75 Pa. Young et al. (2011) found that the fouling rate more or less resumes its original value, albeit at a somewhat lower value of the fouling resistance, indicating that some deposit had been removed. The fast response of the fouling resistance to the step change in stirrer speed also means that any reduction in fouling resistance when the surface shear stress is increased is most likely to be due to a shear removal phenomenon rather than to a mass transfer phenomenon. Nonetheless, it is possible that an underlying and slower mass transfer mechanism is operative as well (Crittenden et al., 1987a).

The effect that surface shear stress has on the removal of deposits was further studied in a different manner for Crude. After a significant fouling had occurred at a stirrer speed of 200 rpm, the stirrer speed was increased to 390 rpm and maintained constant at this value for a reasonably long period with all other conditions held constant. It was then possible to observe deposit removal as manifested by a negative fouling rate as shown in Figure 3.18.

During this experiment, the stirrer speed was monitored closely and found to be fairly constant as shown in Figure 3.18. Similarly the cartridge heating power, which could also affect the surface temperature, was found to be fairly constant during the experiment. When the stirrer speed was stepped down to 300 rpm, the fouling resistance continued to decrease with time although, as might be expected, the gradient was shallower than that at the stirrer speed of 390 rpm on the previous day. The negative fouling rate at 300 rpm was calculated to be -5.28×10^{-9} m^2 K J^{-1} and quantitatively reduced significantly from the value of -8.33×10^{-9} m^2 K J^{-1} at the stirrer speed of 390 rpm on the previous day. This experiment provides further evidence that by increasing the stirrer speed, and thereby increasing the wall shear stress, some of the fouling deposit can be removed from the

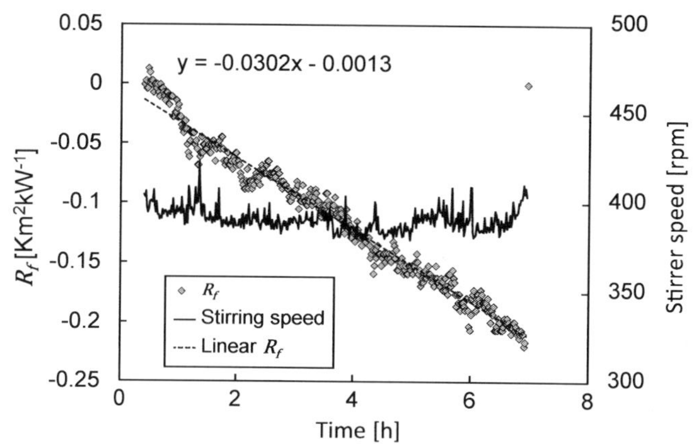

Figure 3.18 Negative fouling rate at increased stirrer speed. Symbols: fouling resistance; black line: stirrer speed. Bulk temperature: 250 °C; stirrer speed: 390 rpm; average heat flux: 37.1 kW m^{-2} (heater power: 175 W) (Yang et al., 2011).

surface by the action of fluid shear. This type of fouling test was repeated more recently for Crude DS, and the same trend was confirmed.

It should be noted that with respect to the data shown in Figure 3.18, the fouling resistance was arbitrarily set to zero when the change in stirrer speed was made. This is why the fouling resistance is shown to become negative when the fouling rate becomes negative. This procedure was adopted for simplicity because, of course, the change in stirrer speed means that the fouling resistance can now no longer be calculated from the situation under the initially clean conditions.

3.2.7 Threshold Conditions

In addition to normal experiments in which the fouling resistance increases with time, experiments on deposit removal (as described in Section 3.2.6) can be made by varying surface temperature at a constant surface shear stress, or by varying surface shear stress at a constant surface temperature. The fouling rate data can, accordingly, be plotted in such a way that it is possible to extrapolate back to find the point at which no fouling occurs. This allows the so-called fouling threshold to be determined (Panchal et al., 1997; Knudsen et al., 1997; Yang et al., 2011). Figure 3.19 shows such a plot. Normally, the fouling threshold would have to be determined by extrapolating plots of fouling rate versus surface temperature (at constant shear stress) back to the origin, that is, back to the point at which no fouling occurs (Panchal et al., 1997; Knudsen et al., 1997). In order to obtain reliable threshold data in

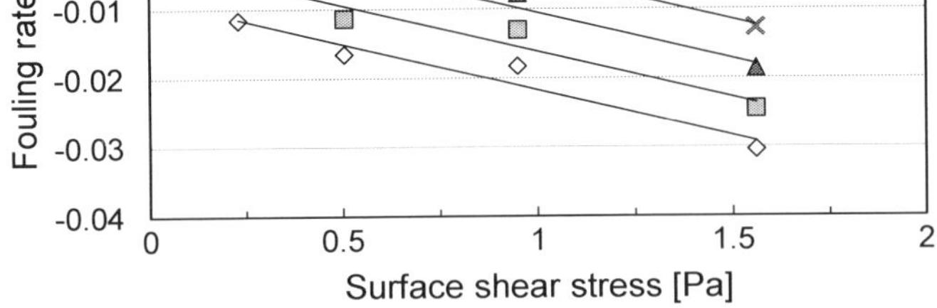

Figure 3.19 Fouling rate against surface shear stress for various initial surface temperatures (Yang et al., 2011).

this way, it would be necessary to carry out a large number of fouling runs, including several tests being run at very low fouling rates in order to improve the accuracy of locating the zero fouling conditions. Experiments in which very low fouling rates need to be studied would, of course, be extremely time-consuming. However, given that negative fouling rates can be found using the batch stirred cell by judicious choice of operating conditions after a deposit has been laid down on the surface, it now becomes possible to identify the fouling threshold conditions by interpolating, rather than by extrapolating, the plots of fouling rate–surface temperature (at constant shear stress). The points at which the fouling rate becomes equals to zero are thereby easily located in Figure 3.19.

By interpolating the plots similar to those shown in Figure 3.19 it is simple to obtain the point at which the fouling rate is equal to zero and, hence, to identify the set of threshold conditions. As an example, Figure 3.20 shows the identification of fouling threshold conditions for four surface temperatures for which the fouling rate–surface shear stress curve crosses from being a positive fouling rate to a negative one: 620 K (347 °C), 630 K (357 °C), 640 K (367 °C), and 650 K (377 °C). For this figure, the shear stress values were obtained by CFD simulations based on the physical properties of the crude oil, the stirrer speed, and the bulk temperature used in the experiments (Yang et al., 2009; Young et al., 2011). It is worth noting from Figure 3.19 that the gradients of the plots of fouling rate against shear stress seem to be the same, regardless of the surface temperature.

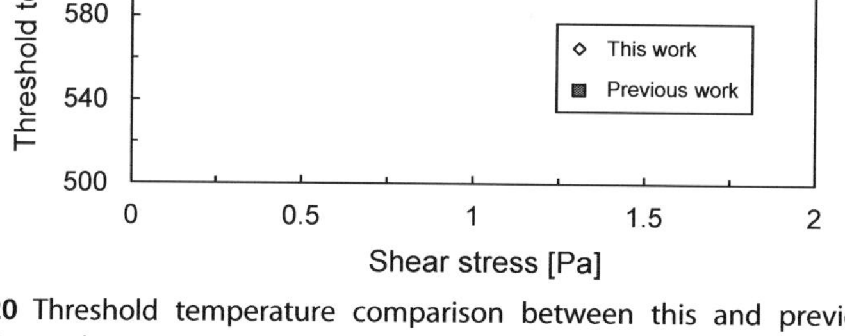

Figure 3.20 Threshold temperature comparison between this and previous work (Yang et al., 2011).

In Figure 3.20 the results from the new interpolation method just described are compared with the threshold conditions of Crude A obtained by extrapolating the data, as reported previously by Young et al. (2011). Agreement of the two methods is very good, despite the extrapolation method potentially suffering from inaccuracy as the fouling rate curves are extrapolated back to the origin. An argument might be raised that threshold conditions obtained using a batch stirred cell might not bear close resemblance to those which prevail in industrial situations. Nevertheless, it should be borne in mind that such threshold conditions would have a significant value in understanding the industrial situation provided that a fouling rate for a particular crude oil was determined solely by the surface temperature and the surface shear stress. That is, the fouling behaviors ought to be similar at the same temperature and under the same surface shear stress, i.e., at the same equivalent Reynolds number regardless of the geometries of the surface (Yang et al., 2009).

3.2.8 Enhanced Heat Transfer Surfaces

A typical fouling resistance curve when the wire nest shown in Figure 3.10 was fitted to the probe is shown in Figure 3.21. When compared with the fouling resistance data found by Young et al. (2011) for the plain probe, it can be seen that there is now much more scatter in the experimental data. This is most probably due to the random removal of the deposit because

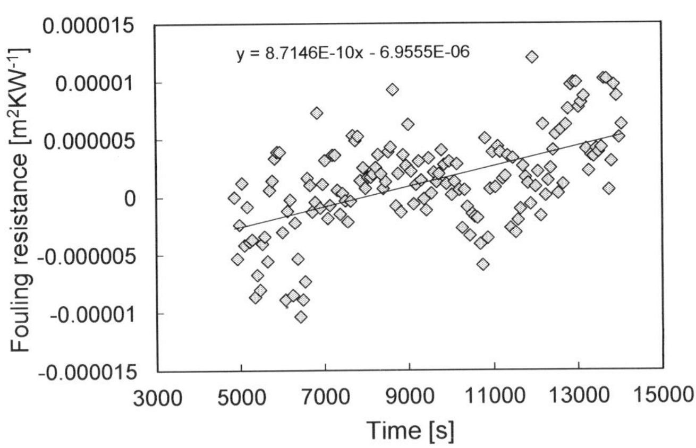

Figure 3.21 Fouling resistance curve on wired probe. Bulk temperature 258 °C; surface temperature 399 °C. Stirrer speed 160 rpm, heat flux 79 kW m^{-2}. *Reproduced with permission from Yang et al.,* Applied Thermal Engineering, *2013, 54, 516–520. Copyright (2013) Elsevier.*

of the higher shear force which is created on the wired probe at any given stirrer speed. The fouling rate for the wired probe, 8.7×10^{-10} m^2 K W^{-1}, was noticeably lower even at a higher surface temperature than that on the plain probe (3.0×10^{-9} m^2 K W^{-1}). The reduction in fouling rate is almost certainly due also to the increased surface shear stress when wires were fitted.

The shear stress around the plain probe is relatively even, whereas the shear stress around the wired probe (Figure 3.13) varies in a periodical manner. With respect to the flow direction, the shear stress reaches a significant peak just ahead of the wire and drops to a low value just behind the wire. The significant difference in shear stress between plain and wired probes clearly demonstrates that the wires significantly enhance the turbulence, resulting in higher shear forces that help to mitigate fouling.

As described earlier, the equivalent Reynolds number is defined as the value for the same fluid flowing in a round tube that gives the same surface shear stress. For the batch stirred cell, data of the average shear stress around plain and wired probes are generated using CFD simulations, taking into account the viscosity as a function of temperature for the crude oil (Yang et al., 2009; et al., 2011). Figure 3.22 shows the correlation between surface shear stress and stirrer speed. These data are then used to generate the correlation between the equivalent Reynolds number and the stirrer speed, as shown in Figure 3.23. It is worth noting that for any given equivalent Reynolds number and surface temperature, the fouling

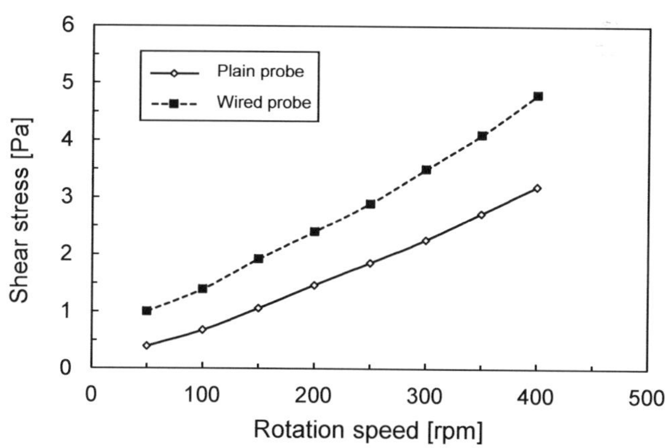

Figure 3.22 Average shear stress over the probe surface with (dashed line) and without wire (plain line). Bulk temperature 250 °C; average heat flux 79 kW m^{-2}. *Reproduced with permission from Yang et al.,* Applied Thermal Engineering, *2013, 54, 516–520. Copyright (2013) Elsevier.*

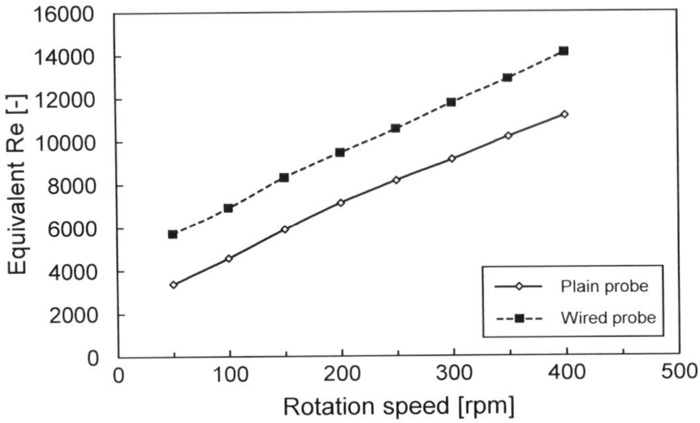

Figure 3.23 Equivalent Reynolds number for the crude fluid flow in the stirred cell with the plain probe (plain line) and with the wired probe (dashed line). Bulk temperature 250 °C; average heat flux 79 kW m^{-2}. *Reproduced with permission from Yang et al.,* Applied Thermal Engineering, *2013, 54, 516–520. Copyright (2013) Elsevier.*

behaviors should be similar regardless of the surface morphology such that the fouling results generated using the batch stirred cell with a wired probe can be compared directly with those generated using a tube fitted with inserts or even using just the stirred cell fitted with a plain probe.

Ebert and Panchal's fouling model (1997), shown as Equation (3.3), has been used to correlate the fouling rate data generated using the stirred cell fitted with both the wired probe and the plain probe (Yang et al., 2011).

$$\frac{dR_f}{dt} = \alpha Re^{\beta} \exp\left(\frac{-E_f}{R_g T_f}\right) - \gamma\tau \tag{3.3}$$

Here R_f is the fouling resistance (m^2 K J^{-1}), Re the Reynolds number, E_A the activation energy (J mol^{-1}), and τ the shear stress (Pa). The parameters used in this model were determined experimentally (Yang et al., 2011; Yang and Crittenden, 2012) for Crude A and found to be 1190 m^2 K J^{-1}, −0.88 (nondimensional), 98.4 kJ mol^{-1}, and 3.85 m^2 K J^{-1} Pa^{-1} for α, β, E_A, and γ, respectively. Figure 3.24 shows the comparison of the predicted fouling rate using the model with the equivalent Reynolds number and the experimental data obtained using the stirred cell with wires attached.

Figure 3.24 also includes predicted and experimental fouling rate data with a Maya crude oil using a tubular test device fitted with hiTRAN® inserts (Phillips, 1999; Yang and Crittenden, 2012). All these data indicate that with the help of the equivalent Reynolds number concept, fouling rate data

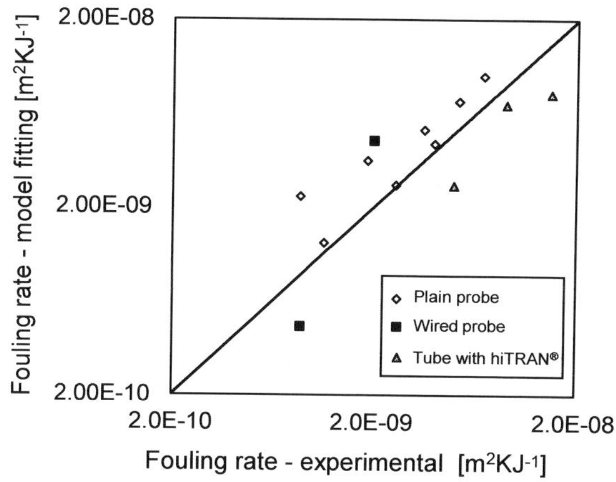

Figure 3.24 Comparison of the experimental data and the model predictions for the fouling rate for the stirred cell with the plain and wired probe and for a tube with hiTRAN® inserts. Note, the fluid tested was Maya crude (Phillips, 1999), and hence the model parameters were different from those used for the above two cases. *Reproduced with permission from Yang et al.,* Applied Thermal Engineering, *2013, 54, 516–520. Copyright (2013) Elsevier.*

become compatible regardless of the morphologies of the test surfaces, such that fouling test results generated using a specific device, e.g., the stirred cell, can be applied to other types of heated surfaces.

3.2.9 Conclusion

Using a batch cell stirred cell system experiments on crude fouling can be carried out relatively quickly using a small volume of crude (about 1 L). It is thereby possible to investigate the effects of key operating parameters on crude oil fouling in a short time period. Crude oil samples can be changed relatively quickly, mixed crudes can be investigated, and deposits can be removed easily for analysis. In addition, the heat transfer surface can be easily modified either by changing the metal and/or by incorporating heat transfer enhancements. Supported by CFD studies, the stirred cell apparatus also allows fouling threshold conditions of surface temperature and shear stress to be identified relatively quickly in the laboratory. With the help of the concept of equivalent Reynolds, fouling rate data become compatible regardless of the nature of the heat transfer surface. In this way, crude oil fouling test results generated using a specific device, such as the stirred cell apparatus, can be applied directly to other types of heated surface.

3.3 LARGE-SCALE EXPERIMENTS IN FLOW SYSTEMS

G.F. Hewitt

The previous two sections of this Chapter deal with micro- and small-scale systems whose aim is to investigate the effects of key parameters on crude oil fouling. By necessity, such systems differ in important ways from the real heat exchanger systems found in preheat trains. It is vital to carry out experimental work on systems more representative of the real case and this has been an aim of a number of studies. Basically, one can classify this work under three headings as follows:

1. **Collection of data from actual plant.** It could be rightly claimed that measurements on actual crude preheat trains are most likely to give meaningful data. This approach has been followed, for instance, by Joshi (2013) who reports analysis of data gleaned from eight crude preheat trains; the measured data were compared with analytical models and provide a level of confidence in such models. Crittenden et al. (1992) report the analysis of data from individual heat exchangers in preheat trains and conclude that the activation energies for fouling in "light" crude oils are much higher than those for "heavy" crudes. The problem with this general approach is that the quality of the data is necessarily limited, often very noisy and that it is not possible to determine the basic parametric effects; the researcher has little or no control over the key variables.
2. **Measurements on simulated heat exchanger systems.** Here, the heat exchanger to be evaluated is placed in a train of heat exchangers and its performance measured. The arrangement used by Ratel et al. (2013) is illustrated in Figure 3.25. Here, the test section is the target heat exchanger and the other exchangers are connected to the heat exchanger under test. Ratel et al. made measurements on shell-and-tube and cross-flow plate exchangers using this approach. A similar approach has been used by Joshi (2013) for single tube heat exchangers and by Gudmundsson et al. (2013) for plate and frame heat exchangers.

Measurements in pilot plant scale flow loops. The use of electrical heating rather than fluid heating has the advantage that the *heat flux* can be controlled and usually accurately measured. Heat Transfer Research Incorporated (HTRI) have constructed a test facility which is based on an annular geometry, with the inside of the annulus being heated using a cartridge heater. The overall layout of the HTRI facility is shown in Figure 3.26 and the detailed test section geometry is shown in Figure 3.27. The HTRI facility uses cartridge heaters inside the center tube of an annulus and this design

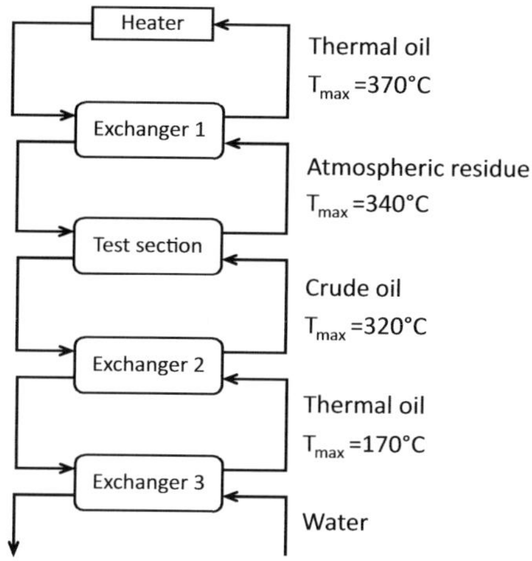

Figure 3.25 Heat exchanger train used in evaluating performance of a test heat exchanger. Reproduced with permission from (Ratel et al., 2013).

was also used in an earlier HTRI probe device which has been widely applied in the study of fouling (Srinavasan and Watkinson, 2005; Saleh et al., 2005; Watkinson et al., 2000; Watkinson, 2007). Radiant heating of the test section tube was used by Crittenden et al. (1987a) and various other electrical heating

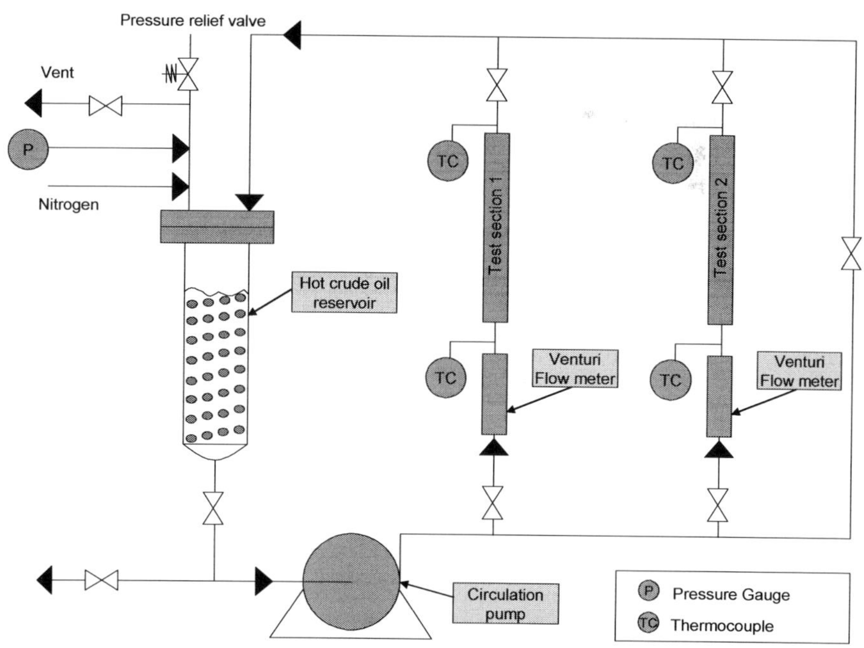

Figure 3.26 General layout of HTRI fouling facility. *Reproduced with permission (Smith, 2013).*

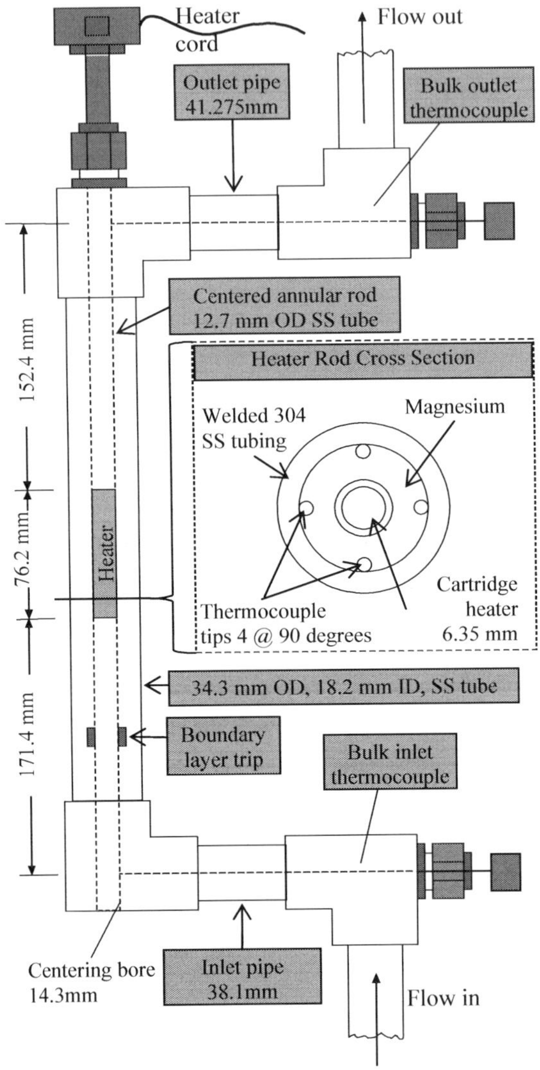

Figure 3.27 Test section in HTRI fouling facility. *Reproduced with permission (Smith, 2013).*

devices have been employed (e.g., band heaters by Taylor and Wallace, 1967, and K-ring heaters by Crittenden et al., 2009). More recently, Coletti et al. (2014) have proposed an alternative heating system based on catalytic heating. Although this system is untested at present, the main advantage of using catalytic heating is that it provides a more uniform heating than traditional furnaces or pipe in pipe heating systems. In contrast to Joule heating, the catalytic system eliminates the need of electrically insulating the flanges and the test section which significantly simplifies the design of construction. Easy setup and dismantling, and safety of the flameless combustion and design, which avoids explosive mixture, are also additional advantages.

During the CROF project (noted in Chapter 1), a major new facility for the study of crude oil fouling—namely the HIgh Pressure Oil fouling Rig (HIPOR)—has been designed and built at Imperial College London and later licensed to Hexxcell Ltd. This rig is designed to operate at temperatures up to 300 °C and pressures up to 30 bar. The basic flow diagram of the HIPOR facility is shown in Figure 3.28.

The rig consists of a storage vessel for the test crude oil and a circuit in which the oil can be preheated, pumped through the test sections, and then cooled before passing back to the storage vessel. The physical arrangement of the rig is illustrated in Figure 3.29. In designing the HIPOR facility, a number of crucial design decisions had to be made as follows:

1. **Type of heating.** It was decided to employ *Joule heating* for this rig. In this type of heating, a large current is passed through the test section wall and this generates a heat flux which can be accurately calculated from the measured values of the current and the voltage drop across the test section. Though this type of heating has been widely applied in heat transfer experiments (for instance, in experiments on heat transfer to water in the

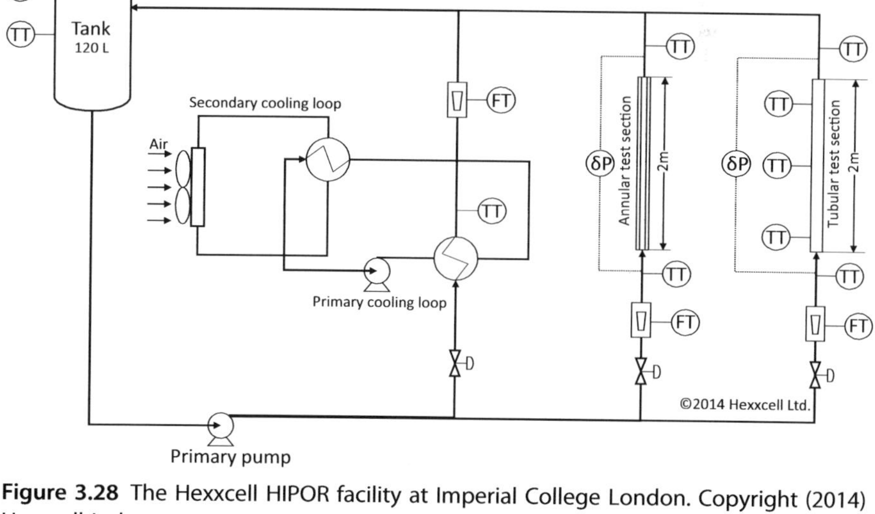

Figure 3.28 The Hexxcell HIPOR facility at Imperial College London. Copyright (2014) Hexxcell Ltd.

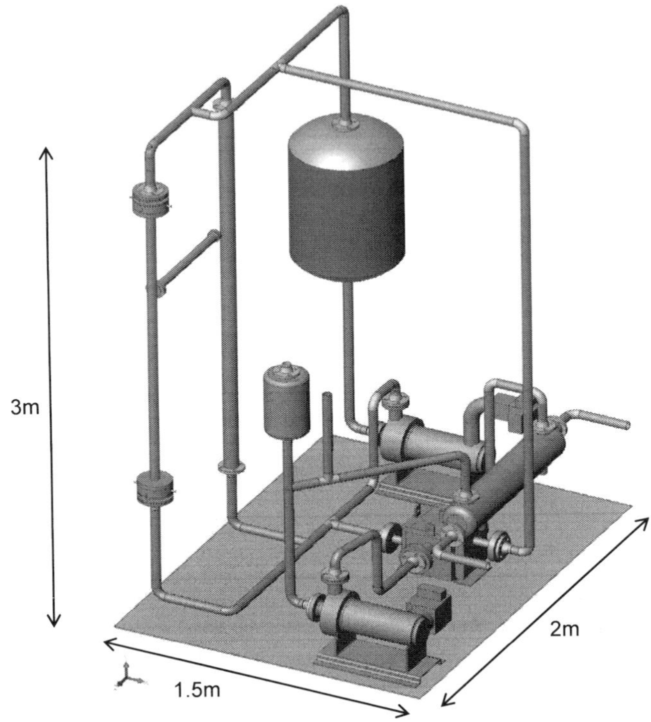

Figure 3.29 Physical arrangement of Hexxcell HIPOR components. Copyright Hexxcell Ltd.

context of nuclear systems), its use for oil fouling experiments is not normal practice (as the discussion above shows). However, the use of Joule heating allows the heat flux to be accurately calculated and controlled. The other variables required to be measured if heat transfer coefficient is to be determined are the temperature of the wall in contact with the fluid and the fluid temperature. The wall temperature can be measured using thermocouples which are electrically separated from the (current-carrying) test section wall and the fluid temperature can be measured using insulated thermocouples or estimated from inlet values using a heat balance. Recent tests have confirmed all flow and temperature quantities can be controlled and measurements obtained with high precision and reproducibility.

2. **Test section geometry.** In most applications, fouling occurs on the inside of heat exchanger tubes and it is thus important to design test sections in which the fluid moves parallel with the heated wall. The simplest approach is to use a *tubular* test section with the (Joule heating) current passing along the wall and with thermocouples measuring the outside

wall temperature from which the inside wall temperature (and hence the prevailing heat transfer coefficient) can be estimated. The thermocouples measuring the outside wall temperature must be electrically insulated from the tube wall and a common way of doing this is to use thin mica strips. A tubular test section has been installed on the HIPOR rig but perhaps the most information can be gleaned by using an *annulus* test section and such a test section has been set up on the HIPOR rig and is illustrated in Figure 3.30.

The annulus test section is much more sophisticated. Though the measurements in the tubular test section are useful and, indeed, closer to the application, there are two main features of the annulus geometry which can give additional information:

a. In contrast to earlier work where annulus sections were used that had cartridge heaters in the center space, the use of Joule heating leaves this space free and it is possible to introduce a radiation equilibrium thermocouple device as illustrated in Figure 3.30. Here, a thermocouple is positioned close to (but not in contact with) the inner wall of the heated tube and the space containing the thermocouple is contained by two ceramic disks as illustrated. The space reaches radiation equilibrium with the tube wall and this is the temperature recorded by the thermocouple. The radiation equilibrium thermocouple can be moved inside the inner tube and a distribution of wall temperature obtained. In the HIPOR facility, this movement is automated so that temperature profiles can be obtained at various times. The build-up of a fouling layer on the outside of the inner tube of the annulus causes this temperature distribution to vary with time and the fouling resistance can be continuously monitored.

b. With the annulus test section, the fouling layer is formed on the *outside* of the heated tube and this fouling layer can be readily examined after removal of the test section on conclusion of the experiment. Furthermore, the thickness of the fouling layer can in principle be monitored continuously using the dynamic thickness gauge described in Section 4.4; the way in which this thickness gauge would be attached is illustrated in Figure 3.30.

Thus, on the one hand, the use of a tubular test section is more representative of the applications in real heat exchangers whereas, on the other hand, the annulus geometry offers the scope for more detailed measurements. In fact, it was considered worthwhile to have the capability of having either type of test section in HIPOR as is illustrated in Figure 3.28.

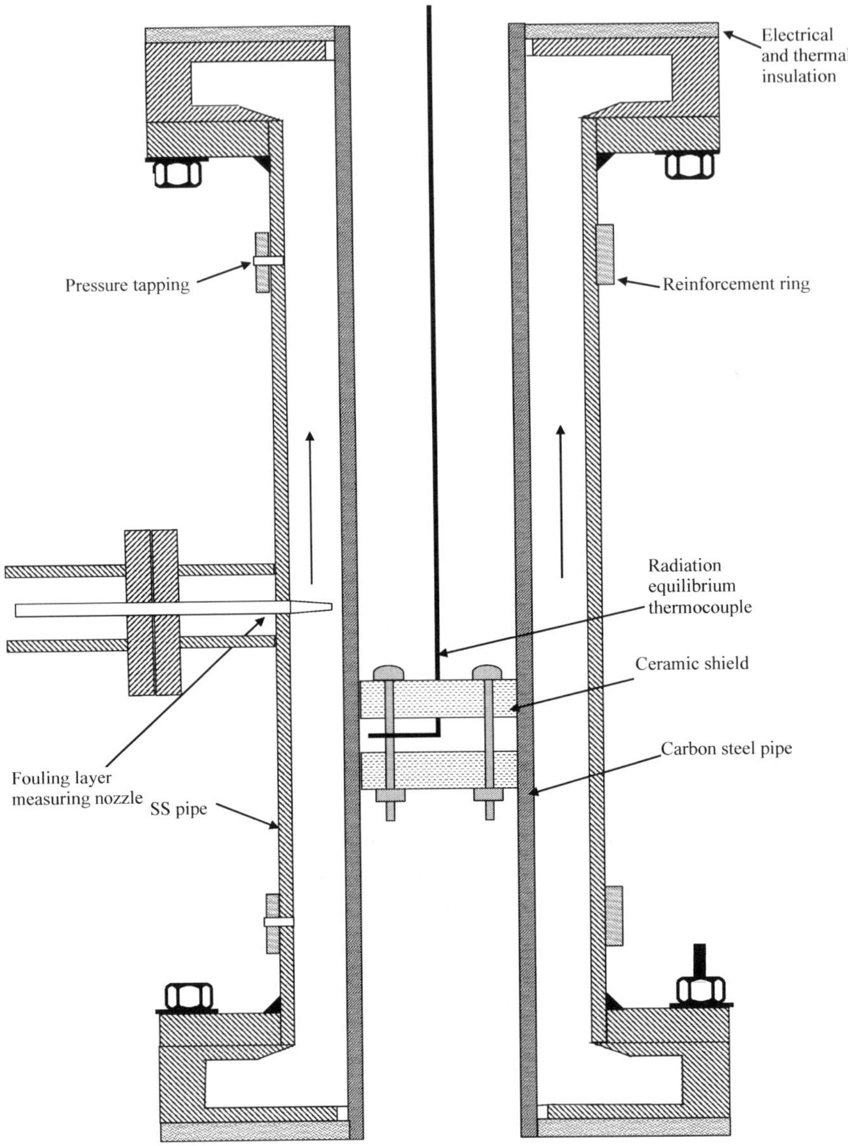

Figure 3.30 Annulus test section on Hexxcell HIPOR facility. Copyright Hexxcell Ltd.

3. **Test section material.** The majority of heat exchangers used for crude oil duty are constructed from carbon steel. However, it is tempting to construct the test sections out of *stainless steel* since this is more corrosion resistant and, furthermore, it has a higher electrical resistance than carbon

steel, which reduces the current required in Joule heating to obtain a given heat flux. However, the results of Wang and Watkinson (2013) show that the fouling propensity of carbon steel is much greater than that of stainless steel in deposition from a sour heavy oil and this is confirmed by the results of Sprague et al. (2005) who studied deposition from a pyrolysis gasoline. Despite the disadvantages in using carbon steel, it was concluded that the test sections in HIPOR (in both the tubular and the annulus inner surface cases) should be constructed from carbon steel tubes which were characteristic of those used in real preheat trains.

4. **Test section length.** In the cases of both the annulus and the tubular test sections, it was considered that a sufficient tube length should be allowed to permit well-established flow—that is, the entrance effects should be of only minor importance.
5. **Safety.** Crude oil requires special handling due to its flammability and the toxicity of some of its potential components (for example, hydrogen sulfide). The HIPOR facility has many flanged and other joints. Although suitable design, construction and maintenance procedures eliminate leakages, safety considerations suggested essential to contain the flow circuit within a sealed enclosure. This enclosure was constructed from steel plates and the enclosure had an upper exit through which the enclosed volume could be vented. The front plates of the enclosure can be removed for access to the circuit between runs and the whole circuit is controlled using a customized software based on the LABVIEW package. Within the enclosure, there are detectors that test for the presence of hydrocarbon vapor and for the presence of hydrogen sulfide. If either is detected, then suitable emergency procedures are initiated. Two approaches are available to deal with the release of potentially flammable materials in the rig enclosure:
   a. The enclosure is filled with nitrogen derived from a pressure swing absorption unit. If the enclosure is filled with nitrogen, then flames cannot occur.
   b. The enclosure can be attached to a ventilation system to provide a large flow of air, replacing the enclosure contents at a high rate. The system is designed so that the concentration of flammables following a potential leak remains below flammability limits at all times. Such ventilation system can be complemented with a suitable fire suppression system, for example, one that floods the enclosure with CO_2 if high level of volatile hydrocarbons or too high a temperature are detected in the enclosure.

Both options above are available on the HIPOR facility.

In conclusions, HIPOR offers an advanced facility for measuring crude oil fouling at scale and conditions directly comparable with those in refineries. Its detailed instrumentation systems will also allow a detailed evaluation and resolution of the phenomena involved, and an assessment of the extrapolability of data obtained from small scale experiments. An industrial testing program is currently underway.

CHAPTER FOUR

Deposit Characterization and Measurements

J. Chew[1], H.M. Joshi[2], S.G. Kazarian[3], M. Millan-Agorio[4], F.H. Tay[5], S. Venditti[6]

[1]Lecturer in Chemical Engineering, University of Bath, UK
[2]Principal Heat Transfer Engineer, Shell, USA
[3]Professor of Physical Chemistry, Imperial College London, UK
[4]Reader in Chemical Engineering, Imperial College London, UK
[5]Head of R&D, EliteTNS, Singapore
[6]R&D Engineer, Resource Centre for Environmental Technologies, Luxembourg

Analytical characterization of crude oil deposits provides useful insights into understanding the underlying mechanisms leading to fouling, to determine its root cause and to find the sources of foulant precursors. Results of fouling deposit characterization can be used to establish the most appropriate method for preventing fouling or cleaning the fouled heat exchanger. For example, if the deposit is predominantly inorganic, coking mechanisms can be ignored and the focus can be shifted to find the source of the inorganic materials (e.g., desalter malfunction).

There is no single analytical technique that can provide a complete characterization of crude oils and crude oil deposits. Instead, their analysis must rely on a combination of several methods, each capable of providing partial information about the sample composition. This limitation in the scope of the characterization methods arises in part from the limitation of the analytical techniques themselves and in part from the complexity of the samples.

This chapter illustrates the use of characterization and measuring techniques of crude oil fouling deposits in refinery heat exchangers. Note that the same techniques can be applied to other fouling deposits such as those generated by experimental rigs (e.g., those described in Chapter 3), in other refining processes, or other industrial equipment.

Section 4.1 is concerned with the industrial practice and provides guidelines on deposit collection and sample preparation and illustrates how the lessons from deposit analysis can be used to effectively solve operating problems in the refinery. Section 4.2 shows the results of a detailed chemical structure and molecular weight characterization of refinery deposits. Section 4.3 presents results for the analysis of crude preheat train deposits using chemical imaging techniques. Finally, Section 4.4 is concerned with the

Crude Oil Fouling
ISBN: 9780128012567
http://dx.doi.org/10.1016/B978-0-12-801256-7.00004-X

measurement of thickness of fouling deposits, an important parameter when considering operations of heat exchangers.

4.1 ANALYSIS OF FIELD FOULING DEPOSITS FROM CRUDE HEAT EXCHANGERS

H. M. Joshi

As discussed in Section 2.1, there are different mechanisms that lead to crude oil fouling in preheat trains (PHTs). Plant experience shows that the most common one is a combination of chemical reaction and corrosion fouling. In this mechanism, a mix of crude and inorganic particles is trapped in the roughness of the tube surface and generates deposits that degrade with time to a coke-like material according to the ageing mechanisms described in Section 2.3.5. The corrosion of the tube wall is believed to lead to the formation of an iron sulfide layer that further increases the roughness of the tube surface and consequently its ability to trap crude oil and other particles. The foulant material in this mechanism typically consists of a mixture of organics, salts, and FeS.

A second, less commonly observed, mechanism is that of deposition of precipitated asphaltenes on the tube wall and their eventual degradation to coke-like material. This typically happens when the crude or crude blend is incompatible (also called unstable) with respect to asphaltenes. The foulant material in this mechanism consists mostly of coke-like (organic) material.

A quantitative analysis of the deposit helps to identify which of the two mechanisms is dominant. If it is the more common fouling mechanism (i.e., with sulfide formation), the amount of different inorganic materials can point to the source of precursors, and the hydrogen and carbon quantities provide information about the temperature relationship of foulant formation. Moreover, regardless of the mechanism, a correct analysis can identify if an unexpected contaminant is responsible for the initiation of fouling.

When analyzing samples of crude oil deposits, to understand and correct operating problems, it is important to know whether fouling is dominated by organic or inorganic deposits, if it is a mix of the two, or if it is driven by a specific reaction mechanism generated, for example, by the use of chemical additives such as those added to prevent corrosion.

To perform an analysis that provides the most useful information, without expensive and time-consuming testing, three major steps are required:

1. Careful deposit collection
2. Correct sample preparation before testing
3. Appropriate sequence of analytical tests.

One such methodology is described by Brons et al. (2010).

This section proposes a recommended methodology to accurately quantify the deposit composition in real-life crude heat exchangers. This includes guidelines for appropriate sample collection, preparation, and analysis. Moreover, indications on practical interpretation of the analysis into actionable recommendations for the plant are provided.

4.1.1 Sample Collection

Sample collection is a relatively simple activity but a few guidelines must be followed to ensure an effective deposit analysis. Two key aspects during sample collection are to preserve the chemical integrity of the sample as much as possible, and to make a clear note of the location of the sample and take photographs of the fouled heat exchanger.

The recommended guidelines for sample collection are as follows:

- *Shutdown procedure.* Safety and environmental considerations require a minimum amount of cleaning, steaming, or flushing (sometimes referred to as decontamination) before a heat exchanger is opened. So that the chemical integrity of the deposit sample is preserved as much as possible, any additional cleaning should be avoided before sample collection. For example, washing with water will dissolve salts, which in some instances could be a large component of the foulant.
- *Collection of the samples.* If it visually appears that the deposits look different at different locations in the heat exchanger, collect multiple samples. Samples from different locations should be clearly marked as to exactly where they were taken, and should not be mixed. Moreover:
  - Tube-side samples should be taken as far inside the tube as possible, and as close to the tube wall as possible. If there is widespread deposit on the tubesheet, or tube inlets are plugged, those deposits should be collected as separate samples.
  - On the shell-side, if it appears that there are different levels and types of fouling in different areas of the exchanger, again collect separate samples. It is important to take at least one sample from the tube outside surface. This is critical for differentiating between a mechanism that is purely a deposition of particles carried in the stream and a mechanism that occurs due to either corrosion or coking at the tube surface. If the samples are collected only from the gap between tubes or the window areas it is possible that the mechanism at the tube surface will be missed at the analytical stage.
- *Quantities.* Fifty or more grams of foulant should be collected for each sample. If this is not possible, as much sample material as possible should be collected. As it will be explained later, a large amount of this

collected material might be lost during sample preparation, thus it is important to have a sufficient quantity remaining to perform all the analyses required.

- *Records*. Keep accurate records of the exact location of the samples. Ideally photographs should be taken during this process to have a record of whether fouling is localized, how much is present in different parts of the exchanger, where the baffles are, etc.
- *Storage*. Store the samples in an airtight jar, to avoid the possibility of oxidation of its inorganic components.

4.1.2 Sample Preparation

Samples collected as per the above guidelines contain the process fluid (crude) which either is trapped within the deposit or is simply coating the samples. In addition, the deposit may be contaminated with flushing solutions or water from a steam-out (i.e., the process of circulating steam to remove contaminants). The trapped fluids or flushing solutions are not part of the foulant and need to be removed before the deposit is analyzed. This is a critical step because leftover process fluid or other contaminants will lead to mischaracterize the chemical makeup of the deposit, especially with respect to the quantification of the organic portion. The measured carbon and hydrogen content will be much higher than what is in the actual foulant (the deposit that creates the heat transfer resistance), leading to the possibly wrong conclusion that the fouling is dominated by organic mechanisms (i.e. asphaltene precipitation).

Solvents such as toluene or methylene chloride are used to dissolve and remove residual feeds and other liquids. An effective sample preparation procedure is described below.

- Add a sufficient quantity (e.g., 10:1) of the solvent in 10–15 g of the collected deposit to dissolve trapped process fluid.
- Leave the mixture at room temperature for a few hours in an airtight jar.
- Filter the mixture and wash with additional solvent a few times until all the trapped fluid is removed—indicated by the filtrate being colorless.
- Dry the remaining deposit for several hours, preferably in a vacuum oven at an elevated temperature, 50 °C or higher.
- Grind and mix the sample as much as possible to obtain a homogeneous powder. Since the interest is in the average composition of the deposit, this step is important to ensure that the final analyses are not conducted on a small and unrepresentative portion of the sample that happens to be rich in one component.

Most samples visually look the same when they are collected in the field. This is usually due to the trapped crude which gives the samples a black color and the appearance of a gooey mixture (often described by plant personnel as "shoe polish"). However, samples will look different when the above preparation is complete and a dry powdery deposit is obtained.

All further analyses detailed below should be conducted on this washed, dried, and homogenized sample. Depending on how much crude was trapped in this deposit when it was removed from the heat exchanger, this final sample might weigh substantially less (<40%) than the original one collected in the field.

4.1.3 Deposit Analysis Philosophy

The sample collection and preparation steps apply to any deposit for which a good quantification is needed for root cause determination or for developing a mitigation strategy. However, subsequent analyses of deposits from crude heat exchangers can be simplified, relative that described by Brons et al. (2010). This is mainly due to the fact that most of the time (about 90%) the fouling mechanism for crudes in operation is one of the two described at the beginning of the section (i.e., (1) chemical reaction/corrosion fouling or (2) precipitation of asphaltenes and their degradation to a coke-like material on the surface). To distinguish between these mechanisms, quantification of only a few elements is critical. This simplified set of tests will be referred to as Level 1 testing (L1). Depending on the results of L1 testing, the need for more specialized and detailed analysis can be established. In these cases (typically 10% of the times), Level 2 testing (L2) will be performed. Figure 4.1 shows a flow chart of the recommended test sequence.

Note that this strategy is aimed at identifying which of the two mechanisms is predominant and providing the refinery with suggestions for corrective actions. Even more detailed analyses (e.g., those described in Sections 4.2 and 4.3) are available to provide a deeper understanding of the fouling mechanisms themselves.

The next sections will describe the recommended analysis in Level 1 and Level 2 testing, and provide examples of analyses performed on deposits collected from oil refineries.

4.1.4 Level 1 Testing

4.1.4.1 Thermogravimetric Analysis

Thermogravimetric analysis (TGA) is a method for characterization of thermophysical properties of materials by probing into thermodegradation

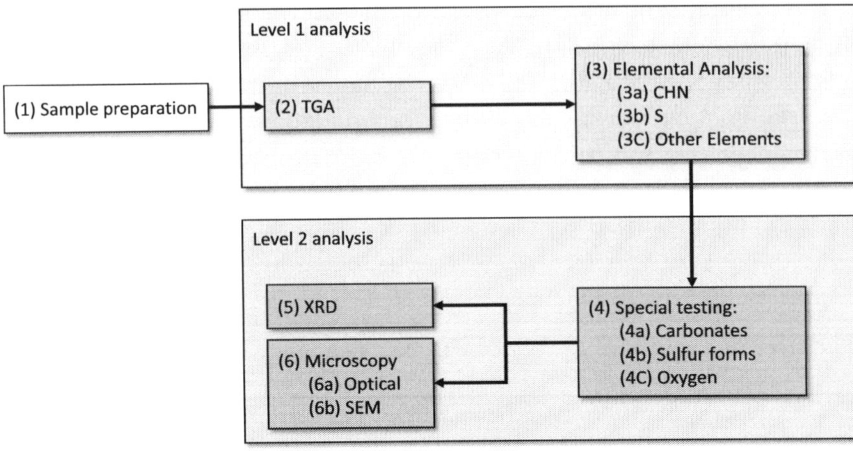

Figure 4.1 Crude fouling deposit analysis test sequence.

reactions at high temperatures. In a TGA test a small amount of sample (10 mg) is placed in a platinum pan under an inert atmosphere inside a glass enclosure. The temperature inside the enclosure is increased from ambient to a maximum temperature (between 600 and 900 °C), at a constant rate (e.g., 10 °C min^{-1}). As the sample is heated, different components of the deposit volatilize at different temperatures, providing information on deposit composition that can be linked to the possible causes of fouling. The response of the sample to heat is determined by accurately measuring the temperature and the decreasing mass inside the pan. Once the maximum temperature is reached, the inert atmosphere in the enclosure is replaced by air or oxygen and the temperature is held constant for a specific amount of time (e.g., 20 min). This results in combustion of the remaining sample. After the combustion phase the experiment is terminated and what remains in the pan is called "ash."

TGA provides a significant amount of useful information in a fast (2–3 h total time) and inexpensive way. For this reason, TGA should be the first test to be performed and it is the one recommended if only one test were to be performed. The main advantage of TGA is that it provides an estimate of the relative quantities of volatile and nonvolatile organic materials, and inorganic materials. However, TGA data do not identify what elements are present, and consequently it cannot be determined, for example, whether the inorganic material is a corrosion product (like FeS) or salts (like carbonates) or something else. Further analysis is needed for confirmation of elements and for quantification.

Another point of interest in TGA is that it allows determination of whether polymeric material such as gums might be involved in the formation of fouling precursors. Polymeric material tends to be emitted at temperatures <370 °C. If a large drop in mass is found in this temperature range, this can be related to the presence of polymers in the sample. This is rare for crude deposit samples and will not be included in the cases studied here.

Two examples of tests on crude oil fouling deposits are presented below, with interpretations of the results. One is for a sample containing a high percentage of organic material (Sample A) and the other for a sample with a large content of inorganic material (Sample B). For samples with compositions in between, some interpolation of the interpretations will apply.

In these examples the key test parameters, such as the rate of temperature increase (10 °C min^{-1}), the time allowed for combustion (20 min), and, particularity, the maximum temperature (800 °C), have been selected on the basis of the author's own experience with crude deposit analysis. Different values can be used in practice.

Sample A has been collected from a heat exchanger toward the hot end of a crude PHT train. The TGA analysis for this sample is shown in Figure 4.2. The sample is heated from ambient temperature (about 30 °C) to 800 °C in ca.75 min. During this heating period about 53% of the mass has volatilized; a further 25% weight loss occurred as a result of combustion when air was introduced at 800 °C. The 53% loss of mass before the combustion step corresponds to easily volatilized material and does not need

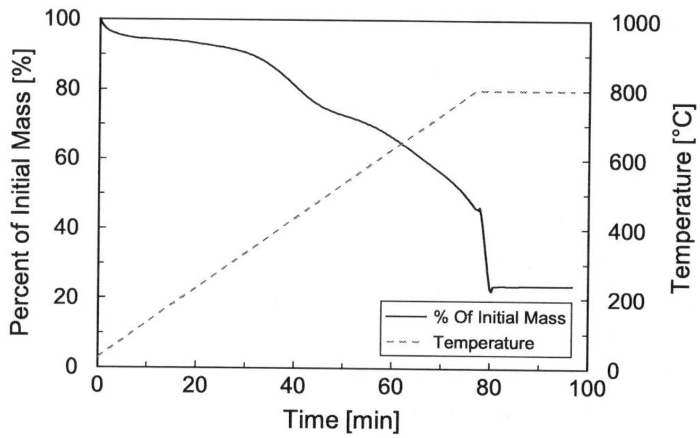

Figure 4.2 Thermogravimetric analysis result for crude heat exchanger deposit for Sample A.

to be combusted. However, the entire 78% mass is organic foulant material. It is foulant formed as a result of the crude being subjected to the high tube wall surface temperatures, and is thermally degraded material that is insoluble in the solvent used to prepare the sample. This is usually referred to as coke or coke-like material. The 25% that combusted is hardened coke, presumably because it has been exposed for a longer time at the high temperature, so as to allow it to harden and become nonvolatile.

The large organic content of Sample A indicates that fouling may have occurred in this heat exchanger because of crude incompatibility. The 22% ash collected at the end of the test is inorganic material and further elemental or other analysis (on the prepared sample, not on the ash) is needed to know what, if any, elements or compounds are dominant.

Sample B is another sample from an atmospheric distillation unit preheat train but it exhibits a completely different behavior when characterized with TGA. Figure 4.3 shows how Sample B contains about 10% of organic material and 90% inorganic material (ash). Inorganic materials flow with crude streams in the form of corrosion products (usually iron sulfide) and salts (typically sodium and calcium salts—carbonates, chlorides, sulfates). For this sample, further elemental analysis is needed to determine which inorganic materials are prevalent. Identifying the exact inorganic species present is key to determining the correct fouling mitigation strategy that will eliminate those components or prevent them from depositing on the tube wall.

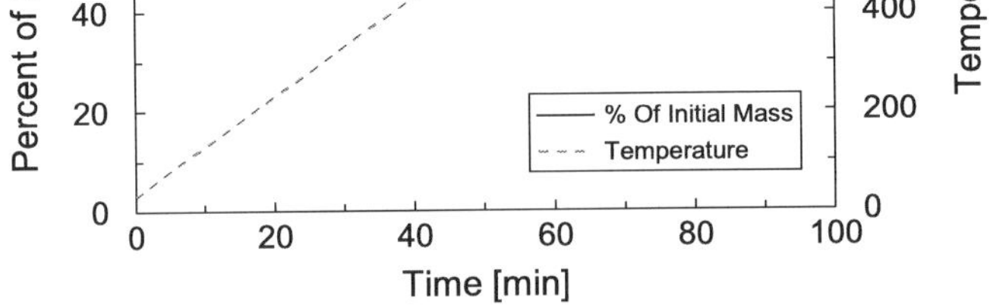

Figure 4.3 Thermogravimetric analysis result for crude heat exchanger deposit for Sample B.

4.1.4.2 Elemental Analysis

Elemental analysis for typical crude deposits involves three types of tests to quantify:

- Carbon, hydrogen, and nitrogen (CHN) content.
- Sulfur content.
- Presence of metals and other inorganic elements.

The determination of which of these tests is necessary should usually be done on the basis of the TGA results. For example, because of the large organic content, for Sample A (Figure 4.2), the CHN test will be sufficient to confirm the asphaltene precipitation mechanism, whereas for Sample B (Figure 4.3), only sulfur and metals need to be measured. All three tests should otherwise be conducted for deposits showing compositions in between, because in these cases both organic and inorganic fouling precursors play a role.

4.1.4.2.1 Carbon, Hydrogen, and Nitrogen Testing

CHN testing focuses on quantifying C, H, and N content in a particular sample. A good method for this type of analysis is provided by the ASTM standard D5373 (ASTM, 2014a). Results are reported as weight percent of the elements in the deposit. There are two useful ways of using data collected from this test. First, CHN tests can be used as a confirmation of the TGA results. The sum of the weight percentages measured with CHN tests should roughly be equal to the organic content inferred from the TGA. For example, in CHN tests for Sample A (Fig. 4.2), the results show C = 70%, H = 5.5%, and N = 1%, for a total of ca. 77%, consistent with the ca. 22% ash remaining from TGA.

The other useful information is the atomic H/C ratio calculated as:

$$H/C = 12 \times \frac{\text{weight\% H}}{\text{weight\% C}}$$

For Sample A, the $H/C = 5.5/77 \times 12 = 0.86$. As the deposit thermally degrades (dehydrogenates) in the operating heat exchanger, over time its H/C ratio decreases. Moreover, the higher the temperature to which the deposit is exposed (the tube wall surface temperature), the faster the rate of degradation. The H/C ratio from a typical deposit is a function of the temperature at which this coke-like material was formed, over a reasonably long time frame, which, for crude heat exchangers, is typically over 6 months. In crude preheat train exchangers the wall surface temperatures in the hotter part of the train are between 200 °C and 330 °C, and the corresponding expected

H/C ratio range is 0.80–0.95.[1] The ratio measured in Sample A falls in this range, which further confirms the results of TGA. If the H/C ratio is higher than 0.95, the deposit has degraded less, which implies that the residence time has not been long enough for the deposit to fully age to a coke-like material. More interesting is the case where this ratio is lower than the above range, ca. 0.70 or lower. In this case, the deposit degraded at a temperature higher than a typical crude preheat. One possible conclusion then is that the foulant was not formed in the heat exchanger but came in with the crude as already formed coke and deposited in the heat exchanger.

4.1.4.2.2 Sulfur Testing

In fouling deposits from crude heat exchangers, sulfur is almost always present, and usually in the form of iron sulfide (FeS). Sometimes it is present in the form of sulfates and sometimes as elemental sulfur with the organic material. The iron sulfide may also be pyrrhotite, a form with a variable iron content: Fe(1−x)S (x = 0–0.2).

A good method for sulfur quantification is the ASTM standard D4239 (ASTM, 2014b). This test gives important information regarding the association of the elements. If the iron and sulfur are in a molar ratio in the range 1 to 0.8 (i.e., with x in the range 0–0.2), it is reasonable to conclude that most of the iron and sulfur are present as iron sulfide (Fe(1−x)S). If the total of Fe and S is a high percentage, e.g., >50%, then at least one of the dominant fouling mechanisms is corrosion based. This includes FeS carried in with the crude or severe sulfur corrosion of the tube surface itself, which contributes to the fouling. High velocities will minimize the deposition of carried-in FeS.

If there is more sulfur in the sample than indicated by the Fe(1−x)S formula, then the excess will be present as metal sulfates or organic sulfur.

4.1.4.2.3 Testing for Other Chemical Elements

Testing for elements other than C, H, N, and S becomes important when the amount of TGA ash is not insignificant, i.e., >30%. The common elements found in crude fouling deposits are:

- *Metals*: Iron (Fe), aluminum (Al), sodium (Na), potassium (K), calcium (Ca), chromium (Cr), cobalt (Co), copper (Cu), magnesium (Mg),

[1] This expected ratio is based on the author's experience. For reference, the starting H/C ratio for crude oils is approximately 1.5–2.0, with asphaltenic material typically between 1.1 and 1.2. Thus the degraded material in crude heat exchangers exhibits a somewhat lower ratio.

manganese (Mn), molybdenum (Mo), zinc (Zn), nickel (Ni), and vanadium (V).

- *Nonmetals*: Chlorine (Cl), silicon (Si), and phosphorus (P).

Among these elements, iron is usually in the largest quantity followed by Cr, Na, and Ca. If sodium is present as a chloride, a stoichiometrically consistent amount of chlorine will usually be detected. The other metals could be in the form of oxides, chlorides, sulfates, or carbonates.

The importance of this elemental testing is twofold: to quantify the amount of iron, as explained in the earlier section on sulfur testing; and to identify if any single element could be responsible for the fouling mechanism, particularly if it is present in large quantities. For example, a large amount of silicon might point to a mechanism where silica is carried in with the incoming crude and deposits due to low velocities. In this case, one possible method of fouling mitigation will be to eliminate the source of the silica.

Similarly, if an unexpected element is detected in a significant amount (>5%), a possible explanation of its presence and a possible cause for fouling is the addition of chemicals either in the crude processing unit itself or upstream all the way to the production facility.

A reliable method for quantifying the elements important to crude fouling deposit analyses is inductively coupled plasma optical emission spectrometry (ICP-OES). Several commercial analytical laboratories provide this test, and sometimes it can be ordered for specific elements only.

4.1.4.2.4 Oxygen Testing and Calculation

A large amount of oxygen (up to ca. 20%) is sometimes found in organic fouling deposits originating from polymers, where oxygen catalyzes the reactions, or where it is a part of the process chemistry. However, in crude oil fouling deposits, this is an exception and only occasionally large quantities of oxides, e.g., silica, will result in a high oxygen content in the deposit.

Oxygen quantification is typically not needed as the amount of oxygen can usually be quantified using the measured silicon. This typically allows bringing the mass balance to a satisfactory closure. However, since some of the metals could be present as oxides, sulfates, or carbonates, sometimes it is necessary to calculate the amount of oxygen that may be present. This helps to close the mass balance of the deposit components, as will be explained in the following section. If oxygen measurement becomes necessary because the mass balance is not closed, a reliable method of quantification that can be used is neutron activation analysis (NAA).

4.1.4.3 Mass Balance

When all the testing is finished, an important step for verifying the quality of the results is to check the mass balance by adding the measured weight percentages of each measured elements. Experience shows that for most crude-related deposits the percentages of the CHN, S, and other metals only rarely add up to 100%. Typically mass balances add up to about 85–95%. Some of this inaccuracy is due to the nonhomogeneity of the samples (despite the care taken to grind and homogenize in the sample preparation step), although some might be due to the accuracy of the measurements.

One step to bring the mass balance closer to 100% is to guess whether oxygen might be associated with certain elements (Si and Al are typical candidates), and calculate how much is needed to have them present as oxides. For example, for silicon to be present as SiO_2 requires the masses of silicon and oxygen to be in the ratio of 28:32. If there is 10% Si in the deposit it is reasonable to expect 11–12% oxygen (which was not part of the measurements).

For the purpose of characterizing a field deposit, an analysis resulting with a mass balance of 90% or better is considered adequate to identify the root cause of fouling and to determine appropriate mitigation or cleaning methods.

To illustrate the concepts above, the analyses of two samples, reported in Table 4.1, are considered. Measured weight percentages for the first sample, Sample C, show an acceptable closure of the mass balance. The second analysis, for Sample D, shows the need to calculate the percentage of oxygen to close the mass balance. In this case it is assumed that silicon is present as SiO_2. Based on experience it can be determined that Si and Al in Sample C are likely to be oxides, and some of the sulfur might be in the form of sulfates. This can be confirmed with Level 2 analysis, and in particular with X-ray diffraction (XRD) and microscopy. If oxygen is calculated with those assumptions the final mass balance will be close to 90% and judged satisfactory.

In cases where it is impossible to close the mass balance to a reasonable accuracy, two possibilities exist. Either an error was made in one of the tests—in which case a repeat is necessary—or there is a missing element for which no testing was done. Experience shows that if a reputable laboratory is used to conduct the testing, it is more likely that an element has been missed. For example, when an ICP-OES test is ordered, the laboratory might provide a standard set of results, or some charge for the test on a per element basis and may not have provided the missing element. Typically, an understanding of the operation of the process unit helps to identify

Table 4.1 Example of two mass balances for different samples. For Sample C, the sum of the measured percentages provide a satisfactory closure of the mass balance. For Sample D oxygen calculations are needed to close the mass balance satisfactorily.

Element	Sample C, wt%	Sample D, wt%
C	35.80	23.00
H	4.61	1.70
N	0.50	0.20
S	16.62	13.00
Cl	2.16	—
Si	0.89	17.00
Fe	23.90	22.00
Na	1.93	0.45
Al	0.46	—
Ca	1.66	—
Zn	1.44	—
V	—	0.35
Total	89.97	77.70
O calculated	—	19.43
Overall	89.97	97.13

why an unexpected chemical is present in the deposit. This includes, for example, whether external additives are injected in the process (e.g., for corrosion control), if temporary streams (such as slops) are being mixed with the crude feed. However, in some cases, this may not be enough to identify a specific fouling mechanism (with related mitigation action) and additional, more detailed testing should be conducted.

4.1.5 Level 2 Tests

At the completion of Level 1 tests, the results of the analysis on a particular sample include the TGA profile and the weight percentage of various elements. For deposits from crude heat exchangers this information is sufficient in 90–95% of the cases to come to usable conclusions and provide guidance to the refinery.

For a few cases, more detailed, Level 2 testing might be required. The intent of Level 2 testing is to gain a deeper understanding of the fouling mechanism, the processes of deposition and coke formation, the exact chemical nature of the compounds, or to identify precursors that otherwise may not be found. Although these types of testing are rarely conducted in industrial practice, they provide key information in nonstandard cases.

4.1.5.1 Carbon as Carbonate Analysis

If a large amount of carbonate is present in a deposit, meaning a portion of the measured carbon is inorganic, the mass balance will be difficult to match because the measured carbon is normally assumed to be organic. In addition, the H/C ratio might look inconsistent with what is expected in a particular deposit. In such a case, a test to quantify the amount of carbon present as carbonate will be useful. One example of such a test is the ASTM standard D6316 (ASTM, 2009).

4.1.5.2 Sulfur Forms Determinations

Total sulfur quantification was described in Section 4.1.4.2. Further analysis of sulfur to identify sulfates or organic sulfur is usually unnecessary but it is a good tool to have when the amount of measured sulfur cannot be explained by association with iron alone.

The ASTM standard D2492 (ASTM, 2012) is a good method for the determination of the percentages of pyritic sulfur and sulfates. Subtracting these from the total sulfur determined from L1 analysis gives the amount of organic sulfur.

4.1.5.3 Neutron Activation Analysis

Neutron activation analysis is a reliable method for measuring oxygen. However, as noted previously under the discussion for oxygen, this test is rarely needed for crude deposits.

4.1.5.4 X-Ray Diffraction

XRD is used to identify specific crystalline compounds or to identify specific phases of crystalline substances such as the type of iron sulfide present in the deposit. This is useful when such specific knowledge is needed, but not necessary for the typical fouling mechanisms of crude oils in heat exchangers.

4.1.5.5 Optical Microscopy

Cross-polarized light optical microscopy is particularly useful in identifying the nature of the coke in the fouling deposit. An expert eye can interpret the results and infer the coke formation mechanism (e.g. whether asphaltenes could have been the precursors), determine the size of inorganic particle in the deposit and obtain other similar information. Figure 4.4 shows an example where inorganic particles (round catalyst fines) serve as nucleation sites for coke growth, and the optical microscopy image clearly reveals this.

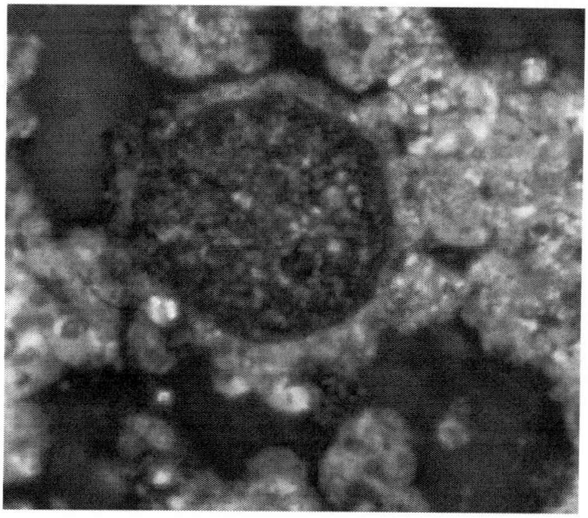

Figure 4.4 Example of optical microscopy.

4.1.5.6 Scanning Electron Microscopy

Scanning electron microscopy (SEM) is another advanced technique that can help to obtain a deeper understanding of the deposit composition as well as the fouling mechanism. For crude fouling deposits, the two most useful features provided by an SEM analysis are element mapping and a deposit profile.

Element mapping shows the association of one or more elements, to confirm the form in which they exist. For example, Figure 4.5 shows SME scans of a deposit where particles are identified with two different elements (Si on the left and O on the right). This confirms that silicon exists as an oxide.

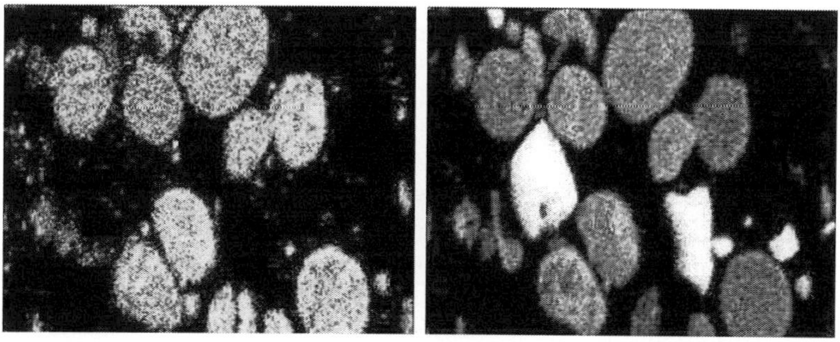

Figure 4.5 Example of scanning electron microscope element map.

If the fouling surface is available for microscopy, e.g., a fouled tube that was removed from the heat exchanger and cross-sectioned, it can be used to study the laydown pattern of the deposit. For example, SEM might show a layer of corrosion (FeS) closest to the surface, and then a layer of salts plus coke-like material.

4.1.6 Guide to the Interpretation of Results

Examples of interpretation of Level 1 analysis for two samples, both collected in crude preheat trains, are shown below. Sample E is very typical of crude preheat trains, in the heat exchangers downstream of the desalter. Sample F is for a specific case of crude oil incompatibility, and will be seen in the hottest part of the crude preheat. Results of TGA and elemental analysis for the two samples are summarized in Table 4.2.

From the TGA test for Sample E it can be seen that about 22% out of the 51% organic portion of the deposit is relatively hardened coke. This is the portion that combusts after air is added during the TGA test. This implies an active, temperature-related coking mechanism occurring in the

Table 4.2 Elemental analysis and TGA summary for Sample E and Sample F.

Element	Sample E wt%	Sample F wt%
C	46.20	77.20
H	3.30	5.00
N	1.00	1.00
S	9.10	2.80
Si	3.10	3.50
Cl	6.90	—
Fe	4.50	4.50
Ca	2.80	—
Na	3.30	—
Al	1.50	—
Mg	0.60	—
Ni	0.20	—
V	0.40	—
Total	82.90	94.00
O calculated	13.06	4.00
Overall	95.96	98.00
TGA	—	—
Volatilized	32.0	50.0
Combusted at 800 °C	22.0	46.0
Ash	46.0	4.0

exchanger. However, given the large inorganic component (45–50%) in the deposit, it can be concluded that this is not the dominant mechanism. From the elemental analysis, it can be seen that the total amount of organic material is ca. 51% (C, H, N, and perhaps some S). This is consistent with 46% ash from TGA and implies that the fouling is roughly an even mixture of coke-like material and inorganic deposits.

The inorganic part of the deposit is dominated by various salts. Since there is not sufficient iron for all the sulfur to be in the FeS form, it is assumed that some of the sulfur was present as sulfate. This is not uncommon in crude deposits, although it is also common to see the inorganic part dominated by iron sulfide. In this example iron sulfide is about 7% of the deposit.

In terms of the quality of the analysis, a close look at the overall mass balance indicated that it is necessary to calculate the oxygen content. In fact, the measured weight percentages added to ca. 83% and, assuming as per above that a large portion is sulfates, the oxygen calculations bring the mass balance to ca. 96%. As noted before, this is satisfactory for practical purposes and also eliminates any doubts that any element has been missed from the tests. If necessary, a sulfur forms test can be used for further confirmation.

The H/C ratio calculated from the HCN test is 0.86, which is consistent with the range expected in such samples (see Section 4.1.4.2 on CHN testing) and confirms that the organic material was formed in the heat exchanger.

From all the observations above, it can be concluded that the overall fouling mechanism for Sample E is a combination of inorganic particle deposition and coking of crude trapped at the surface in these deposits. Experience has shown that preventing deposition, by using high velocities (or shear stress), is the most cost-effective way of dealing with this type of fouling. It might be worth investigating whether the amount of salts can be reduced at their source; however, deposition will still occur at low velocities.

In the case of Sample F, TGA alone provides enough indication to determine the fouling mechanism with a high level of confidence even before conducting elemental analysis. In fact, the large organic (85–90%) material detected via TGA strongly suggests that the formation of the deposits is dominated by incompatibility between the crude oils mixed to form the inlet stream to the heat exchanger. The crude flowing through this heat exchanger has precipitated asphaltenes that agglomerate into large particles and deposit on the tube wall, where they eventually thermally degrade to coke-like material. A test for asphaltene compatibility can be performed to verify if this is the case. Most major oil companies, and several others that provide chemicals to them, have their own, proprietary, tests for asphaltene compatibility (also called asphaltene stability).

Although TGA would have sufficed to identify the fouling mechanism, the elemental analysis provides some more insights on the conditions under which the deposits were formed. The H/C ratio of 0.77 is slightly lower than the values expected in crude heat exchangers. This could be explained by two reasons: high surface temperature in the heat exchanger and/or a longer than usual residence time at the high temperatures.

Experience in dealing with this type of fouling indicates that the most effective way of reducing it is by controlling crude incompatibility, and by eliminating the precursors—precipitated asphaltenes. High velocities have been found not to help as much as they do when inorganic deposition plays a large role in fouling.

4.1.7 Conclusions—Use of Deposit Analysis Results

The previous sections have shown how characterization techniques can be used to postulate a fouling mechanism associated with a specific sample and how to use the results to guide fouling mitigation strategies in oil refineries. These have been discussed in various parts of this chapter—increase velocity, control crude incompatibility, minimize (upstream and in the heat exchanger tubes), remove dominant precursors at their source, and eliminate or control additives that may produce chemicals that promote fouling. Here the typical situations encountered with crude preheat heat exchanger deposits are summarized.

- If the sample composition is dominated by inorganic materials (i.e., corrosion products or salts make up at least 70% of the deposit), the likely fouling mechanism is particulate deposition due to low velocity (or shear stress). In this case, it is commonly believed that increased salt removal in a normally operating desalter has a beneficial impact on heat exchanger fouling. However, it is not proven in practice that this is true and the most effective remedy is likely to be an increase in fluid velocity or shear stress. Other means, e.g., tube inserts, which have the effect of increased shear stress have also proven to be effective in some cases. No economically viable methods (e.g., filtering) are currently known or practiced to mitigate this problem by removal of inorganic particles (which are typically smaller than 30 μm in diameter thus difficult to capture).
- If iron sulfide is dominant in the deposit (>60%), the fouling mechanism is most likely corrosion based and it might be possible to prevent upstream corrosion and minimize its flow in to the heat exchangers (e.g., by

changing metallurgy). However, this may not always be economically feasible as there is a large amount of equipment, piping, etc., upstream that contributes to the creation of these particles.

- If an unexpected element is detected in a relatively significant amount (i.e., more than a few percent for a species that is not expected at all), its presence can often be traced back to a chemical added upstream for purposes such as corrosion control or flow enhancement. Unfortunately, from an operational point of view, it may or may not be possible to control the usage and dosage of these chemicals and their effects.
- A deposit dominant in organic material (>70%) points to an asphaltene precipitation mechanism and to an incompatibility problem with the crude. This should be confirmed by an analysis of the crude. The effective way to deal with this fouling is by making the crudes compatible, but it depends on the economics of buying and blending different crudes, and being able to process them at the given refinery.
- The H/C ratio is significant in determining the nature of the coking process. A ratio >0.95 indicates organic material that is not yet fully degraded and can perhaps be washed away by a solvent. A ratio <0.8 indicates hardened material with long exposure to a hot surface, and a ratio <0.7 indicates exposure temperatures that are higher than those encountered in a crude heat exchanger. The last case is unlikely to be observed in a crude preheat train, and indicates that the organic material was formed elsewhere and not in the heat exchanger.
- If chemical cleaning—circulation of a solvent material—is being considered, the deposit composition will help to determine which solvent will be most suitable. For example, if water-soluble salts are dominant, a steam-out might provide a high degree of cleaning. Or, if the organic portion is dominant with H/C > 0.95, an aromatic solvent might be effective in dislodging the unconverted, partially soluble, material. Note that typical coke-like material (H/C < 0.9) and corrosion products are not soluble in the chemical cleaning agents that are currently in use.

4.2 CHEMICAL STRUCTURE AND MOLECULAR WEIGHT CHARACTERIZATION

M. Millan, S. Venditti

The previous section of this chapter has presented some recommended guidelines for quantitative analysis of heat exchanger deposits in refinery PHTs. This was aimed at identifying the dominant mechanism that leads to fouling and suggesting practical actions to remediate it. This section, based on work originally presented to the 8th Heat Exchanger Fouling and Cleaning Conference in Schladming, Austria (Venditti, 2009b), focuses on developing a more fundamental understanding of the nature of deposits and underlying fouling mechanisms. The development of a combination of techniques used for the characterization of heat exchanger deposits is presented using a case study on the analysis of four refinery samples. It is shown how several methods for the analysis of liquids and solids can be combined to gain a useful insight into the nature of deposits.

The analytical work combined liquid phase analysis of the soluble fractions of the deposits with analysis of the solid samples. For the liquid phase analysis, size-exclusion chromatography (SEC) and synchronous UV–fluorescence spectroscopy (UV-F) were used. For the solid samples, proximate analysis was carried out in a thermogravimetric analyzer and attenuated total reflection-Fourier transform infrared spectroscopy (ATR-FTIR) was used to examine the functional groups of the deposits and to calculate aromaticity indices. Also for the solid samples, scanning electron microscopy coupled to energy dispersive X-ray spectroscopy (SEM-EDX) and XRD have been used to obtain a better understanding of inorganic components in the deposits, which proved to be high in some cases. Despite the limitations of each technique, the results from this combined approach suggest its suitability for the study of crude oil deposits and complex carbonaceous materials.

The main features, advantages, and disadvantages of the analytical techniques whose use is described in the literature on fouling studies are summarized in Table A1 in the Appendix. Some of these techniques (i.e., TGA, elemental analysis (EA), XRD, optical microscopy, and SEM) have already been introduced in Section 4.1; thus this section will only provide, when appropriate, a brief description of SEC, UV-F, and ATR-FTIR.

4.2.1 Methodology

Table 4.3 lists the four heat exchanger deposits analyzed in this work. The origin of the samples is confidential. All samples originate from tube-side locations with exception of the sample DA (shell-side) and were collected by oil refinery operators, following standard shutdown protocols.

Initially, samples of deposits received from the refinery were extracted by a series of solvents, including chloroform, 1-methyl-2-pyrrolidinone (NMP), and toluene, to determine their solubility properties. The extracts were examined by SEC and UV-F, two techniques that have been

Table 4.3 Deposit samples used in this study *VGO (virgin gas oil) is a mixture of coker mid-distillate, light cycle oil, and vacuum gas oil.

Sample name	Unit	Exchanger side	Temperature range (°C)	Feed processed
PHTD	Atmospheric distillation preheat train	Tube	158–166	Desalted oil
KPHTD	Kero preheat train	Tube	240–260	Kerosene
DA	Desulfurizer train	Shell	Up to 315	VGO*
DB	Residue processing	Tube	Up to 370	Oil residue

developed in previous work for the characterization of several heavy hydrocarbon samples, including heavy oil fractions (Álvarez, 2008; Berrueco, 2008; Dabai, 2010; Karaca, 2004).

SEC is a liquid chromatography technique where a sample in solution is injected into a packed column containing a porous stationary phase. The solvent must be strong enough to avoid any interactions between the sample and the column packing, such that the separation takes place purely by molecular size. Small molecules are able to enter the pores and therefore they take a longer path than larger molecules. The latter are excluded from the column pores, eluting at shorter times.

UV-F involves the excitation of molecules in a dilute solution by UV radiation. In molecules presenting fluorescence the electrons return to the ground state, emitting light at a longer wavelength than that exciting the molecule. UV-F spectra can be recorded in three modes, emission, excitation, and synchronous. In emission spectra, excitation is carried out at a single wavelength and the emitted light is recorded in a range of frequencies. Excitation spectra involve excitation over a range of wavelengths and the fluorescence intensity is observed at a single wavelength. In synchronous spectra, both excitation and emission are varied simultaneously, but a constant wavelength difference between them is used. These fluorescence emission spectra reveal information about the abundance and types of functional groups present in fluorescent organic molecules. Of particular importance to the analysis of asphaltenes is the effect of polynuclear aromatic groups on the UV-F spectra. As the size of a fused aromatic ring increases, there is a shift in the fluorescence spectrum toward longer wavelengths, while the fluorescence intensity decreases.

The data collected with SEC and UV-F were combined with analysis of the insoluble material and samples of the whole deposit. The next section describes the characterization of the soluble fraction of the deposits and Section 4.2.3 presents the characterization of the whole deposits and their insoluble fractions.

4.2.2 Analysis of the Soluble Fraction of the Deposits

The first step in the characterization procedure was to test the solubility of the deposits in some common solvents. The results for the four deposits are listed in Table 4.4. DA and DB were less soluble in the solvents used, including toluene, compared to PHTD and KPHTD. This appears to match their exposure to higher temperatures, which contributed to the formation of a coke-like material.

The fractions of the samples soluble in NMP/$CHCl_3$ 6:1 ratio (vol/vol), the solvent mixture used as eluent in SEC, were analyzed by SEC and UV-F. Figure 4.6 presents the SEC chromatogram of the soluble fraction of the deposits. SEC chromatograms of heavy oil fractions and other heavy hydrocarbon samples typically show two peaks. At short elution times (corresponding to larger molecular sizes), between 10 and 13 min, a peak excluded from the column porosity appears. This represents material that due to its large molecular size cannot enter the column pores and therefore elutes in the space between solid phase particles, therefore having shorter paths and elution times. It has been shown in previous work that existing elution time calibrations (based principally on the use of polystyrene standards) overestimate the molecular masses of material eluting in the excluded range. This overestimation was linked to changes in conformation from relatively planar to rigid three-dimensional forms as in fullerenes (Karaca, 2004). The second peak corresponds to material retained by the column eluting at longer elution times. Calibrations have been proven to suitably describe

Table 4.4 Solubility of the deposits used in this study (see Table 4.3 for sources) PHTD are deposits from the atmospheric distillation preheat train, KPHTD are from the kero preheat train, DA are from the desulfurizer train, and DB are from residue processing

	Solubility (% of the Deposit)			
Solvent	**PHTD**	**KPHTD**	**DA**	**DB**
$CHCl_3$	45	80	61	58
NMP	35	38	37	32
NMP:$CHCl_3$	39	42	39	34
Toluene	43	46	33	34

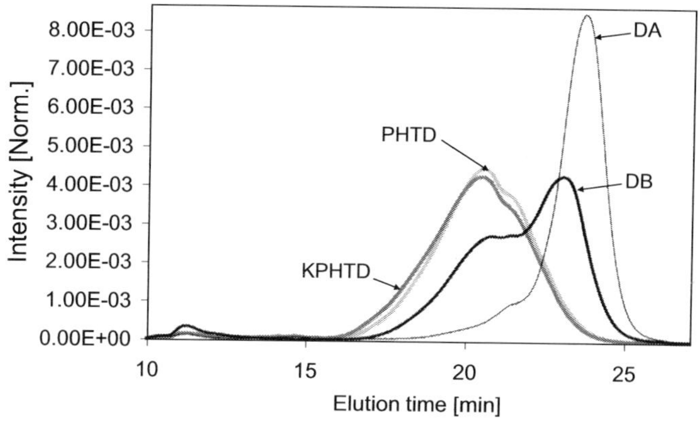

Figure 4.6 Size-exclusion chromatograms of the NMP:$CHCl_3$ (6:1 vol/vol)-soluble fraction of heat exchanger deposits obtained in a mixed-D, size-exclusion chromatography column.

material in the retained region of the chromatogram. The calibration for the column used in this work has been described by Berrueco (2008).

Clear differences were observed between the samples recovered from the four deposits. The retained peak was observed to shift to longer elution times (smaller molecular weights) from KPHTD to DA. This shows the presence of larger molecules in KPHTD compared to DA. The DB chromatogram shows a bimodal retained peak at intermediate values with a maximum intensity for the second peak around 23 min. In terms of distribution, PHTD and KPHTD chromatograms showed the presence of larger molecules in comparison with DB, whereas DA consisted of predominantly light material, likely to have been occluded in the deposit without much alteration.

Intensity-normalized synchronous UV–fluorescence spectra of the deposits are presented in Figure 4.7. UV-F spectra show a shift toward longer wavelengths as the size of the polycyclic aromatic groups increases (Morgan, 2005). The spectra in Figure 4.7 show that DA contains the smaller aromatic chromophores of the four deposits and this can be related to the SEC results that showed that its molecular weight distribution was shifted to low values. The spectra display a shift toward longer wavelengths, from KPHTD to DB and to PHTD, suggesting the largest aromatic cores are in the PHTD sample.

4.2.3 Analysis of the Insoluble Fractions and the Whole Deposits

Heat exchanger deposits and their toluene-insoluble fractions were analyzed by TGA. TGA allowed the estimation of volatiles, ash content, and fixed

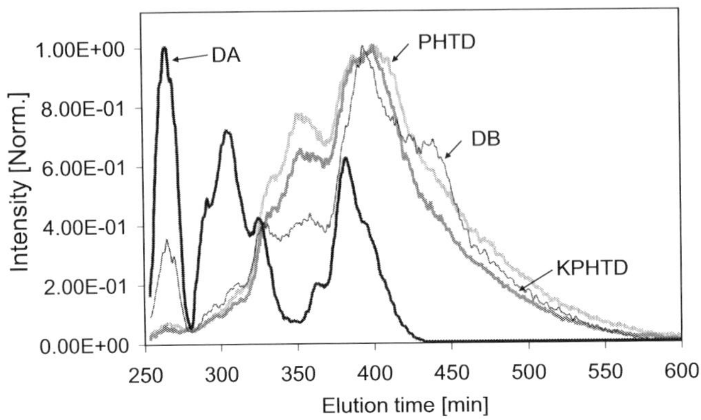

Figure 4.7 Synchronous UV-F spectra of the NMP: $CHCl_3$ (6:1 vol/vol)-soluble fraction of heat exchanger deposits.

Table 4.5 Proximate analysis of the deposits (PHTD, KPHTD, DA, and DB) and their toluene-insoluble fractions (PHTD-TI, KPHTD-TI, DA-TI, and DB-TI) carried out by TGA

Sample	Volatile loss by 370 °C (%)	Total volatiles by 900 °C (%)	Fixed carbon (%)	Ash (%)
PHTD	16.8	46.8	18.0	35.2
KPHTD	7.4	50.3	24.1	25.6
DA	27.0	61.1	33.0	5.9
DB	14.0	40.7	47.9	11.4
PHTD-TI	1.8	29.8	13.8	56.4
KPHTD-TI	2.7	18.9	39.9	41.2
DA-TI	1.0	39.9	57.0	3.1
DB-TI	4.0	21.7	63.3	15.0

carbon in the deposits and their fractions. The results are summarized in Table 4.5. All deposits showed a certain amount of light material that was volatile under the TGA conditions at temperatures up to 370 °C. This was particularly high for DA (27%), which correlates well with the observations made by SEC and UV-F on the soluble fraction. PHTD and KPHTD, on the other hand, presented relatively high ash contents. It is clear that toluene solvent extraction took away most of the light volatiles and therefore the toluene-insoluble fractions showed a marked increase in the fixed carbon.

The high ash content observed in KPHTD and PHTD triggered further analysis to identify the inorganic foulants present by XRD (Table 4.6) and SEM-EDX (Table 4.7). XRD is useful for identifying crystalline phases of

Table 4.6 Inorganic species present in the ash of the deposits measured by X-ray diffraction (XRD). Analyses were carried out in the toluene-insoluble fraction of the deposits

Inorganics present by XRD	PHTD-TI	KPHTD-TI	DA-TI	DB-TI
FeS	Yes	Yes	Yes	Yes
NaCl	Yes	Yes	Yes	No
SiO_2	No	No	No	Yes
$BaSO_4$	No	No	No	Yes
$CaSO_4$	No	No	No	Yes
$FePO_4$	Yes	No	No	No
VS	No	No	No	Yes
MgAlSiO	No	Yes	No	No
AlSiO	No	Yes	No	No
ZnS	No	Yes	No	No
$CaCO_3$	Yes	No	No	No

Table 4.7 Major inorganic elements present in the ash of the deposits measured by energy dispersive X-ray. Analyses were carried out on the toluene-insoluble fraction of the deposits

Inorganics present by XRD	PHTD-TI (wt%)	KPHTD-TI (wt%)	DA-TI (wt%)	DB-TI (wt%)
Na	3.8	2.7	0.1	0.2
Mg	0.3	1.3	—	0.1
Al	0.7	0.5	0.3	1.7
Si	1.5	7.2	0.2	1
K	0.2	—	—	—
P	0.2	—	0.1	0.5
Ca	2.2	1.7	0.2	1.1
V	—	—	—	0.8
Ba	0.9	—	—	1
Fe	10.4	8.8	5.3	4.9
Ni	—	—	0.2	0.7
Zn	1	1.4	—	—

XRD, X-ray diffraction.

inorganic components such as iron sulfide present in corrosion scale. As the toluene-insoluble fraction is regarded to be the main foulant component of the deposit, only those fractions were analyzed. High levels of iron and sulfur are present in all of the samples, especially in PHTD-TI and KPHTD-TI. The high ash, iron, and sulfur contents may indicate corrosion fouling as the main issue for PHTD and KPHTD.

For KPHTD-TI and PHTD-TI, the presence of salts was relevant. High levels of sodium were observed by SEM-EDX, with the presence of sodium chloride confirmed by XRD analysis. This result was somewhat unexpected considering that the feed processed in the PHTD unit was desalted crude. The deposition therefore appears related to incomplete desalting. Analysis of the KPHTD-TI sample also suggested the presence of magnesium silicate; possible sources are the reservoir itself or organic acid salts added during the production process.

The low ash content observed in the analysis of DA and DB suggests that they were mainly related to thermal decomposition due to the high operating temperature. In both cases, corrosion fouling was also observed. Sulfates and sulfides were detected, especially in DB-TI. The presence of Ni and V in DB also suggests a more asphaltenic nature. The DA ash content was found to be lower than that in DB. Originated from a vacuum gas oil feed, DA showed a lighter nature, which would explain the small differences between values for the DA sample and its toluene-insoluble fraction.

Figure 4.8 presents the ATR-FTIR spectra of the set of heat exchanger deposits. In ATR-FTIR, spectra are generated by interaction between the sample and an infrared beam internally reflected in a crystal. The IR beam propagates a few micrometers into the analyte in an evanescent wave and interacts with molecular vibrations in the sample. Each functional group in the sample's compounds has characteristic vibrational modes with frequencies corresponding to a certain energy and wavelength of the incident light. By this technique it is therefore possible to extract information on the functional groups present in the sample.

In this study, the ATR-FTIR spectra showed no bands in the 3900 to 3100 cm^{-1} region, indicating low concentrations of OH or NH groups. The region 3000–2800 cm^{-1}, related to the aromaticity of the samples, presented high intensity peaks.

In the 1750–1600 cm^{-1} region, all samples showed a peak about 1700 cm^{-1} corresponding to carboxylic acids with low intensity. The 1600 cm^{-1} band is known as the coke band and was observed to shift to lower wavelengths in deposits aged for longer, in agreement with observations from the literature (Fan, 2006). These results showed the typical condensed polyaromatic structure of the aged deposits.

The band assignment in the 1300–1500 cm^{-1} region corresponds to CH_3 and CH_2 alkyl chains bending mode. More intense signal in this region was observed in the cases of PHTD and KPHTD.

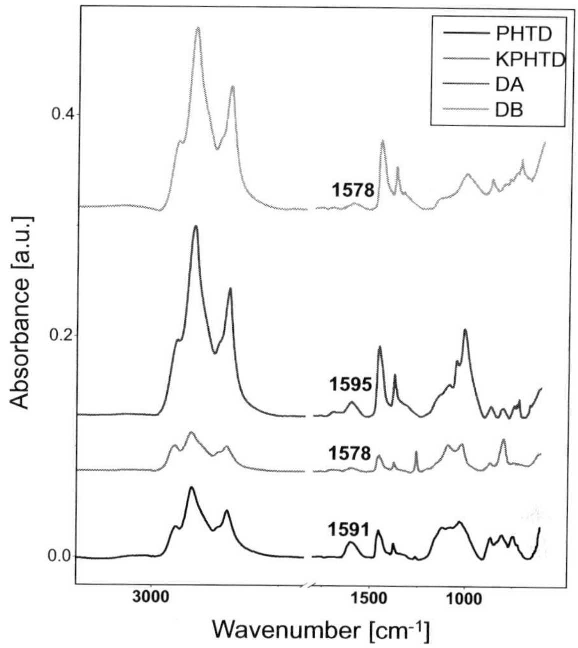

Figure 4.8 Attenuated total reflectance infrared spectroscopy (ATR-FTIR) of heat exchanger deposits.

The band at 1100–950 cm^{-1} is assigned to ethers, alcohols, and sulfoxides. The presence of C=S (1026 cm^{-1}) was observed in all four samples. However, the signal in this band is more evident for the DB and KPHTD samples. This result can be linked to corrosion due to sulfides, in agreement with data from elemental analysis, SEM-EDX, and XRD.

The 900–700 cm^{-1} region shows the aromatic out-of-plane vibrations of aromatic C–H bonds. Three main peaks are observed in all the spectra, and are more important for the DB sample. The band at about 865 cm^{-1} corresponds to isolated aromatic hydrogen. The band in the range 850–800 cm^{-1} may be attributed to systems containing two and/or three adjacent hydrogens. The peak at 750 cm^{-1} is due to the *ortho*-substitution of the aromatic rings.

Combining the information obtained through the different analytical techniques, some possible causes of fouling are proposed for each case. The preheat train deposit (PHTD) seems to be mainly formed through a combination of corrosion and coking products. The deposit had a large toluene-insoluble fraction (57%) and high ash content (35%). The PHTD-TI sample had high values of Fe and S (10.4% and 6.3% respectively) as iron sulfide. Despite PHTD coming from a desalted crude oil, there is evidence of salts and relatively high Na and Ca contents.

Similar conclusions were reached for KPHTD. This deposit was found to be low in asphaltenes and high in ash content (25.6%). The toluene-insoluble fraction contained 8.8% Fe and 3.3% S, respectively. XRD indicated the presence of iron sulfide, sodium chloride, and other salts. The sample was rich in magnesium and other silicates. These findings suggest forms of corrosion fouling associated with the presence of various salts.

DA was different from the other samples. The sample also had low ash content (6%). Metal and salt contents are low. It originates from a relatively light fraction and SEC and UV–fluorescence spectroscopy confirmed the presence of small molecules and small aromatic chromophores in the soluble part of the deposits. This lighter fraction was probably occluded in the deposit without having had time for significant thermal degradation. However, in general, thermal degradation did appear to be the main cause for fouling as the relatively high fixed carbon fraction suggests.

DB also had its origins in thermal degradation. The toluene-insoluble content was high (76%) and the H/C ratio (ca. 0.86) was low. The ash content is low and mostly due to the presence of iron sulfides, a corrosion product. The higher level of organic materials in samples DA and DB compared to samples PHTD and KPHTD seems consistent with exposure to higher temperatures. The presence of Ni and V is in agreement with the more asphaltenic origin of the sample, a vacuum residue in the case of DB.

4.2.4 Conclusions on Chemical Structure and Molecular Weight Characterization

Combining the information obtained through the different analytical techniques, some possible causes of fouling are proposed for each case. The PHTD seems to be mainly formed through a combination of corrosion and coking products. The deposit had a large toluene-insoluble fraction (57%) and high ash content (35%). The PHTD-TI sample had high values of Fe and S (10.4% and 6.3% respectively) as iron sulfide. Despite PHTD coming from a desalted crude oil, there is evidence of salts and relatively high Na and Ca contents.

Similar conclusions were reached for sample KPHTD. This deposit was found to be low in asphaltenes and high in ash content (25.6%). The toluene-insoluble fraction contained 8.8% Fe and 3.3% S, respectively. XRD indicated the presence of iron sulfide, sodium chloride, and other salts. The sample was rich in magnesium and other silicates.

These findings suggest forms of corrosion fouling associated with the presence of various salts.

DA was different from the other samples. The sample also had low ash content (6%). Metal and salt contents are low. It originates from a relatively light fraction and SEC and UV–fluorescence spectroscopy confirmed the presence of small molecules and small aromatic chromophores in the soluble part of the deposits. This lighter fraction was probably occluded in the deposit without having had time for significant thermal degradation. However, in general, thermal degradation did appear to be the main cause for fouling as the relatively high fixed carbon fraction suggests.

DB also had its origins in thermal degradation. The toluene-insoluble content was high (76%). The ash content is low and mostly due to the presence of iron sulfides, a corrosion product. The higher level of organic materials in DA and DB compared to PHTD and KPHTD seems consistent with exposure to higher temperatures. The presence of Ni and V is in agreement with the more asphaltenic origin of the sample, a vacuum residue in the case of sample DB.

The premise of the present study was that characterizing heat exchanger deposits may provide a key to understanding organic and inorganic fouling in these heat exchangers. The work has shown that useful insights into the nature of the deposits can be gained with this approach. The relatively low solubility of these samples requires analyses in solution to be combined with techniques for characterizing solids. Results indicate that this is a suitable method for the study of crude oil deposits and complex carbonaceous materials.

4.3 CHEMICAL IMAGING OF DEPOSITED FOULANTS AND ASPHALTENES

F. H. Tay, S. G. Kazarian

Fourier transform infrared (FTIR) spectroscopy has been one of the most robust techniques in material characterization. The recent emergence of attenuated total reflection–Fourier transform infrared (ATR-FTIR) imaging has permitted the study of heterogeneous systems, providing both chemical and spatial information about the sample. In ATR-FTIR (Attenuated Total Reflection Fourier Transform Infrared spectroscopy), an infrared transparent crystal (for instance a diamond) is brought into close contact with the sample. An infrared beam is fed into the crystal in such a way that the beam is totally reflected at the crystal/sample interface. The interaction of the beam with the interface produces an "evanescence" wave which penetrates into the solid sample (though for a very short distance – hence the capability to measure the near-surface region). The evanescence wave is modulated by the sample in a manner characteristic of the IR absorption by the solid at the IR frequency. The small penetration depth of the evanescence wave of the ATR approach makes it a convenient sampling method with little or no sample preparation and allows it to be applied to highly absorbing materials such as carbonaceous hydrocarbons. One of the advances described in this chapter is the utilization of a movable aperture to control the angle of incidence in the versatile diamond ATR accessory. This is used to correct the distortion of spectral bands due to the dispersion of the refractive index; this allows reliable IR spectral information to be obtained from challenging high refractive index materials such as petroleum deposits and asphaltenes. This development facilitated the acquisition of ATR-FTIR images of actual crude oil deposits from the refinery and laboratory-extracted asphaltenes. The novel applications of combining macro- and micro-ATR modes in FTIR imaging to these materials, with the fields of view of $610 \times 530\ \mu m^2$ and $63 \times 63\ \mu m^2$, respectively, have yielded important information about the spatial distribution of different components in the deposits. The macro-ATR imaging approach provides a larger field of view which can be used to obtain the overall distribution of different components in the measured sample. The enhanced spatial resolution of the micro-ATR approach allows a closer look of the sample and also allows a more representative spectrum of a particular component to be extracted.

4.3.1 Application of FTIR Spectroscopy in the Characterization of Asphaltenes and Fouling Deposits

Yen and Erdman (1962) were among the first to report the application of IR spectroscopy to study the structure of petroleum asphaltenes. Yen and Erdman developed a detailed peak assignment of the IR spectra of asphaltenes; they also proposed methods to determine the size of the aromatic clusters and the aliphaticity of the samples. Their results suggested that as the size of aromatic clusters increases, the number of terminal aliphatic chains decreases.

Coelho et al. (2006) used diffuse reflectance IR Fourier transform spectroscopy (DRIFTS) together with spectral deconvolution modeling to determine the abundances of the carbon chains in asphaltenes and resins by looking at the absorbance of the 2927 and 2957 cm^{-1} bands. Coelho and Hovell (2007) proposed a methodology to determine the functionality and calculate the percentage of single H and paired H atoms, attached to aromatic rings for both asphaltenes and resins. In that study, they compared the absorbance of the spectral bands corresponding to the symmetric and asymmetric aromatic hydrogens in methyl-substituted arenes, in the 3100–2900 cm^{-1} region, and of the bands corresponding to the out-of-plane deformation modes in the 900–700 cm^{-1} region. DRIFTS was also compared with the spectra of asphaltenes acquired from transmission mode and was used to monitor the thermal evolution of asphaltene fractions by Christy et al. (1989).

Siddiqui (2003) studied the hydrogen bonding capabilities of asphaltenes with phenol (OH group) and pyperidine (NH group). It is known that oxygen atoms in asphaltenes can be present as hydroxyl groups and attached to nitrogen atoms in basic pyrrolic forms that may lead to the formation of strong intermolecular hydrogen bonds in asphaltenes. The absorbance of the ν(OH) band of phenol depends on the presence of functional groups in asphaltenes and Siddiqui's results showed that the asphaltenes with high oxygen and low nitrogen contents have poor interaction with phenol, which indicates that oxygen might be incorporated as acidic hydroxyl groups in asphaltenes.

FTIR spectroscopy is a quick and relatively simple technique to give structural and chemical information of samples; thus it has been used in many studies to verify the structure of asphaltenes. Akrami et al. (1997) used this robust technique to characterize pitches from Avgamasya asphaltite prepared by solvent extraction and pyrolysis followed by vacuum distillation of the resulting tar and air blowing of the vacuum-distilled tars. Several chemical transformations were observed in these processes including changes in aromaticity and aliphaticity and the formation of oxygenated groups in the samples. In general, aliphaticity decreases and aromaticity increases with temperature. Huang (2006) used FTIR spectroscopy to study thermally degraded fractions of asphaltenes. The different degrees of oxidation were observed in fractions collected before 450 °C and after 450 °C. The spectrum of the degraded asphaltene fraction (>450 °C) was totally different, thus suggesting that the polycyclic structure of the molecules only decomposes in the temperature range of 450–650 °C.

Methodology using chemometric analysis and deconvolution of FTIR spectra has been developed to predict organic functional groups of asphaltenes. Orrego-Ruiz et al. (2011) use partial least square regression models of spectra obtained using ATR-FTIR spectroscopy of vacuum residues to predict asphaltene content. Coelho et al. (2011) produce theoretical infrared spectra of organic sulfur compounds using first principles and deconvolution of FTIR spectra to predict the organic sulfur constituent of the asphaltene molecule.

Asphaltenes separated from "live" or "dead" oil samples were compared by Aquino-Olivos et al. (2003). As crude oil is normally under pressure in a reservoir, live oil is pressure-preserved oil and dead oil is at atmospheric pressure. FTIR spectroscopy was used to show that there are large differences between asphaltenes extracted in the laboratory and asphaltenes obtained at high pressure. Asphaltenes from the live oil samples appeared to be more polar as revealed by the content of the functional groups and were more aromatic. Aquino-Olivos and coworkers suggested that polarity may be the governing factor rather than size in the precipitation process of asphaltenes in oil reservoirs.

Carbognani and Espidel (2003) compared asphaltenes and resins from stable and unstable crude oil. Oxygenated compounds were observed to be more abundant within fractions isolated from unstable oils, particularly in resins and in the low molecular range fractions.

As described in the previous section of this chapter, Venditti et al. (2009b) combined several analytical techniques such as size-exclusion chromatography, ultraviolet fluorescence spectroscopy, and FTIR to examine the chemical and structural functionalities of petroleum deposits created in a batch microbomb reactor. Their work showed that the mechanism of fouling is not exclusive to asphaltenes and that chemical reactions of different crude fractions could be a larger contributor of deposit formation.

Boukir et al. (1998) studied the photooxidation of asphaltenes. Using structural indices derived from the FTIR spectra, an overall increase in carbonyl groups and a decrease in aliphaticity of the molecules were observed. Juyal et al. (2005) studied the influence of heteroatom groups on molecular interaction that leads to aggregation of asphaltenes. Asphaltene samples were chemically altered by methylation and silylation, and were studied with FTIR spectroscopy. It was found that the silylation reaction was less effective than methylation in reducing the aggregation. The results suggested that the presence of sulfur and nitrogen functional groups has an important role in the aggregation of asphaltenes.

Douda et al. (2004) studied the structure of a Maya asphaltene–resin complex using FTIR. It was found that whole asphaltenes have a higher degree of aromaticity and a higher content of heterocyclic compounds and ketones as compared to maltene fractions.

Calemma et al. (1995), apart from looking at the aromaticity and hydrogen bonding of asphaltenes, also investigated the IR band intensities of carbonyl groups (1750–1600 cm^{-1}). They deconvoluted the spectral zone into four bands centered at 1735 cm^{-1} (esters), 1700 cm^{-1} (ketones, aldehydes, and carboxylic acids), 1650 cm^{-1} (highly conjugated carbonyls such as quinone-type structure and amides), and 1600 cm^{-1} (aromatic C=C stretching). Using an empirical index of carbonyl abundances based on these bands, the contents of the oxygenated groups in different asphaltene samples were compared. Calemma et al. (1995) also investigated the degree of condensation and degree of substitution of different asphaltenes using the out-of-plane aromatic C–H deformation modes in the 900–700 cm^{-1} spectral range.

4.3.2 FTIR Spectroscopic Imaging

A single spectrum provides information about the chemical compositions of the sample in the measured area but it contains no spatial information of the different chemical components. By studying many spectra measured from different locations in a sample, the chemical distribution of various components in the sample can be obtained in an image. Novel approaches to FTIR spectroscopic imaging are needed to fully utilize the power of this chemical imaging technique. The modes of image acquisition are the same as in conventional FTIR spectroscopy, i.e., transmission, reflection, and ATR. The requirements for sample preparation for each mode are similar but in imaging, it is important that the spatial identities of the chemical domains of interest are not altered so as to obtain a true chemical map of the sample. The two most robust and common modes of imaging are in transmission and ATR. The differences between the two FTIR imaging modes have been examined in greater detail by Kazarian et al. (2009) who discuss the advantages and limitations of the particular imaging mode in terms of spatial resolution, the image field of view (FOV), and possible artifacts.

4.3.2.1 ATR-FTIR Imaging

ATR is a robust sampling technique used in infrared spectroscopy which can directly examine gaseous, liquid, and solid samples with minimal preparation. The principle of ATR-FTIR spectroscopy is described in detail by Harrick (1987); it is an extremely versatile sampling technique for surface characterization. In ATR-FTIR imaging, the beam of infrared light enters the high refractive index element, reflects off the internal surface of the crystal in contact with the sample, and goes onto a focal plane array (FPA) which measures thousands of spectra from different regions in a sample. Many areas of research have benefitted from the application of ATR-FTIR imaging, for example, materials and forensic science, pharmaceutical research, conservation science, and biomedical studies.

4.3.2.1.1 Different Capabilities of ATR-FTIR Imaging

Micro-ATR-FTIR imaging offers greatly improved spatial resolution compared to more conventional transmission FTIR spectroscopic imaging. Importantly, ATR-FTIR imaging provides higher spatial resolution than that achieved with FTIR imaging in transmission using a synchrotron source of infrared radiation. Micro-ATR-FTIR spectroscopic imaging opened up many new areas of study, which were previously precluded by inadequate spatial resolution, as discussed by Kazarian and Chan (2010). For example, Kazarian and Chan (2013) used micro-ATR-FTIR imaging to allow more precise analysis of the very small domains in pharmaceutical tablets or biomedical samples. This advanced technique can be a powerful tool for studying petroleum deposits and other samples and allows chemical visualization with enhanced spatial resolution and obtaining valuable chemical information about the compounds present in crude oil and its deposits. This information is crucial for an understanding of crude oil fouling and the ability to control and mitigate this phenomenon. Recent developments in macro-ATR-FTIR imaging, primarily achieved in the Kazarian laboratory at Imperial College London, with the use of inverted prism crystals show good potential with applications to depth profiling of materials, studies of dynamic processes in chemical systems, materials crystallization, and imaging of flows and reactions in microfluidic channels. Macro-ATR-FTIR imaging is a highly versatile technique allowing one to study dynamic processes at different temperatures and pressures, and with different imaging fields of view and spatial resolution. Table 4.8 summarizes the different capabilities of ATR-FTIR imaging.

4.3.2.1.2 Micro-ATR-FTIR Imaging

The main advantage of micro-ATR imaging with the use of a microscope objective is the high spatial resolution images that can be achieved using this method. The high refractive index of the ATR crystal used for this type of ATR imaging (typically a germanium (Ge) crystal with a refractive index of 4) greatly increases the numerical aperture of the system and hence it is possible to achieve spatial resolution beyond the diffraction limit of light in air compared to imaging in transmission mode where the Ge crystal is not used (Chan et al., 2003). High spatial resolution FTIR images up to the diffraction limit were obtained using a bright synchrotron source in the work of Chouparova et al. (2004) and Dumas et al. (2004). However, such images are obtained by rastering and this is usually a relatively slow procedure. Hence, this technique lacks the capability for studying dynamic systems whereas the spatial resolution is still limited by the diffraction of light traveling in air. The signal to noise ratio (SNR) of the spectra collected, on the other hand, is often better than those measured by focal plane array (FPA) detectors and, therefore, it can be a good complementary method. FTIR images obtained with the use of an FPA detector in micro-ATR mode can be obtained within a few minutes of acquisition time and the achieved SNR is often sufficient for most applications. With this advantage, micro-ATR imaging enables measurements of small features which were not attainable before. The high resolving power also enhances the detection limits for heterogeneous materials as reported by Chan and Kazarian (2006). Thus, it opens a range of new opportunities for studying complex materials, polymer blends, and pharmaceutical tablets where the region of interest is often in the micrometer scale as discussed by Chan and Kazarian (2006).

4.3.2.1.3 Macro-ATR-FTIR Imaging

In macro-ATR-FTIR spectroscopy, diamond is widely used as the internal reflection element due to its infrared transparency across the mid-IR spectral range and its relatively high refractive index of 2.4. The hardness of diamond is also valuable property as a high contact pressure between the sample and the crystal is sometimes important to improve the reproducibility of the spectrum as discussed by Everall and Bibby (1997) and Kazarian et al. (1999). This property also allows studies of polymeric materials under a high-pressure and high-temperature environment in situ (Kazarian et al., 2001).

ATR-FTIR spectroscopy is known to be capable of obtaining depth profiles in a nondestructive manner by changing the angle of incidence of the IR radiation. This type of depth profiling has been demonstrated in a number of previous studies such as by Ekgasit and Ishida (1997) and Shick et al. (1996). Chan et al. (2008) presented a simple modification of the diamond imaging accessory by selectively masking different parts of the light exiting the condenser lens to control the average angles of incidence of the IR beam. This new approach allows ATR-FTIR imaging to be carried with high depth resolution and also good lateral resolution (ca. 15–20 μm) due to the high-power optics used in the ATR accessory to focus the IR light onto the small diamond crystal. The introduction of the movable aperture in the diamond ATR accessory to control angles of incidence within a certain range has opened up a range of new possibilities in spectroscopic imaging of different materials. One application is to correct the distortion of

Table 4.8 Different capabilities of ATR-FTIR imaging

	With focal plane array detector					With linear array detector
	Expanded field of view ATR	Variable angle ATR	Macro-ATR (ZnSe)	Macro-ATR (diamond)	Micro-ATR (10 × Ge lens)	Micro-ATR (large Ge lens, Ø ca.12 μm)
Field of view/mm × mm	15.4 × 21.5	3.9 × 5.5	2.6 × 3.6	0.64 × 0.64	0.06 × 0.06	Up to 0.4 × 0.4
Spatial resolution (est.)/μm	500	150	60	15–20	4	4–10
High-throughput applications (No. of samples)	Yes (>100)	Yes (100)	Yes (50)	Yes (<10)	No	No
Depth profiling	No	Yes	No	Yes	No	No

spectral bands due to the dispersion of the refractive index, which allows reliable IR spectral information to be obtained from high refractive index materials. This is described in the next section.

4.3.3 Application of ATR-FTIR Imaging on Fouling Deposits

This section presents the development and application of ATR-FTIR spectroscopic imaging with an FPA detector to characterize carbonaceous materials such as crude oil deposits and asphaltenes. This study utilizes the ATR approach, which has several advantages as compared to other sampling methods such as reflection and transmission. The small penetration depth of the ATR approach makes it a convenient sampling method which requires only relatively simple sample preparation and can be applied to highly absorbing materials such as carbonaceous hydrocarbons.

The chemical maps of crude oil deposits have been reported by Chouparova et al. (2004) using FTIR spectroscopy with a synchrotron source measured in transflection mode. The synchrotron source can be masked into a 5 μm spot using an aperture in the microscope; it is still possible to obtain a quality resultant spectrum due to the much brighter source. However, this facility is expensive and not widely available. Chan et al. (2005) have demonstrated that the approach of obtaining a chemical image of a sample using an FPA can be faster compared to the chemical map generated by raster scanning in the synchrotron approach. They also compared ATR-FTIR imaging with an FPA detector to FTIR microscopy in transmission with a synchrotron source and concluded that while the SNR is much better when using the synchrotron, the spatial resolution is higher with micro-ATR-FTIR imaging due to the use of the ATR objective with a high refractive index (Ge). Thus, the versatility and the opportunities offered by ATR-FTIR imaging are still to be exploited in the characterization of crude oil deposits.

Crude oil deposits and asphaltenes are materials of high refractive indices, this may result in spectral artifacts inherent to the ATR approach. The variable angle diamond ATR accessory, noted in the last section, is used in studies of crude oil deposits and asphaltenes to be able to obtain reliable ATR-FTIR spectra of high refractive index materials. This allows carbonaceous materials to be studied using this technique. The advantages of combining both the macro- and the micro-ATR-FTIR imaging approaches to characterize materials are discussed in Section 4.3.3.4.

4.3.3.1 Conventional Single-Element Detection Challenges

The ATR-FTIR spectrum of a petroleum refinery deposit was measured on a diamond ATR accessory coupled with a single-element detector and is shown in Figure 4.9. The same sample was measured again using a 56° aperture on the lens of the diamond ATR accessory and these spectra were compared. It shows that when no aperture was used (the average angle of incidence was 47°) distortions such as the sharp decrease in absorbance close to a wavenumber of 3000 cm^{-1} and an increase in baseline after a wavenumber of 2800 cm^{-1} can be observed.

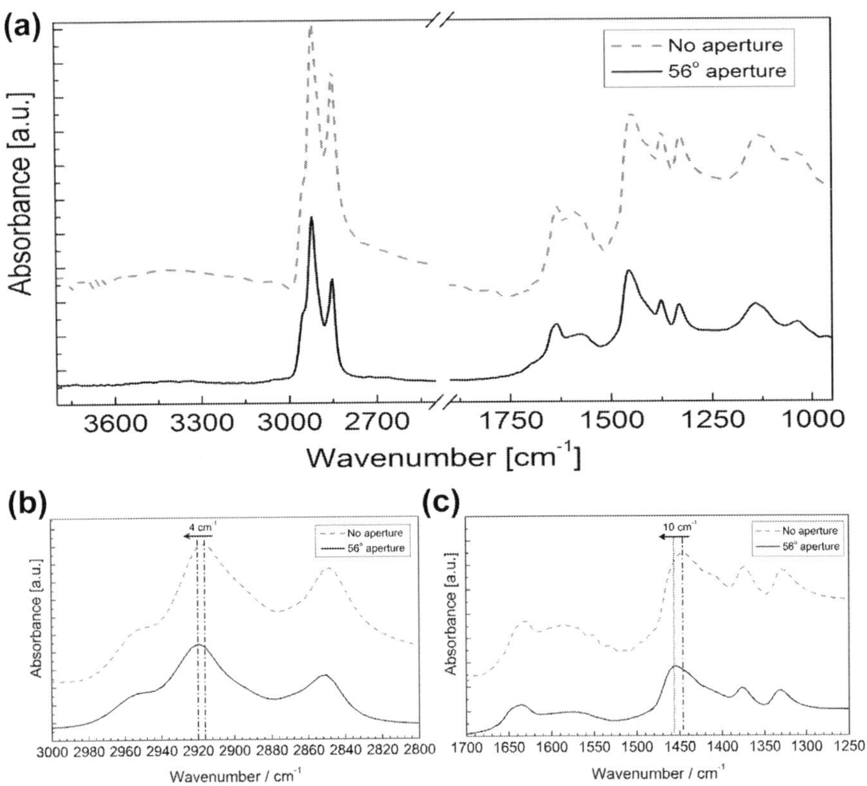

Figure 4.9 (a) Single-element FTIR spectra of petroleum deposit collected using no aperture and using the 56° aperture. (b and c) The distortions of the absorption bands at 3000–2800 cm^{-1} and 1700–1250 cm^{-1}, respectively. *Reprinted with permission from F. H. Tay and S. G. Kazarian, Study of petroleum heat exchanger deposits with ATR-FTIR spectroscopic imaging, Energy Fuels, 2009, 23, 4059–4067. Copyright (2009) American Chemical Society.*

ATR measurements made near the critical angle of the system (the angle the IR beam enters the sample material) would result in spectral bands becoming distorted. Strong bands can become broadened and red-shifted; both effects increase as the angle of incidence is decreased. These effects are due to the dispersion in the refractive index as a function of light frequency. Harrick (1987) demonstrated that the shift in spectral band is small when the angle of incidence is more than a few degrees above the critical angle of the system. Therefore, the important rule of thumb for the practical application of ATR spectroscopy is to keep the measurement angle well above the critical angle of the system in order to avoid distortion of the spectral bands.

The spectrum of the deposit is similar to a typical spectrum of extracted asphaltenes (Figure 4.10), thus indicating that the bulk is asphaltenic. Buenrostro-Gonzalez et al. (2001) and Buckley et al. (1998) have estimated the refractive index of asphaltenes to be around 1.7. Assuming the refractive index of the petroleum deposit to be around 1.7, the critical angle of the system for the diamond and Ge ATR crystals can be calculated to give 44.6° and 25.1°, respectively. Given that the average angle of incidence for the diamond ATR accessory without the aperture is 47°, this angle of incidence of the incoming IR beam is very close to the estimated critical angle for the diamond and the sample, thus explaining the observed spectral distortion.

When an aperture is introduced to the accessory, only the portion of the incoming IR beam that reaches the top surface of the diamond at a greater angle relative to the normal, will pass through, thus the average angle of incidence is increased to 56°. The resultant spectrum obtained, shown in Figure 4.9, has demonstrated that the distortion is removed. This

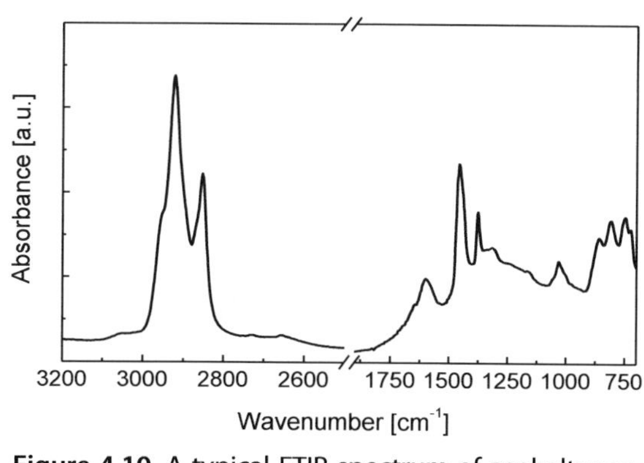

Figure 4.10 A typical FTIR spectrum of asphaltenes.

development allows not only petroleum samples but also other high refractive index materials to be measured with ATR spectroscopy without these optical artifacts. It is important to ensure that spectra are free from significant distortions for a reliable spectral interpretation and comparison of spectra obtained for both qualitative and, more importantly, quantitative analysis of materials.

The interpretation of the aromatic C–H stretching bands at 3030 cm^{-1} serves as an example. This weak band of asphaltenes can be misinterpreted because of the distortion due to dispersion of the refractive index. Figures 4.9(b) and (c) show shifts in the spectral ranges of 3000–2800 and 1700–1250 cm^{-1}, respectively, for spectra acquired with no aperture and with a 56° aperture, respectively. Structural parameters based on deconvolution of certain spectral bands can also be miscalculated due to these band distortions. One example will be the $n\mathrm{CH_2}/m\mathrm{CH_3}$ ratio of aliphatic chains in the samples, which can be correlated to the absorbance ratios of the bands of the asymmetric stretching vibrations of the methylene and methyl group at 2927 and 2957 cm^{-1}, respectively (Calemma et al., 1995). With the distortion as shown in Figure 4.9, the deconvoluted result of these bands can be easily miscalculated. In the fingerprint region, where there may be many overlapping bands, a 10 cm^{-1} shift of the bands may even result in incorrect assignments of functional groups especially in samples where the components are not known.

The concern with the effect of dispersion of the refractive index on the spectrum could be overcome with the utilization of a high refractive index crystal such as Ge or silicon (Si). However, diamond still offers the advantage of being hard, durable, and chemically inert, which allows the use of aggressive solvents for cleaning. A Ge crystal has a high refractive index of 4 but is a softer and more brittle material. For the hard asphaltenes and petroleum deposits samples where much force is needed to press the samples onto the ATR crystal to ensure good contact, repeated measurement can easily damage the crystal. Si crystal has a refractive index of 3.4 and is slightly harder than Ge crystal but it strongly absorbs below 1500 cm^{-1}, thus limiting the information obtained in the fingerprint region. The modification on the optical design of the diamond ATR accessory has made it possible to obtain reliable spectroscopic information on high refractive index materials, such as asphaltenes and crude oil deposits.

4.3.3.2 Macro-ATR-FTIR Imaging

Figure 4.11 shows the image obtained using a diamond ATR accessory with the FTIR spectroscopic imaging system. The set of imaging data is from the

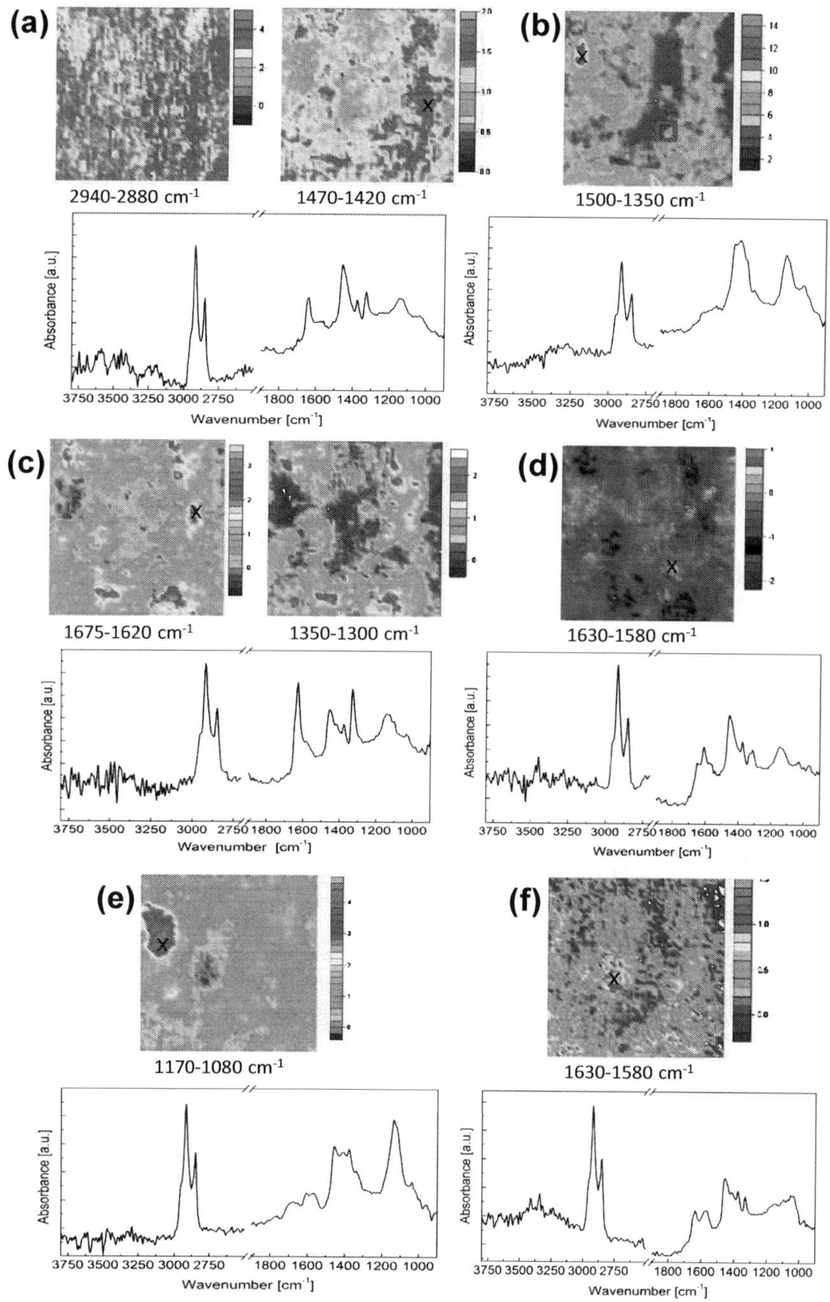

Figure 4.11 Macro-ATR-FTIR (attenuated total reflection–Fourier transform infrared) images of petroleum deposit. Images are based on the integration of the absorption band as indicated below each images. The imaging area is ca. 610 × 530 μm². Spectrum is extracted from point X of the chemical image above it. The red box in (b) shows the relative size of a micro-ATR FTIR imaging measurement. *Reprinted with permission from F. H. Tay and S. G. Kazarian, Study of petroleum heat-exchanger deposits with ATR-FTIR spectroscopic imaging, Energy Fuels, 2009, 23, 4059–4067. Copyright (2009) American Chemical Society*. (For interpretation of the color in this figure legend, the reader is referred to the online version of this book.)

same petroleum deposit sample measured using a conventional single-element detector in Figure 4.9. Each pixel of the 64 × 64 FPA detector measures a full mid-IR spectrum. The macro-ATR-FTIR images are generated by allocating a color to each pixel based on the integrated absorbance of the particular spectral band as indicated below each image in Figure 4.11. The red in the scale denotes a high value and blue denotes a low value. This color coding represents the concentration of the component and this can be quantified with a calibration graph of the known component. The univariate analysis of the FTIR images has revealed the distribution of several chemically different components. The image size of this setup is about 610 × 530 μm^2 and the spatial resolution of this approach is ca. 12 μm as described by Chan et al. (2008).

Good contact of the sample with the ATR crystal had been achieved as shown by the homogenous distribution of the absorbance of the asymmetrical methylene stretching mode at about 2920 cm^{-1} in Figure 4.11(a). It also shows an image based on the integration range of 1470–1420 cm^{-1} and this is assigned to the bending modes of CH_2 and CH_3 groups. The reason it does not correspond to the image of the stretching modes of C–H in the 2940–2880 cm^{-1} region is that there are also other functional groups such as carbonates that appear in the 1470–1420 cm^{-1} range.

The images based on the 1675–1620 and 1350–1300 cm^{-1} ranges show similar distributions in the measured area, therefore implying that the two bands are likely to belong to the same component. It was found from a spectral library search that these two bands can be assigned to the oxalate functional group. Using conventional FTIR spectroscopy, the band at 1635 cm^{-1} can be seen but the band at 1330 cm^{-1} is not obvious. It can be difficult to assign a specific compound for an unknown sample as the 1635 cm^{-1} band can also be assigned to different conjugated carbonyls. In an FTIR image, domains of high concentration of this component can be located thus a more representative spectrum of the component can be extracted and spectral assignment can be made with higher confidence.

The image in Figure 4.11(b) is based on the integrated absorbance of 1500–1350 cm^{-1} and it shows the distribution of carbonate compounds in the sample. This characteristic carbonate band overlaps with the band assigned to the bending modes of CH_2 and CH_3 groups at 1470–1420 cm^{-1}; thus the assignment of the carbonate compounds would not have been obvious if just based on the interpretation of the FTIR spectrum measured with a single-element detector for petroleum deposit in Figure 4.9(a). Figure 4.11(e) and (f) are based on the distribution of the integrated absorbance in the range of 1180–1100 and 1080–1000 cm^{-1}. They can be assigned to the sulfate and sulfoxide functional groups, respectively.

Although conventional single-element FTIR spectroscopy may have a better SNR and a faster acquisition time, its sensitivity is inferior to the more advanced imaging technique. In single-element measurements, chemical information about the sample is averaged across the whole measured area of the ATR crystal; thus spectral bands of a component in a heterogeneous sample will be diminished by the spectral bands of other components depending on their concentration and absorptivities. Interpretations of the set of chemical images are that the crude oil deposits from the heat exchanger are extremely heterogeneous. The clusters of distinct compounds ranging in size from ca. 40 μm for the oxalate to ca. 150 μm for the sulfate can be observed. This emphasizes how the enhanced sensitivity of the imaging approach can provide more accurate chemical information about a sample, especially for heterogeneous materials. When components are resolved spatially, spectroscopic information that better represents the components can be obtained.

The fouling mechanism of crude oil heat exchangers is highly complex. It is well accepted that fouling involves the process of crystallization of inorganics, deposition of particulates, corrosion and chemical reactions of organics such as oxidative polymerization, asphaltenes precipitation, and coke formation (Crittenden et al., 1992; Watkinson and Wilson, 1997). The mitigation strategy is therefore plant specific and largely dependent on the feedstock composition of the crude oil. The FTIR images have shown both chemical and spatial information about the organic and inorganic fractions in petroleum deposits. From the measurements with a single-element detector and the distribution of the methylene asymmetrical stretching mode measured by the macro-ATR-FTIR imaging approach, the bulk of the samples were observed to be mainly asphaltenic. Detailed analysis of the organics such as aromaticity, length of the aliphatic side chains, and carbonyl abundances can be carried across the large data set obtained from each imaging measurement. The chemical images also reveal chemical heterogeneities in petroleum deposits especially for the mineral compounds. This may help in indicating the main mechanisms that contribute to fouling, thus providing information to the heat exchanger specialist in deciding mitigation strategies. For example, the oxalates detected are not commonly found in crude oil but are known to exist in the form of an organic acid salt in sedimentary rocks (Jehlicka and Edwards, 2008). This particular compound is present in significant amounts in the deposit obtained with this particular crude oil; it could therefore be an initiator or a cofactor to the fouling process.

4.3.3.3 Micro-ATR-FTIR Imaging

A higher spatial resolution ATR imaging approach has been used to examine the petroleum deposit. The micro-ATR-FTIR imaging technique uses a microscope objective with a Ge crystal. The Ge crystal is also hard, 6 on the Mohs scale, but it is brittle compared to the diamond crystal. It has a high refractive index of 4 that provides the high spatial resolution. The image size of the micro-ATR measurements is about 63 × 63 μm^2 and the spatial resolution is about 2–4 μm as described by Kazarian et al. (2009). A red square in Figure 4.11(b) shows the relative image size of a micro-ATR measurement, which is almost 100 times smaller compared to the size of a macro-image obtained with the diamond ATR accessory. As shown, the sample is chemically dispersed; thus a single micro-ATR-FTIR image is not a good representation of the sample. Multiple measurements of the sample are required to better understand the distribution of the chemical components.

The micro-ATR-FTIR image in Figure 4.12(a) represents the distribution of a component based on 1630–1580 cm^{-1}. Figure 4.12(a) shows a cluster of domain size of ca. 7 μm and this chemical component was not detected in the initial analysis from the macro-ATR-FTIR images. According to the spectrum extracted, this component has two strong bands at 1605 and 1313 cm^{-1}. The oxalates component observed from the macro-ATR measurement in Figure 4.11(c) also has two similar bands at 1635 and 1330 cm^{-1}. It is speculated from these results that this unknown component may be a different form of complex of the oxalate. With the information of a new chemical component, the integration range of 1630–1580 cm^{-1} is applied to the macro-ATR dataset to show the distribution for this compound and the image generated is presented in Figure 4.11(d). The domain size of the component is large enough to observe a distribution in the image, but the spectral information of this component is still masked by the surrounding contributions. By contrast, the spectrum from the micro-ATR measurement is able to show stronger bands of this component due to its enhanced spatial resolution.

The distribution of the carbonates shown in Figure 4.12(b) is based on the integrated absorbance in the range of 1500–1350 cm^{-1}. The spectrum extracted from the area of high concentration of the component shows a

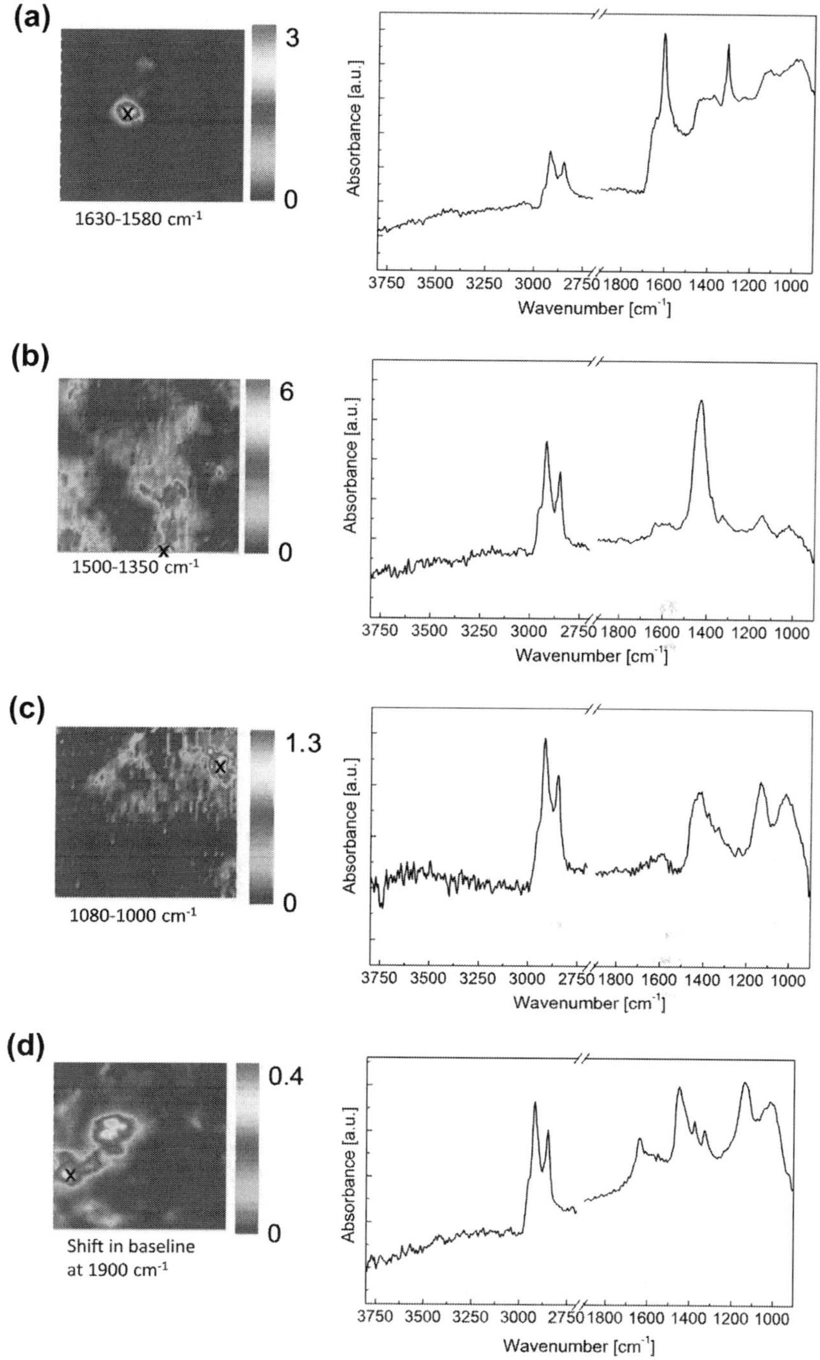

Figure 4.12 Micro-ATR-FTIR images of the petroleum deposits. Each of the images is from different measurements. Images (a) to (c) are based on the integration of the absorption band as indicated below each images. Image (d) is based on the shift of absorbance of baseline at 1900 cm^{-1}. Spectrum is extracted from point X of the chemical image beside it. *Reprinted with permission from F. H. Tay and S. G. Kazarian, Study of petroleum heat-exchanger deposits with ATR-FTIR spectroscopic imaging, Energy Fuels, 2009, 23, 4059–4067. Copyright (2009) American Chemical Society.*

single band at 1430 cm^{-1}. This "purer" spectrum of the carbonate component, compared to the extracted spectrum from the macro-ATR-FTIR image, can be obtained with the enhanced spatial resolution, thus allowing better chemical analysis and spectral interpretation. Figure 4.12(c) shows the distribution of the sulfoxide functional group based on the integrated absorbance in the range of 1080–1000 cm^{-1}. The spectrum extracted from this micro-ATR image also shows a much stronger sulfoxide band when compared to the spectrum of the component extracted from the macro-ATR measurement.

Figure 4.12(d) shows the image based on the shift of absorbance of baseline at 1900 cm^{-1}. The presence of carbonaceous materials such as graphite and also petroleum deposits can result in an increase of the absorbance baseline in the whole measured IR spectrum. The image in Figure4.12(d) shows clusters of ca. 30 μm in diameter, corresponding to the materials that result in an increase of the absorbance baseline. Coke formation is one of the possible mechanisms of fouling, and it is correlated to the thermal degradation of crude oil fractions. The reaction that results in coke formation is still debatable but is closely associated with the asphaltenes as reported by Rahmani et al. (2002). The chemical structure of coke consists mainly of large poly aromatic hydrocarbons with few aliphatic side chains. These aromatic hexagonal sheets have structures similar to graphite; thus they behave similarly in the IR region of the spectrum. The coke material was only observed in the micro-ATR approach and only appeared occasionally. Although the bulk temperature of the processing line where this deposit originated from was between 155 and 166 °C, the temperature of the heat transfer surface will have been higher. Fan and Watkinson (2006) compared the DRIFTS spectra of graphite and industrial samples from a bitumen coker unit and deduced that the industrial deposits are mainly graphitic. Although the sample discussed here was from a lower temperature stream compared to the 500 °C of a coker, the ageing effect of deposits is not well studied. The sample was in the heat exchanger unit for an undetermined time, and it is possible that it has aged, thus resulting in materials of a more graphitic nature. Such compounds may be formed especially on hotspots of the heat exchanger surface. These highly absorbing blackbody materials would not have been observed in techniques used before and FTIR imaging may provide a way to monitor this ageing effect of the deposits.

With a higher spatial resolution from the micro-ATR imaging, smaller domains are magnified; thus the sensitivity of the ATR-FTIR imaging effectively increases. In-depth analysis of the different components in the deposits can be carried out by using the better represented spectrum obtained using the micro-ATR approach. On the other hand, the macro-ATR approach, with a larger field of view, can give a better overview picture of the whole sample. Quantitative analysis of the deposits, such as the particle size analysis of the different components, can be obtained from the FTIR images. Although multiple measurements will still be needed to obtain a statistically significant representation of the sample, this technique is relatively fast where each measurement only takes less than 2 min.

4.3.3.4 Comparison of Petroleum Deposit and Laboratory-Generated Asphaltenes with ATR-FTIR Imaging

Asphaltenes have been closely linked and related to the level of fouling in heat exchangers; thus it is important to characterize these materials and their association with crude oil fouling. In this section, the developed methodology of combining macro- and micro-ATR-FTIR spectroscopic imaging has been applied to real petroleum deposits from the refinery and to extracted asphaltenes from three different crude oils. The deposits were obtained from the preheat train of a distillation column in a refinery with inlet and outlet temperatures at around 158 °C and 166 °C, respectively. The shutdown procedure was standard; thus the deposits were only exposed to nitrogen and steam. The deposits were removed from the process unit and sent to the laboratory. The typical crude oils received and blended at the refinery originate from different geographical regions and are known to foul when processed. The asphaltenes were extracted following the ASTM-D3279 (ASTM, 2007). The asphaltene samples have been labeled as ASP-A, ASP-M, and ASP-K. Three macro-ATR-FTIR imaging measurements were made and analyzed for each sample. The higher spatial resolution micro-ATR-FTIR images were also obtained.

The macro-ATR-FTIR images of the deposits in Figure 4.13 show the distribution of different chemical groups based on the integrated absorbance of the spectral bands as indicated on the left side of the images. The images along the first row, which were based on the integrated absorbance in the range of 3000–2800 cm^{-1}, show the distribution of the alkyl C–H stretching modes. Since the sample is dominated by hydrocarbon compounds, the high absorbance of this band throughout the images can be used as an indication that good contact between sample and ATR crystal has been established. The result has shown that ATR-FTIR imaging is suitable for studying this type of sample, despite that the samples are hard.

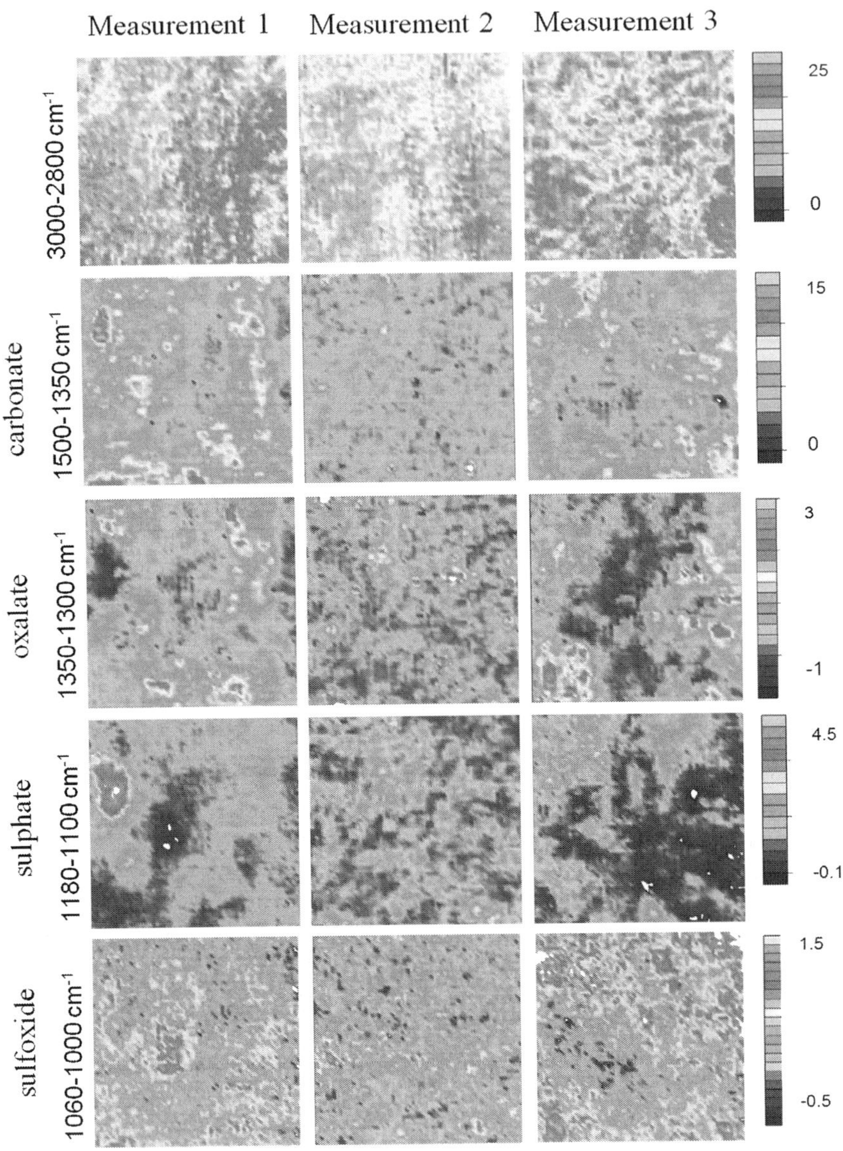

Figure 4.13 Macro-ATR-FTIR images of petroleum deposits.

All three measurements have shown the presence of the following chemical components: carbonates, oxalates, sulfates, and sulfoxides. The distribution of these components is not the same in each of the images, despite the fact that they were all measured from the same batch of petroleum deposit sample. In measurements 1 and 3, there was a higher presence of carbonates as compared to measurement 2. The distribution and particle domain size of the oxalate compounds were similar for all three measurements, but the distributions of sulfates and sulfoxide compounds were different in each image. This shows that a single macro-ATR-FTIR measurement with a sampled area of ca. $610 \times 530\ \mu m^2$ is not enough to provide representative data concerning the distributions of the chemical components in the petroleum deposits. More chemical images need to be measured in order for reliable parameters of the deposited foulant to be derived. Nevertheless, it is possible to analyze the sample qualitatively, which would still provide valuable information about the chemical nature of the sample.

The data produced by chemical imaging offer a far greater level of detail about the sample studied compared to the averaged information obtained from typical single-element measurements. This is clearly demonstrated in Figure 4.14 where the averaged spectra of the three imaging measurements of petroleum deposit are shown. Little difference can be observed among the three spectra even though the spatial distribution and the cluster size of the different components vary significantly. Due to the overlapping characteristic spectral bands of the different chemical components, the distribution of asphaltenic components in the deposits is difficult to present as an image using univariate analysis. The averaged spectra resemble the spectrum of asphaltenes (Figure 4.9); however, it can be observed that there are still contributions from other chemical

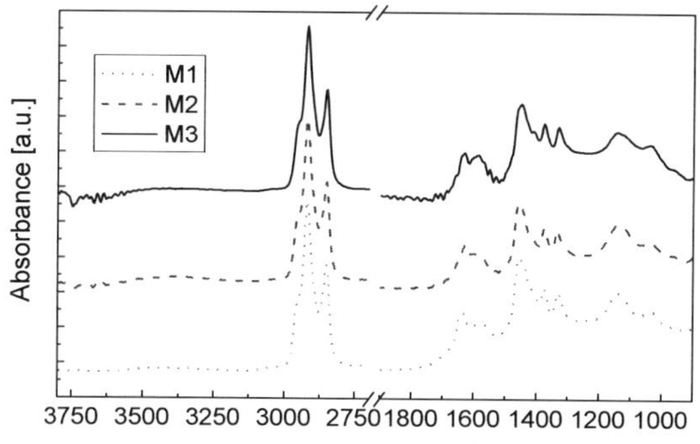

Figure 4.14 Averaged spectrum of deposit across the macro-ATR-FTIR image obtained in three repeated measurements.

components, such as carbonates, oxalates, sulfates, and sulfoxides, to the overall spectrum of the deposits.

The petroleum deposit sample was also measured using the micro-ATR-FTIR imaging and the result is shown in Figure 4.15. The measured area of this approach is ca. 63 × 63 μm^2, which is about 100 times smaller than the measured area of the macro-ATR approach. The higher spatial resolution of the micro-ATR imaging approach reveals even finer details of the sample, allowing smaller domains of components to be observed and, therefore, a more representative spectrum of a component can be obtained.

The image showing the distribution of the alkyl C–H stretching modes in Figure 4.15(a) indicated that good contact between the sample and the ATR crystal was established for the measurement. The distributions of carbonates, oxalates, and sulfates in the deposit sample are presented in Figure 4.15(b)–(d), respectively. In order to obtain a representative spectrum

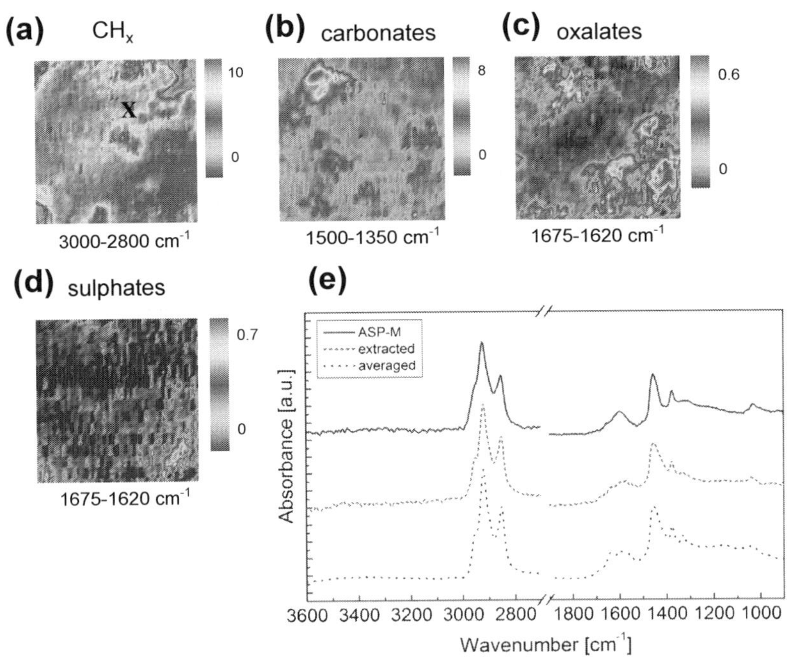

Figure 4.15 Micro-ATR-FTIR images of the deposit sample. Each of the images corresponds to the same measurement. Images (a) to (d) are based on the integration of the absorption band as indicated below each image. Spectrum is extracted from point X of the chemical image.

of the asphaltene component, a spectrum (red) was extracted from the micro-ATR-FTIR measurement where the contributions from the inorganic species are at their lowest. This point is marked X as shown in Figure 4.15(a) and the spectrum is presented in Figure 4.15(e). The extracted spectrum is compared to the spectrum (blue) of the laboratory-extracted asphaltenes, ASP-M. It can be observed that the extracted spectrum of the deposits sample has almost no contribution from the absorption bands of the other inorganic components and is very similar to the spectrum of the extracted asphaltenes. Therefore, with micro-ATR-FTIR imaging, a more representative spectrum of the asphaltene component in deposited foulants can be obtained. This result has demonstrated that spectroscopic information of the asphaltenic component in the heterogeneous deposits can be acquired easily and analyzed with the imaging system. Otherwise, routine extraction processes, such as Sohxlet extraction, would have to be carried out in order for the asphaltene fraction in the deposit to be analyzed through conventional analytical techniques.

From the discussion in Section 4.3.3.2, six chemically different compounds, namely the asphaltenic component, carbonates, sulfates, sulfoxides, and two different types of oxalates, were observed in the petroleum deposit sample. Five micro-ATR-FTIR images were selected from a set of 13 measurements of petroleum deposits. The representative spectra for each of the different components were extracted and presented in Figure 4.16. The assignment of these components was largely facilitated by the capability of the micro-ATR imaging approach to extract purer spectra of the components.

The objective of the work presented here is to develop and demonstrate an analytical methodology using ATR-FTIR spectroscopic imaging to investigate the extent of asphaltene deposition in real crude oil fouling scenarios. Ideally, a relationship between the feed crude oils and the deposited foulants they form should be found. The asphaltenic content of the deposit should be compared to the composition of the feed and its asphaltene fraction. Unfortunately, limited information, such as the type of crude oil, was available due to confidentiality issues with the refinery. Also, during the operating lifetime of the refinery, from start up to shutdown of the plant for maintenance, it is likely that more than one type of crude oil was processed. Hence, the crude oil that resulted in the formation of the deposit was not available for the analysis. Nevertheless, asphaltenes from three different regions were analyzed using macro- and micro-ATR-FTIR imaging to gather typical chemical information about asphaltenes.

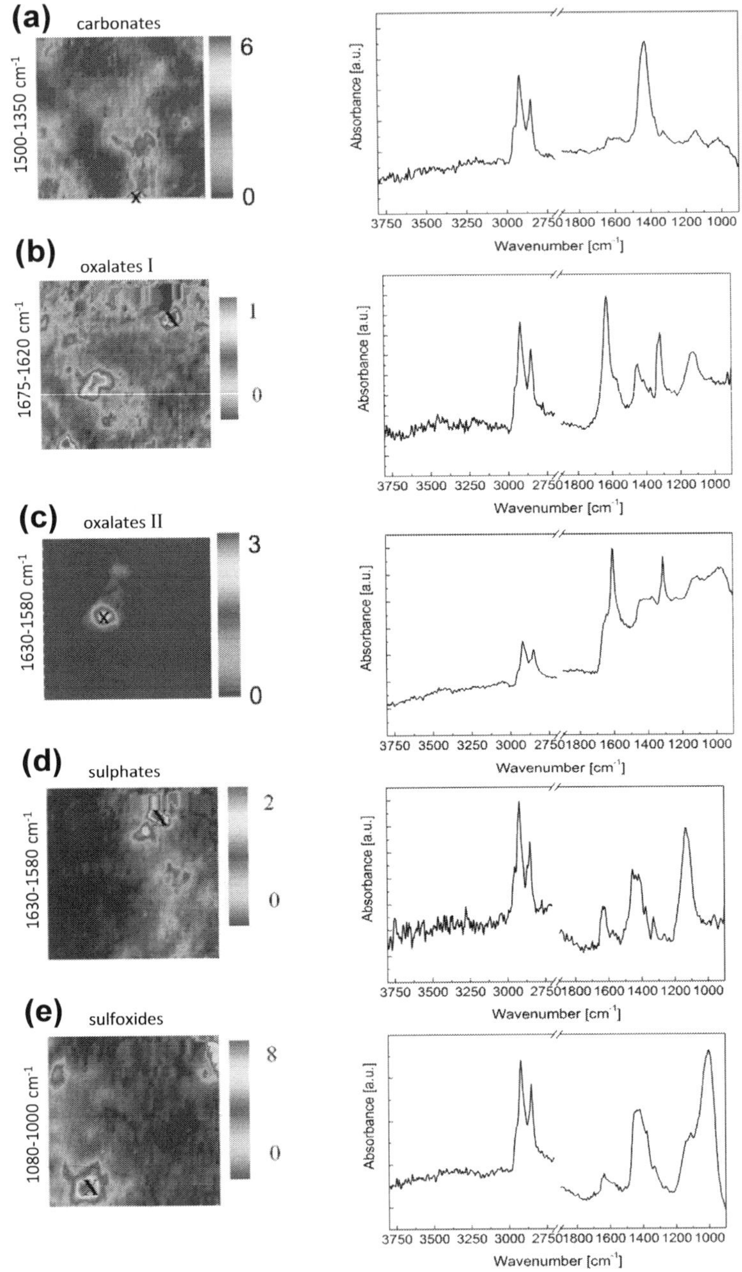

Figure 4.16 Micro-ATR-FTIR images of petroleum deposits. Each of the images corresponds to a different measurement. Images (a) to (e) are based on the integration of the absorption band as indicated at the side of each image. Images (a) and (c) are results from Figure 4.12. Spectrum is extracted from point X of the chemical image beside it.

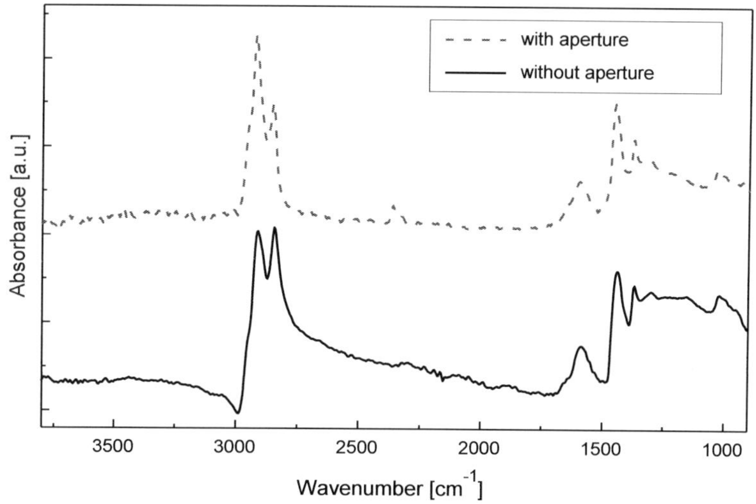

Figure 4.17 Single-element fourier transform infrared spectra of ASP-M collected with and without the use of the 56° aperture.

This demonstrates the potential of this methodology for extracting the information from asphaltenes in the laboratory using ATR-FTIR imaging.

Figure 4.17 shows the single-element spectra of ASP-M collected with and without the use of the 56° angle of incidence aperture in the diamond ATR accessory. The spectral distortion observed in the spectrum without the use of the aperture is due to the dispersion in the refractive index as a function of light frequency. The distortion is greater than the example shown in Figure 4.10 for the real crude oil deposits. This is probably due to the refractive index of ASP-M being higher than the refractive index of the crude oil deposit, which resulted in the average angle of incidence in the ATR accessory being closer to the critical angle. This highlights the importance of using the aperture to increase the average angle of incidence of the IR beam to avoid the optical artifacts, so that a reliable ATR-FTIR spectrum can be obtained.

Figure 4.18 shows the spectra of ASP-M, ASP-K, and ASP-A measured with a single-element detector. It can be observed that the general spectral features of the asphaltenes are similar. ASP-K is shown to have a larger carbonyl band at 1700 cm^{-1} whereas ASP-A has the smallest band at 1030 cm^{-1} assigned for the sulfoxide functional group. Figure 4.19 shows the macro-ATR-FTIR images of ASP-M. Each row of images was based on the integration range stated on the left of the images. Various integration

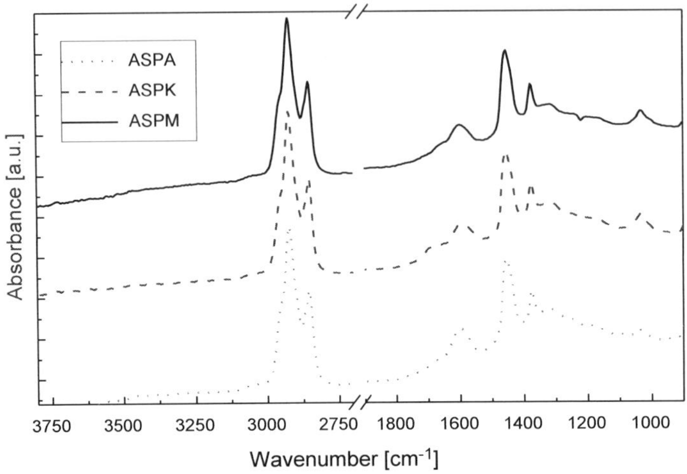

Figure 4.18 Averaged fourier transform infrared spectra of ASP-M, ASP-K, and ASP-A.

ranges were used to generate the images and they mostly show a homogeneous distribution. However, at least two chemically different components were observed. The row of images based on shift of absorbance of baseline at 1900 cm^{-1}, as shown in Figure 4.19(c), revealed the distribution of "coke-like" materials in the asphaltenes. This suggests that ordered graphitic structures exist in asphaltenes and are found in clusters of ca. 20–30 μm diameter. The top middle sections of the images, where they are boxed in Figure 4.19(c), are image artifacts and they should be ignored.

Interestingly, a small area of the sample has shown a strong absorbance band in the 1175–960 cm^{-1} region. Figure 4.19(d) shows the images based on this band. A spectrum is extracted from the red region of the image as indicated by the red arrow. This chemical species can also be observed in other repeat measurements using the macro-ATR approach and thus is not a contamination that occurred during the sampling process. The band at 1175–960 cm^{-1} is a characteristic band of cellulose which is a linear polymer of D-glucose units linked by glycosidic bonds. One possible material that bears similar spectral features is the vitrinite compound, which is one of the main constituents of kerogens (Lin and Patrick, 1993; Schenk et al., 1990; Sun, 2005). Vitrinite compounds are derived from plants and thus contain the cellulose functional group. They have been found in crude oil and are common in organic source rocks (Yao et al., 1997; Ye et al., 2007).

Two other spectra were extracted from the green and blue regions of the image at locations indicated by the green and blue arrows, respectively. The

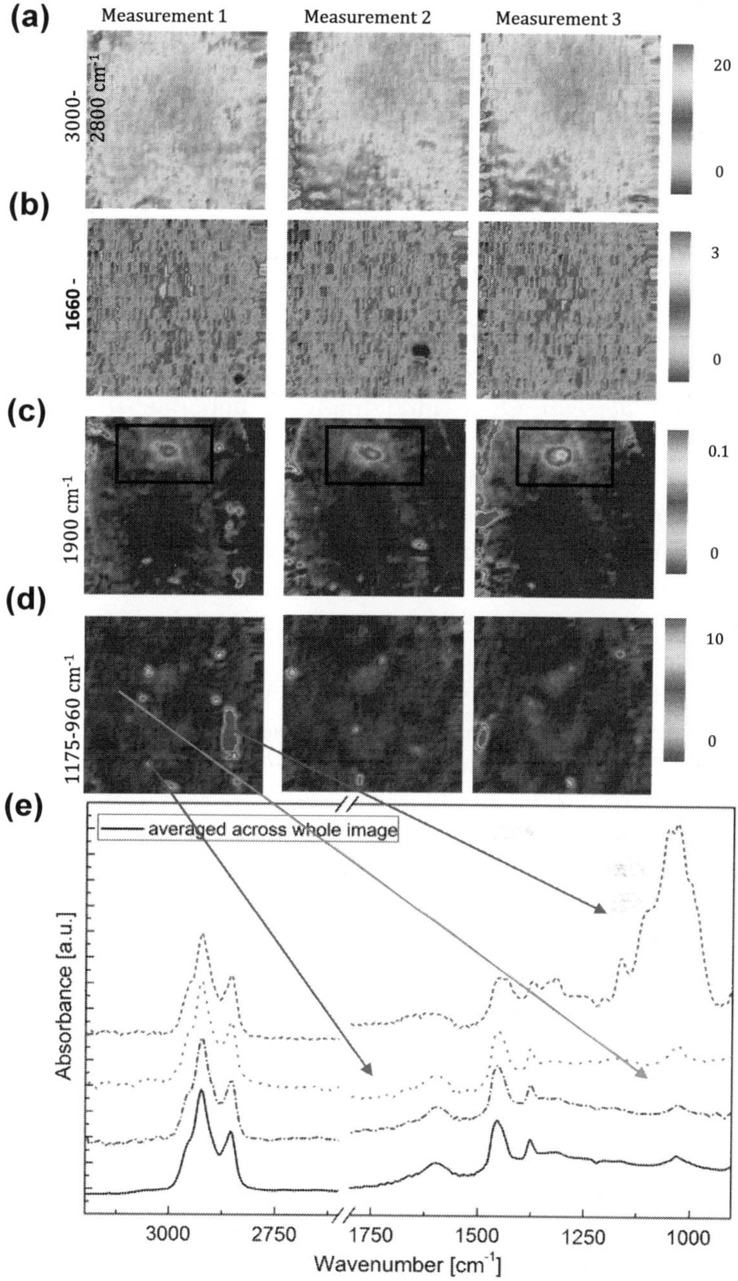

Figure 4.19 Macro-ATR-FTIR images of ASP-M. Spectra are extracted from the location of the image as denoted by the arrow.

shapes of the spectral band in the 1100–1000 cm^{-1} region extracted from the red and green regions are different, suggesting that the two spectra were extracted from areas with two chemically different species. The green spectrum in the 1100–1000 cm^{-1} region is still representative of the sulfoxide functional group. As the image generated is based on the integration range set, the distribution shown in the image may represent chemical components with overlapping bands as in this case. Therefore, it is always important to refer to the spectra data to validate the image generated. The averaged spectrum of the spectral data set of the measurement was also presented in the plot. The spectrum shows that the vitrinite compounds would not have been easily identifiable by conventional FTIR spectroscopy using a single-element detector. One would simply attribute the band observed in the 1060–1000 cm^{-1} entirely to the sulfoxide functional group.

Figure 4.20 and Figure 4.21 show the macro-ATR-FTIR images of ASP-A and ASP-K, respectively. Similar to ASP-M, the main functional groups have shown homogeneous distributions with the exception of the coke-like materials and vitrinite compounds. Both ASP-A and ASP-K show the presence of vitrinite compounds whereas only ASP-K shows the presence of the coke-like material.

Micro-ATR-FTIR imaging measurements were carried on the extracted asphaltenes. Figures 4.22–4.24 show the micro-ATR-FTIR images of ASP-M, ASP-A, and ASP-K, respectively. ASP-M and ASP-K show the presence of the vitrinite compounds but not ASP-A. In macro-ATR imaging mode, ASP-M shows more coke-like materials compared to ASP-A and ASP-K. However, in micro-ATR imaging mode, all three asphaltenes show the presence of this material. This suggests that coke-like materials exist in ASP-A and ASP-K in smaller domains compared to ASP-M.

No other distinct distributions of chemical components were observed in the micro-ATR approach, providing a FOV of $63 \times 63\ \mu m^2$ and spatial resolution of 2–4 μm. This shows that asphaltenes extracted in the laboratory are relatively homogeneous up to the spatial resolution achieved in this imaging setup.

4.3.4 Conclusions on Chemical Imaging of Deposited Foulants and Asphaltenes

A detailed chemical analysis of petroleum deposit samples was made using ATR-FTIR imaging. The visualization of the different chemical components in the deposit has shown that the sample is extremely heterogeneous

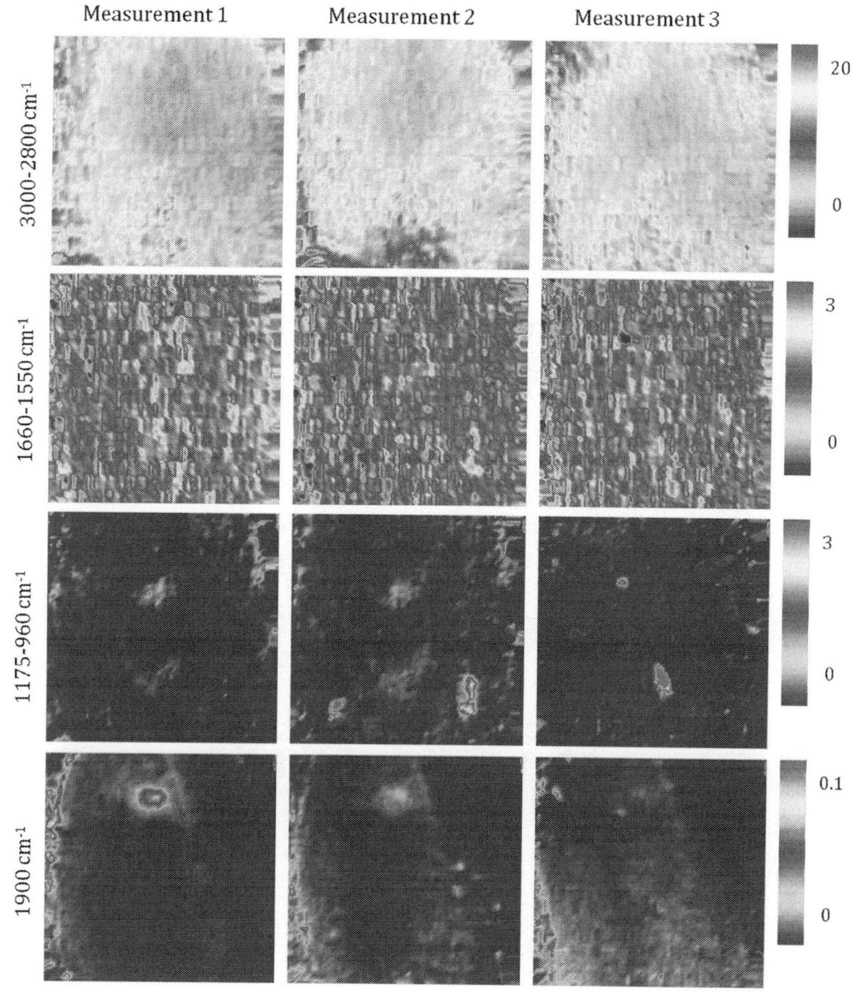

Figure 4.20 Macro-ATR-FTIR images of ASP-A.

and the FOV of a single macro-ATR-FTIR imaging measurement is not sufficient to illustrate the true distribution of the different chemical components of the measured sample. From the repeated measurements, seven chemically different components were found. The representative spectra of asphaltenes, carbonates, two different forms of oxalates, sulfates, and sulfoxides were extracted and presented using the micro-ATR-FTIR imaging approach. Macro- and micro-ATR-FTIR images of laboratory-extracted asphaltenes were obtained for the first time. The ATR-FTIR images of asphaltenes extracted from three different crude oils were analyzed and

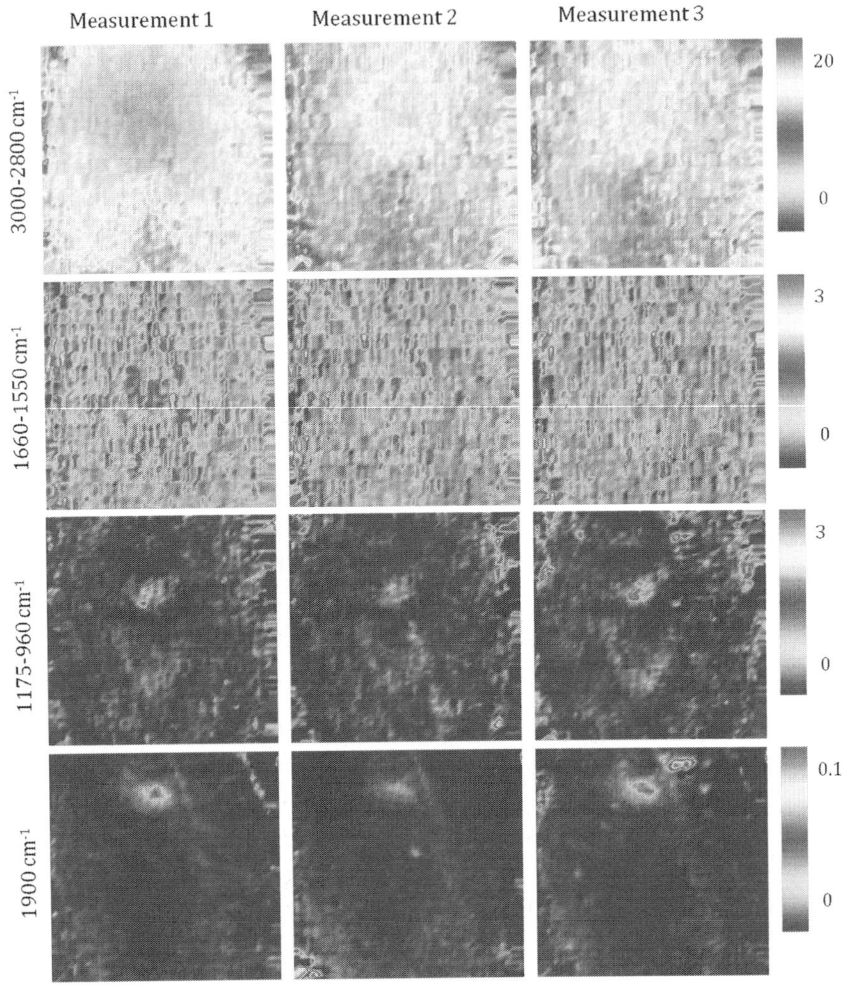

Figure 4.21 Macro-ATR-FTIR images of ASP-K.

compared. The clusters of two chemically different compounds, namely coke-like materials and, possibly, vitrinite compounds, were observed in these samples. This work has also demonstrated a methodology for comparing, spectroscopically, the asphaltenic component in a heterogeneous deposit with laboratory-extracted asphaltenes using the ATR-FTIR imaging approach, so that the significance of asphaltene deposition in crude oil deposits can be investigated.

The analytical approach of ATR-FTIR imaging using the diamond accessory can be extended to monitor the heating of crude oil in situ and

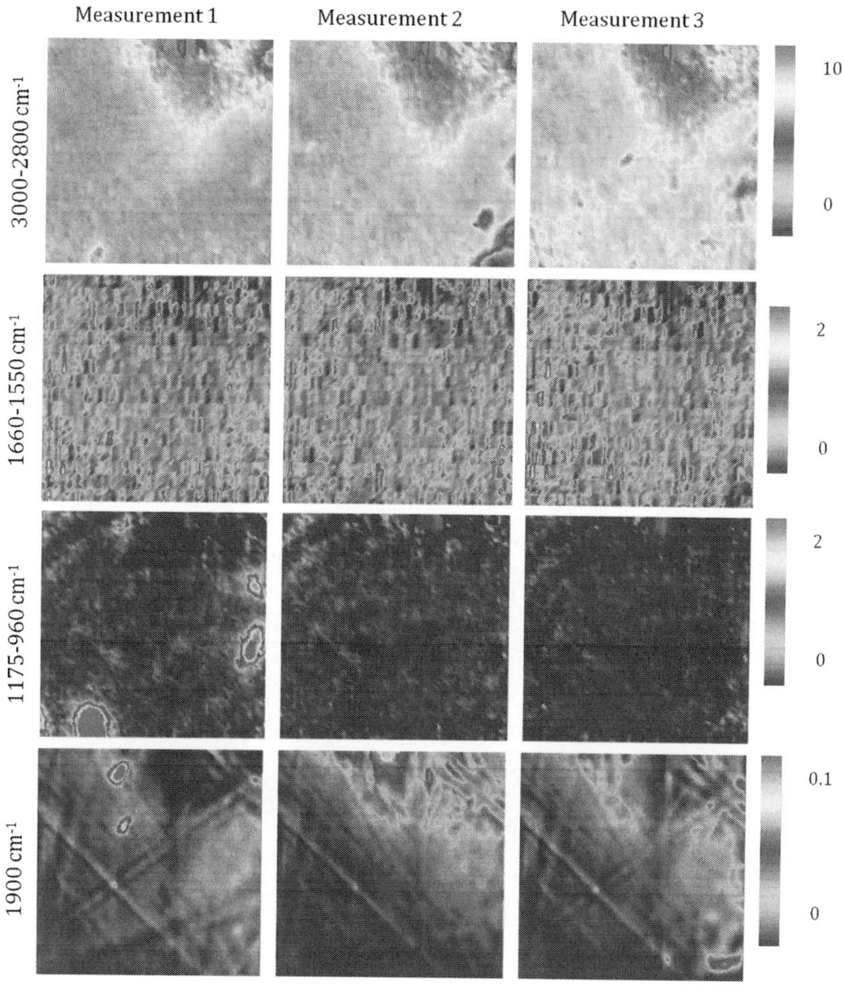

Figure 4.22 Micro-ATR-FTIR images of ASP-M.

the onset of asphaltene deposition. It is possible to determine the onset of crude oil fouling and dominant fouling mechanism at different time–temperature–pressure conditions. Gabrienko et al. (2014) used ATR-FTIR imaging to monitor the precipitation of asphaltenes in situ from Tatarstan crude oil induced by *n*-heptane. It was demonstrated that this approach can be used to detect the onset of precipitation and the dynamics of deposit particles formation and growth and to obtain chemical information of the deposition process.

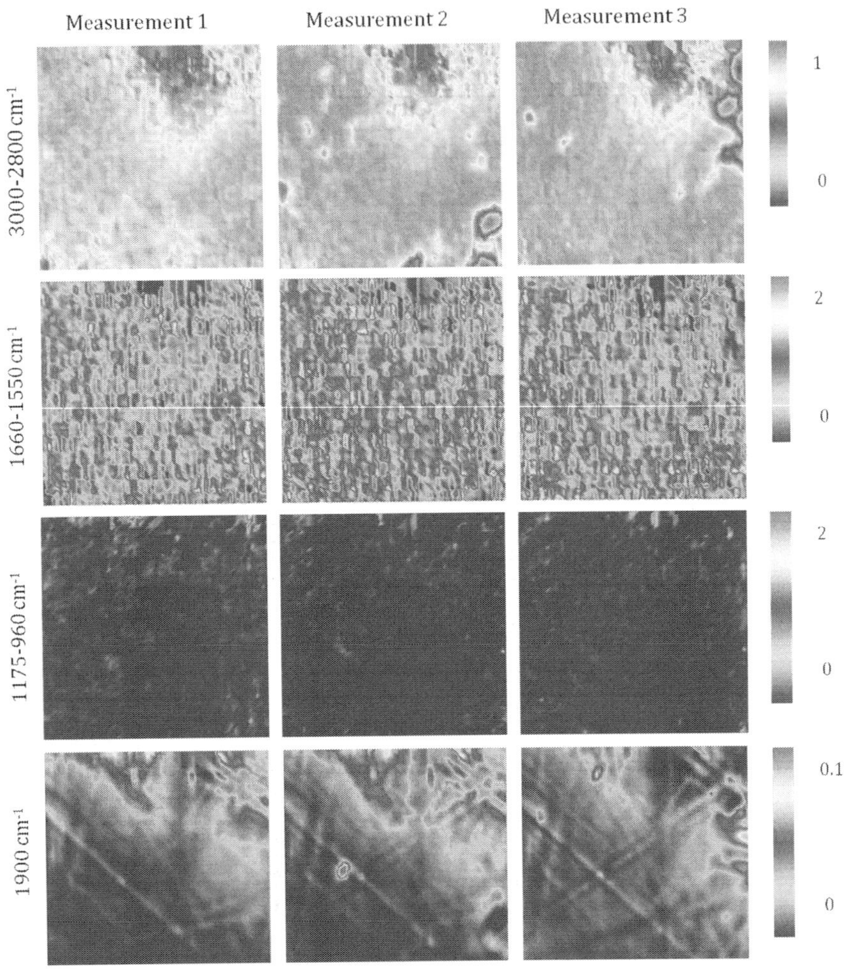

Figure 4.23 Micro-ATR-FTIR images of ASP-A.

ATR-FTIR imaging allows the visualization of different chemical components in a heterogeneous sample. By combining both macro-ATR and micro-ATR-FTIR spectroscopic imaging, the complex crude oil deposits and extracted asphaltenes can be better characterized. Relevant spectroscopic information, representing the chemical species of interest, can be extracted and analyzed from the chemical images. Hence, ATR-FTIR imaging provides an important tool in the chemical characterization of fouling materials which will aid in understanding the fundamentals of crude oil fouling.

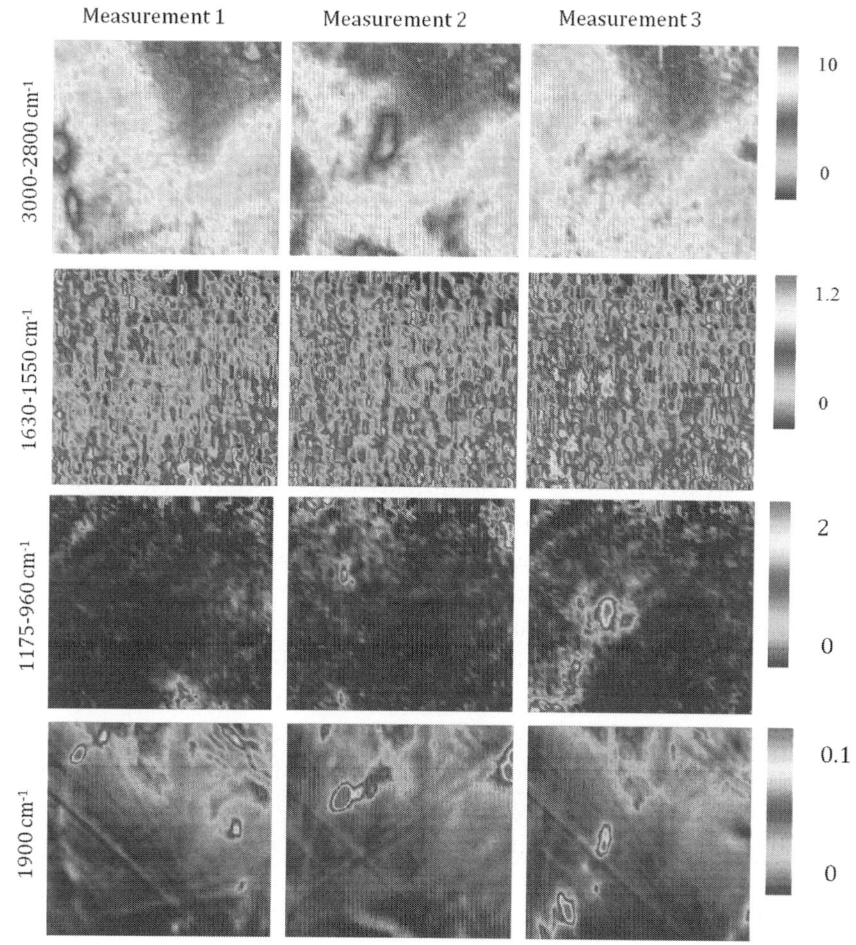

Figure 4.24 Micro-ATR-FTIR images of ASP-K.

4.4 FLUID DYNAMIC GAUGING: THICKNESS AND STRENGTH MEASUREMENTS

J. Chew.

Monitoring of deposit buildup and removal under different operating conditions is fundamental to understanding the underlying mechanisms of fouling and cleaning. When fouling develops, the thermal and hydraulic performance of a heat exchanger goes off design and decreases; thus efficient energy usage can no longer be ensured and extra pumping power is required to maintain fluid flow. This leads to product loss and the necessity for cleaning of the heat exchanger equipment using a fixed cleaning cycle. Cleaning is often time-consuming and costly, and the cleaning cycles used are seldom adaptable to different fouling conditions. Hence, fouling sensors are imperative and various methods have been developed to detect foulants.

One of the fundamental needs for fouling and cleaning studies is the development of methods for measuring the thickness of deposits and its variation with time. Deposit thickness, or its consequences, is usually an indicator for operation shutdown and cleaning periods. Strength (adhesive and cohesive) of deposits is also one of the key parameters in cleaning and removal, and it changes over time and is also often difficult to quantify. Furthermore, the deposit strength is usually a function of the degree of ageing in the layer; Müller-Steinhagen (2000) described ageing as the most poorly understood aspect of fouling. Ageing usually affects the deposits by making them stronger as a result of extended reaction between organic components, diffusion to reduce deposit voidage, or generation of adhesive extracellular material. Thus, the development of a satisfactory deposit thickness and strength measurement device would be welcomed in fouling, cleaning, and scheduling investigations.

This section is organized into three parts. The first part gives a short account of current fouling (and cleaning) detection methods. The second part introduces the principles of fluid dynamic gauging (FDG) as a fouling sensor and the different modes of its operation. The third part describes how FDG can be developed to measure fouling thickness and strengths in situ and in real time under high-temperature and high-pressure conditions.

4.4.1 Fouling Sensors

Any measurements made with respect to fouling and cleaning need to be reliable, reproducible, and accurate as far as is possible so they may be applied to process plant operation with confidence. The quality of the test data is hugely affected by the quality of the experimental design and sensors employed. The main factors that influence the accuracy of test data in fouling and cleaning of heat exchanger studies include (Glen et al., 1997; Withers, 1996):

- The geometric, dynamic, and kinematic similarities and the operating conditions used for measurements should be as close as possible to the systems they model.
- The measurements should be made in situ where possible and at a location that gives a representative impression of the degree of fouling at other locations.

- For integrated techniques, e.g., heat transfer coefficients or pressure drop methods, measurements should be performed at various sections of the heat exchanger.
- Regardless of the techniques employed, an assessment of the errors of measurement should be undertaken.
- The choice of measurement technique should be selected according to its response and sensitivity to the controllable variables, e.g., pressure drop measurements are much more sensitive to fouling of small channels than are measurements of heat transfer coefficient.
- The sensor must be capable of online application; thus, its sampling and signal processing rates need to be several orders of magnitude faster than the rate at which fouling occurs in the process.

It is almost always impossible to replicate or simulate the fouling conditions of industrial operations. Glen et al. (1997) recommended that it is important to establish the relationship between the extent of fouling and cleaning observed and assessed under well-defined but limited test conditions and the wider range of operating conditions that prevail in the full-scale industrial equipment. Furthermore, it would be of considerable benefit if the same sensor could be used to monitor the cleaning process, so as to indicate when the fouling film had been entirely removed.

There are various approaches to classify sensors, such as *intrusive* and *nonintrusive, in situ* and *ex situ*, or *integrated* and *localized*. These divisions are, however, not mutually exclusive, e.g., a sensor can be *nonintrusive, in situ*, and *localized*. Figure 4.25 illustrates these classifications. Whatever the classification, all sensors face the challenge of measuring foulants that are rarely uniformly distributed in space. Crattelet et al. (2013) and Prakash et al. (2005) presented extensive summaries of different sensor techniques.

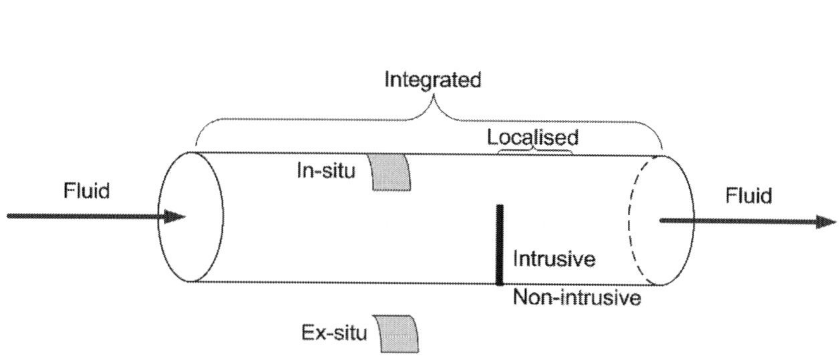

Figure 4.25 Illustration of sensor terminology.

Here, a brief review on recent sensor developments in the area of heat exchanger fouling and cleaning is presented.

4.4.1.1 Intrusive Sensors

Ideally, any fouling or cleaning sensor should be noninvasive, although this is not always possible. In the event that the sensor is invasive, its design should obviously be such as to minimize any effects it might have on the process. An example of an intrusive measurement is the conductivity technique reported by Shanrang et al. (2009). In this technique, conductivity electrodes were inserted inside a heat exchanger tube to measure calcium carbonate scaling. An AC signal was imposed on the electrodes (in electrolyte solution) and the impedance of the electrodes was compared between clean and fouled states to obtain the fouling mass. Then, by assuming a uniform deposit density, the fouling thickness is calculated.

4.4.1.2 Nonintrusive Sensors

The pressure drop (Bott, 1995) and heat transfer coefficient measurement techniques (Burton, 1968) are examples of nonintrusive sensors. Both techniques are commonly employed for heat exchanger fouling studies where fouling is usually not visible from outside the processing equipment. The sensors/gauges are usually mounted externally such that they do not interfere with the process operations. Such techniques tend to be relatively insensitive and indicate the presence of the fouling film only when it is already well established.

Another example is the mechatronic surface sensor developed by Pereira et al. (2006). The technique has been employed to monitor biological and inorganic deposits and milk fouling deposit (Pereira et al., 2009). The principle is that of ultrasonics; i.e., a sinusoidal wave is used to excite the monitored surface using a piezoelectric actuator. The amplitude and the damping factor of the excitation are related to the amount and type/nature of the deposit. Throughout the measurements, the deposit layer was assumed not to be disrupted by the actuator excitement.

Radioactivity measurements represent another type of nonintrusive sensor. These have been used by researchers since 1950 (Corrieu, 1981), having been widely adopted due to their sensitivity, reproducibility, and relative ease of adaptation to laboratory research. Such measurements require the deposit material to contain a radioisotope and the integration of a detector. The use of radioactivity measurements is obviously limited to laboratory studies and infeasible for industrial applications.

4.4.1.3 In Situ Sensors

In situ optical methods are commonly employed to evaluate the cleanliness of transparent surfaces such as glass, by comparing their light transmittance before and after fouling/cleaning. Examples include the work by Schöler et al. (2009), who used phosphorescent crystals as optical tracers to monitor the efficiency of pulsed-flow cleaning of food soils in complex pipe geometries, including gradual or sudden expansions/contractions and elbows. The sensor should not influence to any significant degree the process or the phenomenon that it is monitoring; more specifically, the presence of the sensor in the system should not promote atypical fouling.

Other online optical methods such as turbidimetry (Hardy et al., 1985), refractometry (Payne et al., 1993), and backscattered light (Tamachkiarowa and Flemming, 1999) are also widely used due to their nonintrusive nature. However, the main drawback of these methods is that they are not always adaptable to nontransparent equipment.

4.4.1.4 Ex Situ Sensors

Ex situ sensing involves dismantling the fouled section and analyzing the deposit outside its native environment. This is typically due to the incompatibility of sensing techniques with process equipment; i.e., the lens/probes/needles used cannot be fitted into tubes or ducts. Deposit from the dismantled section can be used for microscopic examination (Visser et al., 1997) or for chemical analysis. Microscopic and topographic techniques such as atomic force microscopy, scanning electron microscopy, confocal laser scanning microscopy, and Fourier transform infrared spectroscopy (see Section 4.3) can be used to characterize deposits. However, as described in previous sections, these methods provide very useful insights when used to examine deposits ex situ.

4.4.1.5 Integrated Sensors

Integrated techniques involve measuring the overall relevant parameter of interest, such as rate of heat transfer or pressure drop, over a test section (Visser et al., 1997). The interpretations usually assume uniform spatial distribution of the deposit, which is seldom the case. There are also errors associated with interpretation of the data in terms of the mechanistic models (e.g., correlations for heat transfer and pressure drop). However, integrated techniques are still commonly used in industry owing to their ease of installation.

4.4.1.6 Localized Sensors

A key point to consider in fouling/cleaning detection is the location of the sensor in the plant. The ideal would be a portable sensor that could be used at various locations throughout the plant. Localized sensors, such as heat flux sensors and thermocouples, are capable of making measurements at specific points within the test unit. Heat flux sensors (Davies et al., 1997) offer local measurements of heat flux across a surface. Temperature sensors (often thermocouples) are often incorporated into apparatus to follow changes in local fouling resistance related to growth or removal of a fouling layer. An example is the stirred vessel at the University of Bath (Young et al., 2009) described in Section 3.2. From these local measurements, the deposit thickness may be inferred by assuming a value for the thermal conductivity of the deposit.

However, given the nature of the sensing techniques listed above, it seems unlikely that a portable device would be feasible owing to the need for the sensor to have some form of access to the internal surfaces of the pipe or heat exchanger.

4.4.1.7 Soft Sensors

In addition to the aforementioned physical sensors, other modeling based approaches, the so-called "soft" sensors, have been derived to detect fouling in cross-flow heat exchangers off-line (Gudmundsson et al., 2009). These models are usually mechanistic and require no extra equipment. However, the data will need to be verified by experimental studies.

A summary of published measurement techniques potentially suitable for heat exchangers is given in Table 4.9. These are just some of the existing methods relevant to heat exchanger fouling.

It is clear that each technique has its unique applicability and suitability in laboratory and industrial studies. In general, various assumptions must be introduced for most sensors in order to interpret the measurements to give the parameter of interest. In the author's opinion, there is no universal sensor capable of yielding all the desirable (physical, chemical) information for every existing fouling layer. Rather, a combination of different methods and switching between the techniques are more practical. The important thing is to choose, from the existing methods, a method suitable to the problem at hand.

Experimental studies of crude oil fouling have been performed worldwide (e.g., Asomaning and Watkinson, 1999; Srinivasan and Watkinson, 2005; Watkinson and Wilson, 2007). An important aspect in the study of

Table 4.9 Summary of laboratory and industrial techniques to monitor fouling and cleaning in heat exchangers Basis, principle of measurement; Issues, limitations of the technique; Level, existing laboratory research work or industrial applications; Detectable range, range of measureable scale

Methods and sensors	Basis	Issues	Level [Detectable range]	Examples
Intrusive:				
Electric resistance or conductivity	Electrical impedance of deposit as a function of an alternating current	Solution dependent. Inapplicable to nonelectrolytic solutions. Distortion of the deposit due to contact.	Industrial/laboratory [μm−mm]	Shanrang et al. (2009)
Nonintrusive:				
Pressure drop	Overall pressure drop	Instrument sensitivity limitations. Assumes uniform deposition in space. Assumes knowledge of surface roughness	Industrial/laboratory [mm−cm]	Bott (1995)
Heat transfer coefficient	Local temperature, overall heat transfer coefficient	Accuracy of thermocouple, accuracy of heat transfer correlations. Assumes uniform deposition in space.	Industrial/laboratory [mm−cm]	Burton (1968)

(*Continued*)

Table 4.9 Summary of laboratory and industrial techniques to monitor fouling and cleaning in heat exchangers Basis, principle of measurement; Issues, limitations of the technique; Level, existing laboratory research work or industrial applications; Detectable range, range of measureable scale—cont'd

Methods and sensors	Basis	Issues	Level [Detectable range]	Examples
Mechatronic surface sensor	The amplitude and damping factor of the excitation are related to the amount and type/nature of the deposit.	Assumes the deposit layer is not disrupted by the actuator excitement. Sensitive to temperature. Space limitation. Applicable to flat surfaces.	Laboratory [μm–mm]	Pereira et al. (2006, 2009)
Radioactivity measurement	Detecting and measuring ionizing radiation	Requires the deposit material to contain a radioisotope. Limited to laboratory studies and infeasible for industrial applications.	Laboratory [μm–mm]	Corrieu (1981)
In situ:				
Optical methods (e.g., light transmission or reflectance turbidity, refractometry, backscattered light)	Scattering of light (optical or near infrared) beam or pulses during transmission through deposit. Optical contrast between deposit and substrate surface.	Introduction of dye into the deposit. Use of sophisticated (expensive) imaging acquisition equipment and software hardware and software. Requires transparency of test section. Thin films only.	Industrial/laboratory [μm–mm]	Schöler et al. (2009), Hardy et al. (1985), Payne et al. (1993), Tamachkiarowa and Flemming (1999)

Ex situ:				
Microscopic and topographic techniques	Examination of the surface characteristics of the deposit	Limited to laboratory studies and infeasible for industrial applications due to opaque pipes and fittings.	Laboratory [μm−mm]	Visser et al. (1997)
Integrated:				
Pressure drop	As noted above			
Heat transfer coefficient	As noted above			
Localized:				
Heat flux and temperature sensors	Measurements of local heat flux and/or temperature on a surface	Multiple measurements at various locations are needed. Difficulties for the sensors to have some form of access to the internal surfaces of the pipe or heat exchanger.	Laboratory [mm−cm]	Davies et al. (1997), Young et al. (2009)
Soft:				
Computational or mathematical modeling	Modeling of momentum, mass and energy transport	Assumptions made in the models. Data will need to be verified by experimental studies	Laboratory/industry	Gudmundsson et al. (2009)

crude oil fouling is the development of surface sensors. Surface sensors are essential for monitoring the severity of crude oil fouling and also to determine the end point of cleaning. Industry requires sensors that can measure relevant parameters accurately, which are robust, reliable, sensitive, easy to install and maintain, and compatible with other equipment on the plant (Wilson, 2005).

The key challenge facing all sensors is that deposits are rarely uniformly distributed; a large amount of deposit may be found in local "dead spots." The dead spots are places in the crude oil fouling unit that are difficult to access, for example, in pipes with complex geometries or in bends, where the velocity is low and little or no circulation exists. If crude oil fouling measurements are taken over the whole fouling unit it may be difficult to detect these local dead spots. There is therefore a need for local sensors.

4.4.2 Fluid Dynamic Gauging

4.4.2.1 Origin of FDG

The challenges in fouling and cleaning studies demand a noncontact method capable of providing measurements in situ, in real time, and requiring no physical contact with the materials being studied. FDG is such a technique; it was developed at the Department of Chemical Engineering, Cambridge, in 1997. The inspiration for FDG came from pneumatic gauging, which is an earlier example of a noncontact gauging technique. The operation of pneumatic gauging involves a jet of air moving outward from a nozzle mounted perpendicular to the surface to be gauged (Gale, 1995). The pressure profile within the nozzle will be affected by the proximity of the deposit surface, whence the thickness of a deposit on a substrate can be determined from the pressure profile.

The concept of FDG arose from adapting pneumatic gauging to using the process fluid, rather than air, as the working fluid; and of using "suction" rather than "ejection" action for the jet. The operation of FDG was first proved and reported in Tuladhar et al. (2000). Since then, various configurations of FDG have been built and employed in both academic and industrial applications. The next section reports the main operating principle of the technique.

4.4.2.2 Principles of FDG

FDG offers a robust, simple, and effective method for estimating the thickness of fouling layers in liquid environments in situ and in real time. Although it uses a probe and is not strictly noninvasive, it offers unparalleled operability in a range of media with minimal deformation of the foulant.

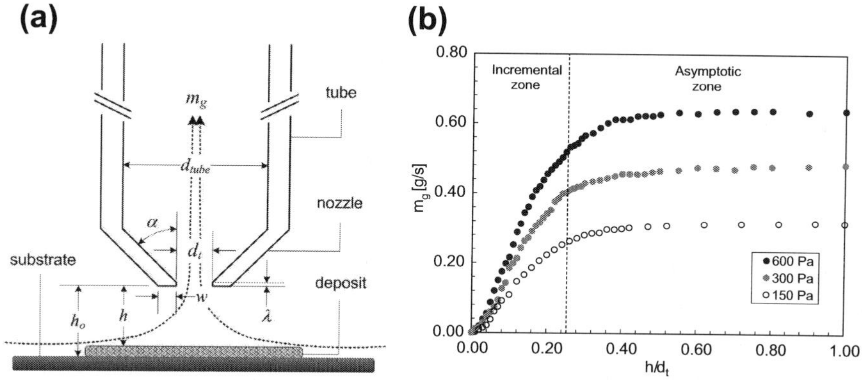

Figure 4.26 Operating principle of FDG (a) where d_t is the inner diameter of the nozzle throat, d_{tube} is the inner diameter of the tube, h is the clearance between the nozzle and the deposit, h_o is the clearance between nozzle and substrate, w is the width of the nozzle rim, λ is the length of nozzle exit, α is the angle of nozzle entry, and m_g is the mass flow rate; and (b) typical calibration profiles of m_g versus h/d_t, for water at different pressure driving force, showing the transition from incremental to asymptotic zones where discharge through the nozzle becomes less sensitive to h/d_t. Nozzle dimensions: $d_t = 1.0$ mm, $d_{tube} = 4.0$ mm, $w = 0.5$ mm, $\lambda = 0.1$ mm and $\alpha = 45°$.

It can also be used to infer useful information on the sensitivity of fouling layers to applied fluid shear stress.

Figure 4.26(a) illustrates the key concepts behind the FDG technique. Flow behavior is exploited to locate the layer; it is a noncontact technique, requiring that the layer is flat and relatively stiff over the time and scale of the measurement. A small conical nozzle, with a throat diameter d_t, is connected to a tube of diameter d_{tube}. The nozzle is positioned close to the deposit, while a pressure difference between the bulk liquid and the nozzle discharge draws liquid into the nozzle. The flow rate into the nozzle, m_g, is sensitive to the clearance between nozzle and layer, h, provided that the clearance is small, typically when $h/d_t \leq 0.25$. Figure 4.26(b) shows a typical relationship between m_g and h/d_t, where it demonstrates the transition between the incremental (or operating) and the asymptotic zones. While operating FDG within the incremental zone, the clearance value, h, at any instant can be determined from the registered mass flow rate, m_g. When the deposit grows or shrinks on the surface, the clearance value will change to a new value of h. If the initial location of the nozzle is fixed, the clearance value from the clean surface, h_o, can be measured. The thickness of the deposit, δ, can then be calculated from:

$$\delta = h_o - h \tag{4.1}$$

In the incremental zone, i.e., $h/d_t \leq 0.25$, the flow pattern through the nozzle is expected to be complex and to be affected by the proximity of the gauging surface. Figure 4.27 shows a schematic of the liquid flow path divided into a series of regions with numbered stations (1 to 7). The flow converges as it approaches the suction region (1 to 2), in which the flow is predominantly radial (2 to 3). The flow then redirects and enters the nozzle (3 to 4). The flow complexities between 3 and 4 are due to the change in flow direction from radial to axial and flow expansion. Downstream from the nozzle tip, the flow diverges in the nozzle throat, generating recirculation (4 to 5). The flow separates downstream of the nozzle entry and then reattaches to the inner wall of the tube further downstream (5 to 6). The flow within the tube (6 to 7) is relatively well defined and laminar; hence the Hagen-Poiseuille flow can be applied. It is worth noting that Chew et al. (2004b) reported the nonexistence of flow recirculation and separation (4 to 6) when the mass flow rate is small; i.e., the flow has entered the creeping flow regime. This type of flow exists for a very viscous fluid such as concentrated sucrose solution and mineral oil.

The total pressure drop across the nozzle and the tube can be written as:

$$\Delta p_{17} = \Delta p_{12} + \Delta p_{23} + \Delta p_{34} + \Delta p_{45} + \Delta p_{56} + \Delta p_{67} \tag{4.2}$$

or

$$\Delta p_{17} = \Delta p_{15} + \Delta p_{56} + \Delta p_{67} \tag{4.3}$$

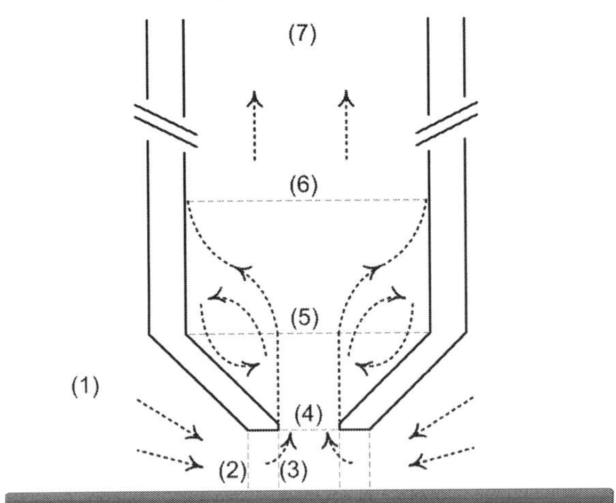

Figure 4.27 Schematic of flow through the FDG nozzle, with stations numbers (1) to (7) between regimes, and the approximate direction of liquid flow indicated by dashed arrows. Evidence for the flow recirculation shown within the nozzle is discussed in Chew et al. (2004b).

where Δp_{ij} is the pressure drop from stations i to j. Assuming the flow through the tube is fully developed, the pressure change in the tube is governed by the familiar Hagen-Poiseuille result:

$$\Delta p_{67} = \frac{128\mu m_g l_{eff}}{\rho\pi d_{tube}^4} = \frac{\mu m_g l_{eff}}{2\rho\pi d_t^4} \tag{4.4}$$

where μ is the dynamic viscosity of the process fluid, and l_{eff} is the effective length of the tube, defined as the length of the hypothetical straight tube that would support the same resistance to flow as the real tube at the same flow rate, measured from stations 6 to 7. The value of l_{eff} is determined using independent tests performed at high values of h, for which the nozzle had been removed from the tube. The total pressure drop is then equal to the pressure drop over the real tube, including losses due to the bends and fittings. Hence, for fully developed laminar flow:

$$l_{eff} = \frac{2\rho\pi d_t^4}{\mu m_g}\Delta p_{17} \tag{4.5}$$

The pressure change between stations 5 and 6 can be described by assuming conservation of momentum, and neglecting any flow contraction such that:

$$\Delta p_{56} = \frac{16 m_g^2}{\rho\pi^2 d_{tube}^2}\left(\frac{1}{d_t^2} - \frac{1}{d_{tube}^2}\right) \tag{4.6}$$

In practice, there are irreversible energy losses due to the complexity of the flow near the nozzle tip and clearance region. The resistances to flows in and about the nozzle are described by a discharge coefficient, C_{d}, i.e., the ratio of the measured flow rate (m_g) to the ideal flow rate, which could be estimated from the Bernoulli's equation in the absence of losses:

$$C_d = \frac{m_g}{\frac{\pi d_t^2}{4}\sqrt{2\rho\Delta p_{15}}} \tag{4.7}$$

The experimental results presented in Figure 4.26(b) can be replotted in dimensionless form i.e., C_{d} against h/d_{t} (Figure 4.28).

Figure 4.28 indicates that the calibrations profiles collapse to a single, generalized form. The complete profile of the dimensionless calibration plot can be described by the following empirical model:

$$C_d = A\left(1 - e^{-B[(h/d_t)+C]}\right) \tag{4.8}$$

where A, B, and C are fitted constants. This generalized model allows an explicit relationship between h and m_g to be expressed, which enables real

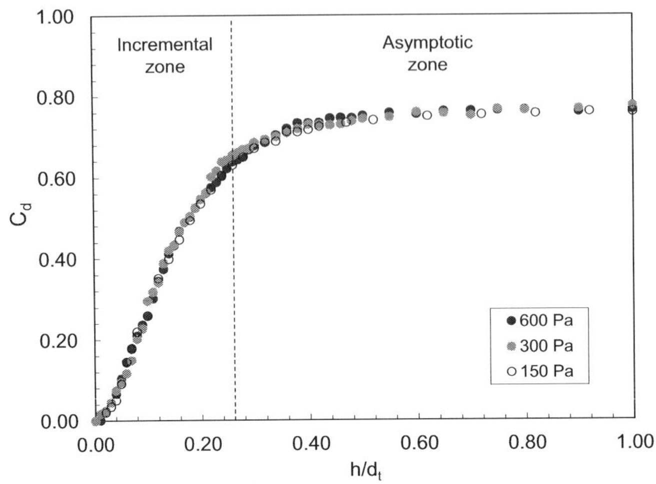

Figure 4.28 A dimensionless representation of the calibration profiles plotted in Figure 4.26(b).

time calculation of the deposit thickness for similar flow conditions. Chew et al. (2004b) and Peralta et al. (2011a,b) presented a comprehensive study of the dependence of C_d on the nozzle geometry and flow conditions. It is worth noting that when the existing generalized model is found to be inaccurate an alternative set of calibration parameters (A, B, and C) can be specified.

An alternative method for obtaining a generalized model for C_d against h/d_t is reported by Gordon (2013) where the author performed an analytical approximation for the fluid flow around the gauge. In most cases, the model shown in Equation (4.8) is sufficient for thickness measurements.

4.4.2.3 Estimation of Deposit Strengths Using FDG

Experimental studies using FDG (Tuladhar et al., 2002) indicated that the forces imposed by gauging flows on weak deposit layers could cause significant deformation of the surface layer. This is undesirable for thickness measurement, but the onset of deformation—which could be recorded by gauging—is related to deposit strength. Knowledge of the stresses imposed by the gauging flows on the surface would therefore afford a method for measuring film strength in situ.

The forces exerted by the gauging flows on the gauged surface can be estimated using computational fluid dynamics (CFD). Chew et al. (2004b) reported the first CFD studies of FDG operation and demonstrated how shear and normal stress distributions on the gauged surface could be quantified. Chew et al. have also demonstrated the use of the combined FDG and CFD techniques for fouling and cleaning studies of food soils (Chew et al., 2004a), polymeric films (Chew et al., 2006), and inorganic fouling systems (Chew et al., 2005a). In later work, Gu et al. (2009), Peralta et al. (2011a,b) Lewis et al. (2012), and Ali et al. (2013) performed a series of CFD studies of FDG operations under various controlled flow conditions.

The shear stress acting on the surface under the gauge nozzle, τ_w, at a radial distance, r, from the center of the nozzle, can also be estimated from the analytical solution for converging radial flow between parallel disks (Middleman, 1998):

$$\tau_w = \frac{3\mu m_g}{\pi \rho h^2 r} \tag{4.9}$$

CFD simulations showed that the maximum shear stress typically occurs underneath the inner rim of the nozzle (Chew et al., 2004b), i.e., when $r = d_t/2$. It is assumed that deformation of deposit is caused by the maximum shear stress exerted on the surface. Thus, knowledge of the registered mass flow rate, m_g, and clearance, h, will afford a method for estimating the cohesive and adhesive strengths of the deposit using Equation (4.9).

4.4.2.4 FDG on Curved Surfaces

Gu et al. (2009) reported the first application of FDG to curved surfaces with and without the presence of bulk flow using an annular geometry. Figure 4.29(a) shows the FDG nozzle, where the gauged surface is curved. $D_i = 21.3$ mm and $D_o = 35.1$ mm are the inner and outer diameters of the annulus, respectively. They performed experimental studies for two different operating cases: *quasi-static FDG*, where the only substantial fluid movement is the gauging liquid and *flow FDG*, where the bulk liquid is flowing in the turbulent regime. In both configurations, the measured mass flow rate through the nozzle is driven by a fixed pressure diving force. They also showed that experimental results for the quasi-static case showed very good agreement with simulations of the configuration using CFD. Later in Gu et al. (2011a,b), CFD studies of FDG in the presence of bulk flow are reported, which also showed excellent agreement with experiments.

Figure 4.29(b) shows the calibration profile of curved surfaces for different annulus Reynolds number, Re_a, defined by:

$$Re_a = \frac{\rho u_m D_H}{\mu} \tag{4.10}$$

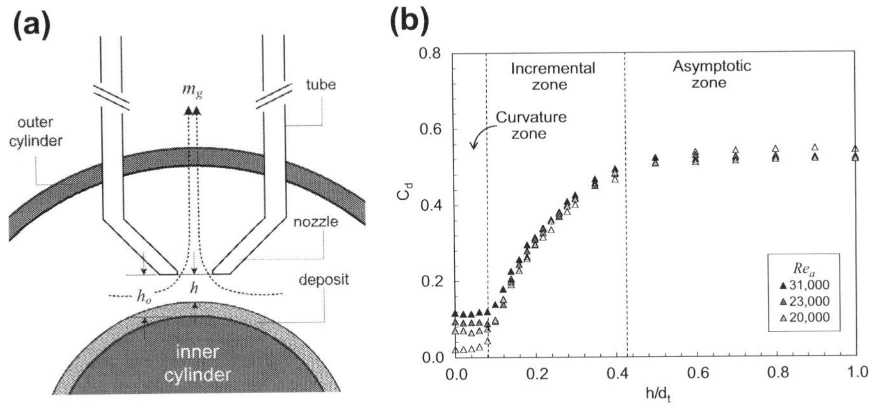

Figure 4.29 Schematic of FDG nozzle on curved surface; and (b) calibration profiles with varying bulk flow conditions, showing the curvature, incremental and asymptotic zones. Nozzle dimensions: $d_t = 1.0$ mm, $d_{tube} = 4.0$ mm, $w = 0.5$ mm, $\lambda = 1.5$ mm and $\alpha = 45°$.

where μm is the mean velocity and D_H is the hydraulic diameter of the annulus (i.e., $D_H = D_o - D_i$). The $m_g - h/d_t$ profiles exhibit three zones. The first zone (at $0 < h/d_t < 0.06$) is described as the curvature zone, where the mass flow rate remains approximately constant, independent of the gauging position. The curvature zone arises from the presence of a curved surface. m_g never reaches zero because there will always be a gap for liquid to flow through between the flat nozzle tip and the curved substrate surface. The second zone (between $0.06 < h/d_t < 0.4$) is labeled the incremental zone, where the flow increases almost linearly with increasing clearance; this is the useful working range of the instrument, to be exploited for thickness measurement. The third zone is the asymptotic zone, at $h/d_t > 0.4$, where the mass flow rate is relatively insensitive to the clearance. With the exception of the initial curvature zone, the behavior of the gauge in the annular geometry and the resulting discharge mass profile resemble those for flat surfaces (Figure 4.26(d)).

4.4.2.5 Modes of Operation of FDG

The FDG technique was initially developed to study surface layers on flat surfaces, where the bulk liquid was still, apart from the flow generated by the gauging action. Equation (4.7) shows that the discharge coefficient, C_d, is uniquely related to m_g and Δp_{15} (hence Δp_{17}) for a given nozzle dimension. Initial variants of FDG employed the *mass flow mode*, whereby the pressure driving force, Δp_{17}, for suction flow is held constant and m_g measured. The mass flow mode is ideally suited to operation at near-ambient pressure. This FDG configuration has been successfully applied to fouling and cleaning studies on a diverse range of foulant materials and surfaces (reviewed by Saikhwan et al., 2007). Accurate control of the pressure driving force to small values (typically a few kPa) at higher operating pressures (>2 bar) is challenging, and the associated variation in flow rate of a hazardous liquid (and the risk of spillage if pressure control is lost) means that mass flow mode operation is unattractive. FDG can be operated in the *pressure mode*, in which the flow rate is instead maintained at a constant level, and the pressure difference around the nozzle is recorded. This is particularly useful for high-pressure systems or where a consistent gauging flow is preferred. Other advantages offered by pressure mode include the following: constant bulk flow conditions can be maintained by minimizing fluid removal from the bulk system by the gauge, and monitoring of the gauging flow allows for better control of applied shear stress. This new mode of operation has been demonstrated by Gu et al. (2011a) in annular and duct flow apparatuses, and by Lewis et al. (2012) and Jones et al. (2012) in membrane fouling studies. Figure 4.30 shows a typical calibration profile obtained using the pressure mode FDG on a flat surface. The profiles are broadly similar to those illustrated in Figure 4.28 (flat surface) and Figure 4.29(b) (curved surface) showing transition between the different zones.

4.4.3 Development of FDG for HIPOR Apparatus

Section 3.3 described the development of a high-temperature and high-pressure oil rig (HIPOR) also reported by Macchietto et al. (2011). This

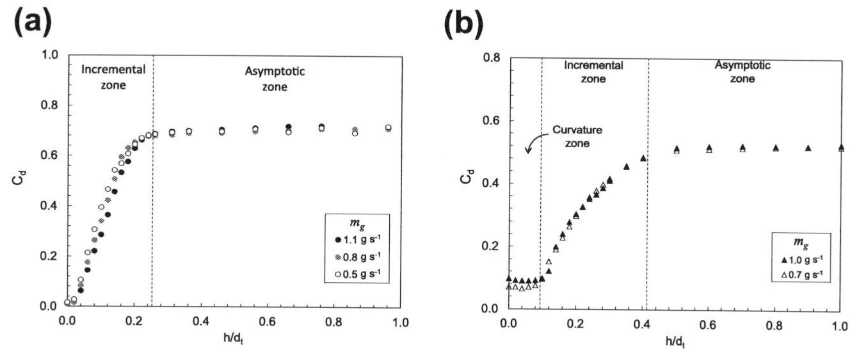

Figure 4.30 Calibration profiles for water using pressure mode at different gauging mass flow rates for (a) flat surface, $Re_a = 1100$; and (b) curved surface, $D_i = 21.3$ mm and $D_o = 35.1$ mm $Re_a = 1000$. Nozzle dimensions: $d_t = 1.0$ mm, $d_{tube} = 4.0$ mm, $w = 0.5$ mm, $\lambda = 1.5$ mm, and $\alpha = 45°$.

fouling unit consists of a flow loop with an annular and a tubular test section (3.33). The inner rod of the annular test section is heated and it was built to study crude oil fouling with simultaneous measurement of pressure drop and heat transfer coefficient. Thermal resistances and pressure drop were monitored over time and from these parameters an indirect estimate of the fouling thickness may be obtained.

This section concerns the development of FDG for measuring crude oil fouling layers on the inner rod of the HIPOR annular flow test section at elevated temperature and pressure in situ and in real time. Apart from measuring deposit thickness, FDG can also be used to measure the yielding strength of the deposit, which is useful for cleaning studies. The challenges faced here are the high-temperature and high-pressure operating conditions of the crude oil fouling apparatus (up to 300 °C and 30 bar). These operating conditions require substantial engineering effort on the design of the FDG technique, with meticulous consideration of the mechanics, seals, instrumentation, safety, tolerance, and operability of the gauge.

In addition to design and experimentation, CFD modeling is also required to elucidate the fluid mechanics of FDG operation in annular geometries in the presence of a bulk flow. CFD evaluates the flow and pressure fields in the modeled domain and the shear stresses caused by the gauging fluid on the substrate surface can then be extracted. Knowledge of the shear stress is important because it allows estimation of the forces involved in deposit removal, which is directly relevant to cleaning.

4.4.3.1 FDG on Heated Curved Surfaces

An important feature of the HIPOR test unit is its curved surface, in particular the heated inner convex surface of the annulus. Gu et al. (2011) demonstrated the application of mass flow mode FDG to study fouling of whey protein on curved surfaces in the presence of bulk flow in the laminar, transitional, or turbulent regime at surface temperatures up to 110 °C. The FDG nozzle is fitted to the apparatus for online thickness fouling measurements. The calibration profiles for the different flow regimes are presented in Figure 4.31. The wall temperatures were maintained at between 20 and 110 °C for Reynolds number in the annulus, Re_a, of 1700, 3000, and 10,000. The working range for the apparatus was between $0.03 < h/d_t < 0.28$. The profiles show that there was little effect of surface temperature on the incremental zone over the range of heat fluxes and flow rates tested. The curvature zone was smaller and this difference is attributed to the differences in the annular geometry.

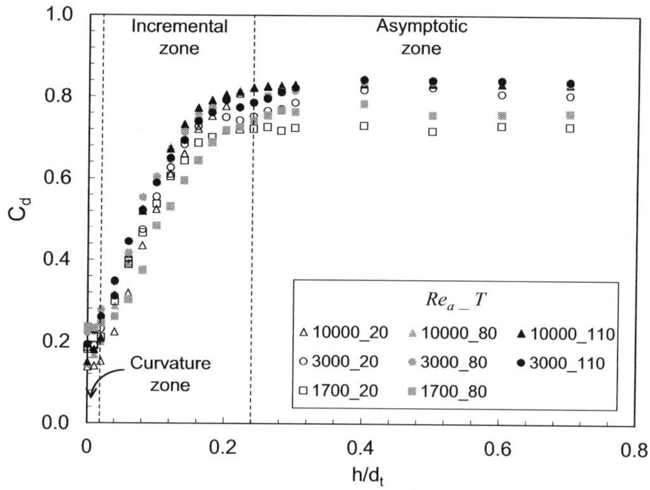

Figure 4.31 Effect of inner wall temperature on discharge coefficient using mass flow mode. Annulus dimensions: $D_i = 12.0$ mm and $D_o = 30.0$ mm. Nozzle dimensions: $d_t = 1.0$ mm, $d_{tube} = 4.0$ mm, $w = 0.12$ mm, $\lambda = 1.5$ mm, and $\alpha = 30°$.

Figure 4.32 shows the shear stress imposed on the inner surface of the annulus at different positions along the arc, *a*, from the central axis of the nozzle under different temperatures flow conditions. The gap between the nozzle and the surface increases along the arc. The shear stress is approximately zero at the centerline ($a = 0$) of the nozzle, reaches a maximum beneath the nozzle lip (*a* ca. 0.5 mm), and approaches zero asymptotically for large values of $a > 0.8$ mm. The values of shear stress are different for different wall surface temperatures. There is a clear decline in the peak shear stress as the temperature of the surface increases from 20 to 50 and 80 °C. This could be explained by the fact that as the wall surface temperature increases, the fluid near the heated surface becomes less viscous. The fluid flows more readily, as a result the shear stress caused by the gauging flow decreases. The figure also indicates that Middleman's (1998) analytical approximation (Equation (4.10)) deviates from the CFD simulation significantly when $a > 0.8$ mm. So Equation (4.10) is less applicable for FDG for curved surfaces.

4.4.3.2 FDG Design for HIPOR Apparatus

The previous section discussed the operability of FDG on heated surfaces up to 110 °C. The mass flow mode design is, however, unsuitable for application in the HIPOR test apparatus, mainly because the operating temperatures and pressures during fouling studies in the HIPOR test unit are high

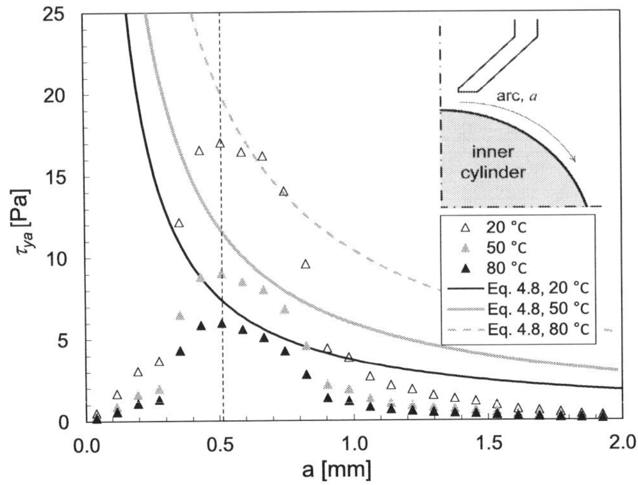

Figure 4.32 Shear stress, τ_{ya}, imposed by the gauging flow on the inner surface of the annulus at different positions along the arc, *a*, as illustrated in the inset. The vertical dashed line indicates the position of the inner radius of the nozzle rim. $Re_a = 1700$ and $h/d_t = 0.18$. Annulus dimensions: $D_i = 12.0$ mm and $D_i = 30.0$ mm. Nozzle dimensions: $d_t = 1.0$ mm, $d_{tube} = 4.0$ mm, $w = 0.12$ mm, $\lambda = 1.5$ mm, and $\alpha = 30°$.

(up to 300 °C and 30 bar). It is, therefore, desirable to operate FDG in the pressure mode (Section 4.4.2.5). The principal reason is to withdraw only a controllable small amount of liquid from the bulk flow, to avoid significant disturbance of the flow regime in the annulus, as well as to avoid turbulent flow in the gauge. Another benefit of this mode of operation is that one could limit the shear stress exerted by the gauge during deposit thickness measurements. On the other hand, for strength studies, mass flow rate and thus the shear stress exerted on the gauged surface could be increased to remove the deposit.

Ali et al. (2013) reported the proof of concept of the operation of pressure mode FDG for measuring the thickness and strength of soft solid fouling layers immersed in an opaque liquid in situ and in real time at elevated pressures and temperatures. Tests were performed under quasi-stagnant conditions using mineral oil at temperatures and pressures up to 140 °C and 10 bar, respectively. Although the hardware limitations of the apparatus prevented testing up to the design limits of the HIPOR apparatus, Ali et al.'s studies presented a significant advancement of the FDG technique where most applications reported to date have involved liquids at temperatures near ambient (10–80 °C) and pressures ranging from 1 to 3 bar.

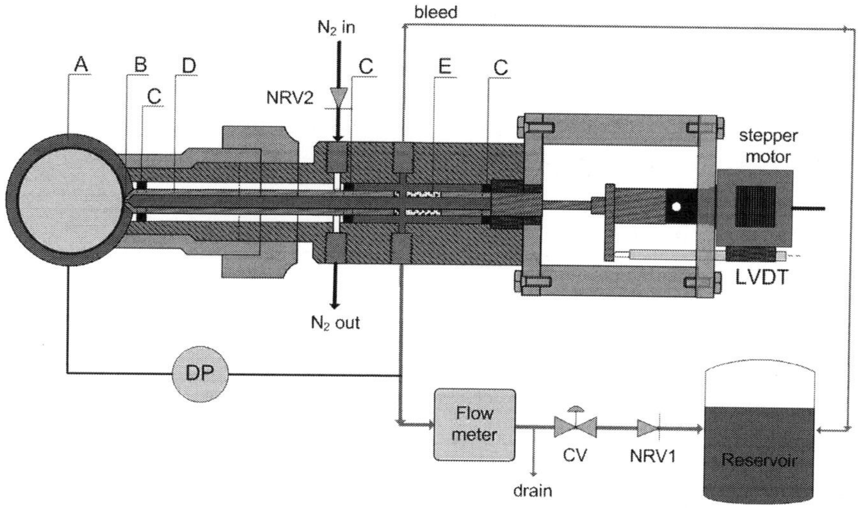

Figure 4.33 Schematic of HIPOR FDG. A, annular cross-section; B, gauging nozzle; C, PTFE seals; D, gauging tube; DP, differential pressure transducer; E, perforated; E, perforated gauging tube; CV, control valve; LVDT, linear variable displacement transducer; NRV, nonreturn valve; and T, thermocouple.

Figure 4.33 shows a schematic of the high-pressure FDG prototype. The FDG is mounted horizontally and perpendicular to the annulus axis. The annulus dimensions are those of the HIPOR annular fouling test rig reported by Macchietto et al. (2011), i.e., inner, D_i, and outer, D_o, diameters of annulus are 24.3 mm and 35.1 mm, respectively. The FDG nozzle design (Figure 4.29(a)) was similar to that reported by Gu et al. (2011), namely $d_t = 1$ mm, $w = 0.12$ mm, $\lambda = 1.5$ mm, with an internal nozzle angle, α of 45°. The gauging tube internal diameter, d_{tube}, was 4 mm with a wall thickness of 2 mm. The device was constructed from 316L stainless steel.

The mineral oil flows from the annulus (A) into the gauging nozzle (B), along the gauging tube (D) and into the perforated wall section (E). The three high-temperature polytetrafluorethylene (PTFE) spring-loaded seals (labeled C in Figure 4.33) created two internal chambers. The first seal isolated the FDG device from the oil in the annular test section. The space adjacent to the outer annulus pipe was a vapor chamber that was swept continuously by nitrogen gas to remove any leaked oil. The second chamber was the receiver for the gauging flow: the tube wall in this section was drilled with many 1-mm-diameter holes to allow the oil to pass through with minimal pressure drop. The flow rate through the gauge is regulated by a control valve (CV) to allow FDG operating in pressure mode (fixed mass flow rate and measure pressure differential). The pressure drop across the FDG nozzle is measured by a differential pressure transducer (DP).

The displacement of the FDG nozzle relative to the annulus is driven by an automated stepper motor. A linear variable displacement transducer (LVDT) records the movement of the stepper motor to provide an independent measure of the nozzle position. The stepper motor movement fittings included a mechanical stop to prevent the nozzle contacting the inner rod surface.

The annular test section and FDG probe were heated using electric heating tape. The temperature was estimated using a thermocouple (not shown) that measured the outer surface temperature of the nozzle, close to the outer annulus wall. The prototype did not allow the temperature of the oil in the nozzle or flow path to be measured directly. The inner annulus wall, gauging tube, HIPOR FDG, and outer annulus wall are expected to expand when oil at higher temperature passes through the device. The expansion of the first two parts brings the gauging nozzle closer to the surface, whereas the latter two move it away. The linear expansion of each element can be estimated from the known coefficients of thermal expansion using Equation (4.11) (Bejan and Kraus, 2003).

$$x_e = \beta l_e \Delta T_e \tag{4.11}$$

where x_e is the distance expanded, β is the linear expansion coefficient, l_e is the length of the element, and ΔT_e is the temperature increase. Thermal expansion calculations indicated that the nozzle position could change by 470 and 640 μm when the temperature was raised to 140°and 300 °C, respectively. The time taken for the stainless steel to expand was approximately 5 s, which is small compared to the time required to heat the oil to its test temperature (several minutes).

Figure 4.34 shows calibration plots obtained at ambient and high temperatures and pressures. All the plots exhibited three distinct zones, namely the curvature zone, incremental zone, and asymptotic zone, i.e., similar to those obtained using the mass flow mode (Figures 4.30(b) and 4.31). The useful range for FDG measurement is in the region of $0.05 > h/d_t > 0.25$. For temperature 19 °C, C_d is independent of the absolute pressure below 10 bar. The values collected at 10 bar for $h/d_t > 0.25$ are noticeably lower than the other data sets. This could be attributed to slow leaks in the system: a leak will lower the recorded values of m_g and thus reduce C_d (see Eqn. (4.7)). This effect is only evident at $h/d_t > 0.25$ because at small clearances the value of C_d is more sensitive to the large pressure drop across the

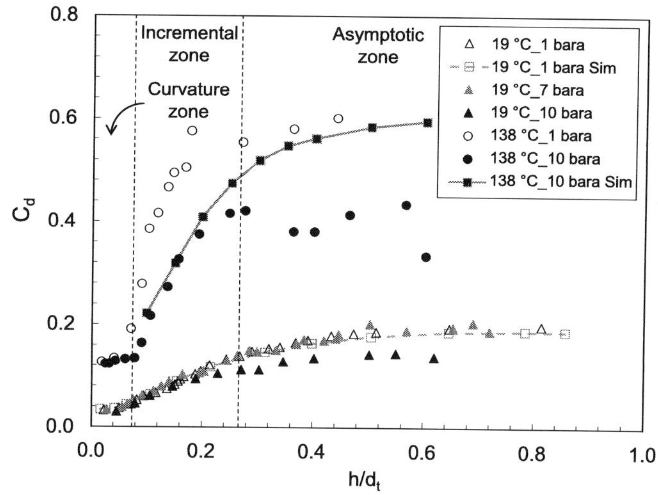

Figure 4.34 Calibration profiles for mineral oil using pressure mode at different temperatures and pressures. Fixed mass flow rate, $m_g = 0.4$ g s^{-1}. Annulus dimensions: $D_i = 24.3$ mm and $D_o = 35.1$ mm. Nozzle dimensions: $d_t = 1.0$ mm, $d_{tube} = 4.0$ mm, $w = 0.12$ mm, $\lambda = 1.5$ mm, and $\alpha = 45°$. *Data replotted from Ali et al. (2013)*

nozzle than a small difference in flow rate. These results confirm that some aspects of the hardware require improvement in order for the device to operate reliably at 30 bar and 300 °C.

CFD studies are employed to model the fluid flow and stress profiles exerted by gauging flows on the gauged surface. The $C_d - h/d_t$ profiles for 19 °C and 1 bar and 138 °C and 10 bar generated by the CFD simulations are plotted in Figure 4.34. The agreement with the experimental data is very good for 19 °C and 1 bar. There is also good agreement for 138 °C and 10 bar up to h/d_t ca. 0.25, while there is a noticeable difference between the measured C_d values and the simulations at larger values of $h/d_t > 0.25$. This difference arises from the uncertainty in the gauging mass flow rate owing to the leaks noted above. The shear stress exerted on the inner surface of the annulus by the gauging flow (not shown) shows profiles similar to those illustrated in Figure 4.32. The main difference is the magnitude of the peak shear stress exerted by mineral oil, which is in the range between 400 and 1000 Pa. The large value in this case arises from the high viscosity of the oil at 19 °C and will decrease significantly with temperature. Ali et al. (2013) also showed that when dealing with deformable deposit layers, Equation (4.9) can be employed to estimate the largest shear stress imposed on the inner rod surface.

4.4.4 Concluding Remarks

This section described the operating principle and development of the FDG technique to measure fouling thickness and strengths in situ and in real time under high-temperature and high-pressure conditions, and for a heated annular surface. This gauging technique is believed to be superior to existing methods for the following reasons: it works in situ and in real time; it avoids introduction of foreign matter into the process fluid, thus avoiding contamination; it does not involve complex or costly sensors which require expensive maintenance; deposit thickness is determined from the discharge flow rate and differential pressure measurements, which can be measured with high accuracy; and, lastly, no prior knowledge of the physical properties of the deposit is required. Moreover, initial commissioning tests showed that the FDG device can potentially be applied to elevated temperatures and pressures similar to those in crude oil fouling. The main remaining challenge is the leakage problems experienced at pressures >10 bar. The limitations of the current design can be eliminated by careful design considerations such as improved connections and sealing at FDG device and annulus pipe connections.

CHAPTER FIVE

Modeling of Fouling from Molecular to Plant Scale

F. Coletti[1], B.D. Crittenden[2], A.J. Haslam[3], G.F. Hewitt[4], G. Jackson[5], G. Jimenez-Serratos[6], S. Macchietto[7], O.K. Matar[8], E.A. Müller[9], D. Sileri[10], J. Yang[11]

[1]Chief Technology Officer, Hexxcell Ltd, UK
[2]Professor of Chemical Engineering, University of Bath, UK
[3]Research Fellow, Imperial College London, UK
[4]Professor of Chemical Engineering, Imperial College London, UK
[5]Professor of Chemical Physics, Imperial College London, UK
[6]Research Associate, Imperial College London, UK
[7]Professor of Process Systems Engineering, Imperial College London, UK
[8]Professor of Fluid Mechanics, Imperial College London, UK
[9]Professor of Thermodynamics, Imperial College London, UK
[10]Senior Business Consultant, Misys, Hong Kong
[11]Research Associate, Imperial College London, UK

Chapter 2 discussed the basic science of fouling and highlighted the complex chemical and physical aspects of fouling deposition, which are highly dependent on local conditions (e.g., temperature, velocity, etc.). These variables are determined by the design of the heat exchangers and by the arrangement of these heat exchangers in the preheat train. However, as will be seen in Section 5.1, existing methodologies used for the design of heat exchangers do not incorporate fouling as a design variable and lump together all the interactions between the complex phenomena involved with crude oil fouling, summarized in Figure 2.6, into a fixed fouling factor.

To overcome limitations of existing heat exchanger design methodologies, it is necessary to increase the level of description of existing fouling models. The multiscale modeling approach proposed in this chapter relies on the detailed description of all the main phenomena involved with crude oil fouling at different levels of investigation, from the nano- and microscale to the plant level, using various modeling techniques, from molecular modeling to computational fluid dynamics.

Figure 5.1 illustrates the overall framework and interactions between modeling activities involved with this approach. Experimental data, collected with techniques such as those described in Chapter 4, can be used to develop and test thermodynamic equations of state and molecular models. In turn, these are capable of providing information on the bulk

Crude Oil Fouling
ISBN: 9780128012567
http://dx.doi.org/10.1016/B978-0-12-801256-7.00005-1

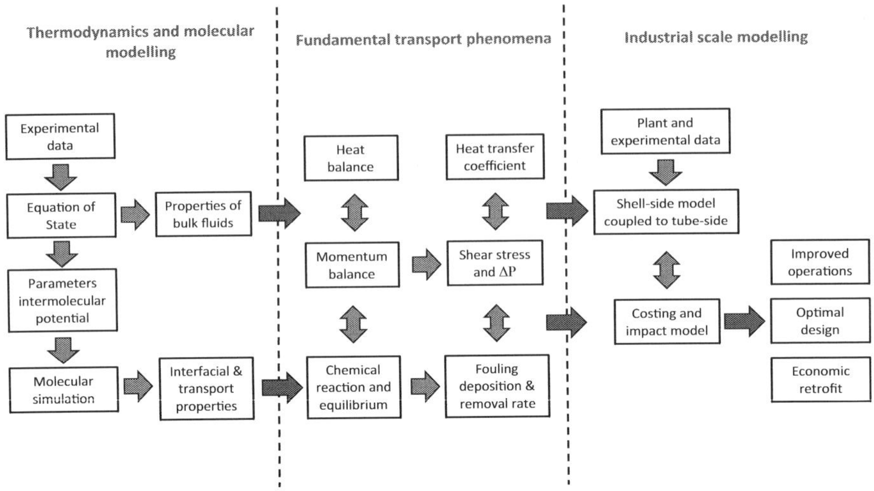

Figure 5.1 Framework of multiscale modeling of crude oil fouling in heat exchangers.

properties and the interfacial and transport properties that can be used by the fundamental models of transport phenomena to determine (via the solution of detailed heat, momentum, and mass balances) deposition and removal of fouling, heat transfer rate, wall shear stress values and pressure drops across individual tubes. Such information can then be coupled with shell-side models to obtain a full-scale heat exchanger model that can be replicated as many times as needed to simulate the behavior of entire networks of heat exchangers. As a result, the effects of oil composition, phase precipitation, chemical reactions, and interactions with fluid flow and heat transfer are integrated to obtain a model that features fully predictive capabilities at the plant scale. The only information needed by such a model would be the oil composition and the operating conditions under which a refinery would like to process it. The multiscale model would then be capable of predicting the fouling behavior of that specific oil—or oil blend—and use this information to assess the economic impact of fouling on the refinery's bottom line, devise retrofit options of individual heat exchangers or change the arrangements of the network that minimizes fouling, and suggest optimal cleaning schedules. Macroscopic experimental data, collected with techniques such as those described in Chapter 3 or directly from plant-scale equipment, can also be used to fit directly the required parameters for the industrial-scale model discussed in Section 5.4.

After a review of heat exchanger design methodologies and available fouling rate models in Section 5.1, the following sections describe the key ingredients involved in the approach proposed: a model capable of describing crude oil fouling physical properties and phase separation of fouling species (Section 5.2), a kinetic model that captures the dynamics of the fouling process as a function of local conditions (Section 5.4), and a thermohydraulic model for heat exchangers capable of providing accurately such conditions (Section 5.4).

5.1 REVIEW OF HEAT EXCHANGER DESIGN METHODOLOGIES AND MATHEMATICAL MODELS OF FOULING

F. Coletti, B. D. Crittenden, G. F. Hewitt, and S. Macchietto

As noted in Chapter 1, the most commonly used type in crude preheat trains is the shell-and-tube heat exchanger. In some cases, other types such as the spiral (Wilhelmsson, 2005; Jones, 2008) and the plate heat exchangers (Tornqvist, 2007; Tamakloe et al., 2013) are used. The claimed advantage of these two types is the higher shear stress achievable for a given flow rate comparable to the shell-and-tube type. However, this latter type is typically perceived in industry as a safer option because of the wider cross-sectional area, which provides more allowance for fouling buildup. Given that the vast majority of heat exchangers are shell-and-tube types, this section will focus only on this type.

The basic design of shell-and-tube heat exchangers was introduced in the early 1900s (Taborek, 2008a), but it was not until 1941 that the Tubular Exchangers Manufacturers Association (TEMA) produced a document to standardize the mechanical design, thus improving safety, quality control, communication between parties, and also lowering costs. Although attention has been paid over the years to improving some aspects of the design (e.g., baffle spacing: Li and Kottke, 1998; Saffaravval and Damangir, 1995; Eryener, 2006), the practice of design of shell-and-tube heat exchangers has not changed much and the use of the latest version (9th edition, 2007) of the original TEMA document is considered a standard practice in industry all over the world. A key element of this standardized design methodology is the representation of fouling by fixed numbers (the TEMA *fouling factors*). As will be realized, fouling is a complex and *dynamic* process and it has been recognized for a long time that representation of fouling by simple fixed resistances cannot really be appropriate. In this section, the application of *models* for fouling will be discussed. Though, at the present time, such models are relatively unsophisticated, their use has a profound effect on the design and performance evaluation of heat exchangers.

This section begins with an overview of the basic design methodologies of heat exchangers (Section 5.1.1) and heat exchanger networks (Section 5.1.2) with particular focus on how to account for fouling. Section 5.1.3 discusses the importance of monitoring fouling in heat exchangers and describes commonly used methods for this activity in oil refineries. In Section 5.1.4 the focus is shifted to review the mathematical models developed over the years to capture the fouling rate as a function of time and process conditions. These are typically developed by fitting data of fouling resistance obtained experimentally in simplified systems (e.g., fouling rigs as described in Chapter 3). Section 5.1.5 presents a review of integrated models of heat transfer and fouling in industrial-scale heat exchangers.

5.1.1 Heat Exchanger Design Methodologies

The local heat transfer rate, Q, between a hot and a cold stream of local temperature respectively T_h and T_c is given by

$$\mathrm{d}Q = U(T_h - T_c)\mathrm{d}A \tag{5.1}$$

where U is the overall heat transfer coefficient as defined in Equation (5.6). A typical design problem is to find the required surface area, A, for a given total heat duty, Q_T and for known stream inlet and outlet temperatures. Modern computer software commercially available for the design of heat exchangers obtains A by integrating Equation (5.1):

$$A = \int_0^{Q_T} \frac{\mathrm{d}Q}{U(T_h - T_c)} \tag{5.2}$$

Equation (5.1) and thus Equation (5.2) are rigorously valid only at steady state (i.e., they do not apply when process conditions vary with time), but this is generally considered a good approximation because in normal operations the heat being transferred is vastly larger than the storage capacity of the exchanger (Bell, 2008).

This incremental temperature/heat load methodology, which can be graphically represented in an enthalpy space, requires the solution of Equation (5.2) and cannot be easily performed by hand. If the assumption is made that there is no variation of physical properties with time or temperature across the heat exchanger (i.e., the heat transfer coefficient does not vary), Equation (5.2) is easily integrated analytically to give

$$A = \frac{Q_T}{U\Delta T_{lm}} \tag{5.3}$$

where ΔT_{lm} is the mean temperature driving force defined as

$$\Delta T_{lm} = \frac{\Delta T_1 - \Delta T_2}{ln(\Delta T_1/\Delta T_2)} \tag{5.4}$$

In Equation (5.4), ΔT_1 is the temperature difference between the two streams at one end of the heat exchanger and ΔT_2 is the temperature difference between the two streams at the other end. For countercurrent arrangements Equation (5.4) becomes

$$\Delta T_{lm} = \frac{\left(T_h^{in} - T_c^{out}\right) - \left(T_h^{out} - T_c^{in}\right)}{\ln\left(T_h^{in} - T_c^{out}\right)/\left(T_h^{out} - T_c^{in}\right)} \tag{5.5}$$

where T_h^{in} and T_c^{in} denote, respectively, the inlet and outlet temperature for the two fluids and T_h^{out} and T_c^{out} denote the outlet temperatures for the two fluids.

An alternative to the log mean temperature difference (LMTD) method described above is the so-called ε-NTU approach by Kays and London (1964), which relies on the definition of thermal efficiency and number of transfer units. Although this latter approach has several benefits over the LMTD approach (e.g., it does not require initial guesses of inlet temperatures), it is subject to the same assumptions (i.e., constant U), which may lead to gross errors in the estimation of the area required for a specific duty. Hewitt et al. (1994) have shown an example for a feed-effluent exchanger for a chemical reactor, fed with a multicomponent feedstock, in which the use of the LMTD method (i.e., Equation (5.3)) led to a design with a 36% larger area than that calculated with the incremental method (i.e., Equation (5.2)).

Further details on available methodologies can be found in several publications (e.g., Taborek, 1978; Smith, 1979; Hewitt et al., 1994; Shah and Sekulic, 2003; Hewitt, 2007, 2008; Coletti, 2014). What is key in the context of this discussion is that any of the traditional methodologies, including the incremental temperature/heat load methodology typically implemented in commercial software packages, account for fouling by adding fixed values for fouling resistances in the calculations of the overall heat transfer coefficient.

Under clean conditions the overall heat transfer coefficient, U_c, is given by the sum of the individual thermal resistances,

$$\frac{1}{U_c} = \frac{D_o}{D_i}\frac{1}{\alpha_i} + R_w + \frac{1}{\alpha_o} \tag{5.6}$$

where α_i and α_o are, respectively, the inside and outside tube convective heat transfer coefficient and R_w is the thermal resistance of the metal wall. In Equation (5.6) the ratio between the inner tube diameter, D_i, and the outside one, D_o, is used to adjust the change in surface area from the outside to the inside of the tube wall. Acknowledging that heat transfer is impaired by the presence of fouling layers on both the inside and the outside of the tube, the respective heat transfer resistances, namely $R_{f,i}$ and $R_{f,o}$, are added to Equation (5.6) (Hewitt, 1994):

$$\frac{1}{U} = \frac{D_o}{D_i}\frac{1}{\alpha_i} + \frac{D_o}{D_i}R_{f,i} + \frac{D_o}{2\lambda_w}ln\left(\frac{D_o}{D_i}\right) + \frac{1}{\alpha_o} + R_{f,o} \tag{5.7}$$

Here R_w is rewritten in terms of λ_w, the thermal conductivity of the metal wall. The physical significance of the nomenclature is summarized in Figure 5.2.

A direct consequence of including fouling resistances in the calculation of the overall heat transfer coefficient in Equation (5.7) is that the surface area required for the given duty is increased. Thus the selection of the values for the fouling resistances $R_{f,i}$ and $R_{f,o}$ is a key factor influencing the design of the heat exchangers (Gram, 1960; Hewitt, 2008). Unfortunately, as seen in previous parts of this book, different fluids, under different process conditions, foul in different ways, making such a selection a challenging activity for design engineers who relied for the past 70 years on values provided by TEMA (1941).[1] These are reported in the form of a table listing values of fouling resistances for various fluids under various conditions. Values for the TEMA fouling resistances of typical PHT streams are reported in Table 5.1.

It is important to point out that once these values have been selected and the exchanger has been designed, constructed, and commissioned, the user must live with the economic impact of its operation and maintenance during its useful life that typically spans 20–30 years (Sheikh et al., 2000). For this reason several authors have proposed guidelines to deal with fouling (Taborek, 1979; Chenoweth, 1988a; Bott, 1995a,b; Hewitt, 2008; Nesta and Bennett, 2004; Bennett et al., 2007). There have been a number of case studies on means of fouling mitigation (for example, that of Al-Bagawi and Said, 2001), but these are based on heuristics and not on a sound systematic approach.[2]

The TEMA *fouling factors* are perceived in industry as *safety factors* that prevent the heat exchanger from losing performance over time. However,

[1] It should be noted that the TEMA are not the only fouling factor values available. Other tables exist in the literature (Nelson, 1958) and often companies use values based on in-house experience.

[2] Chenoweth (1988a) suggested that other considerations, often overlooked, can make the difference in the fouling behavior of the unit. These involve the procedures of construction, storage and erection, shakedown, and start-up.

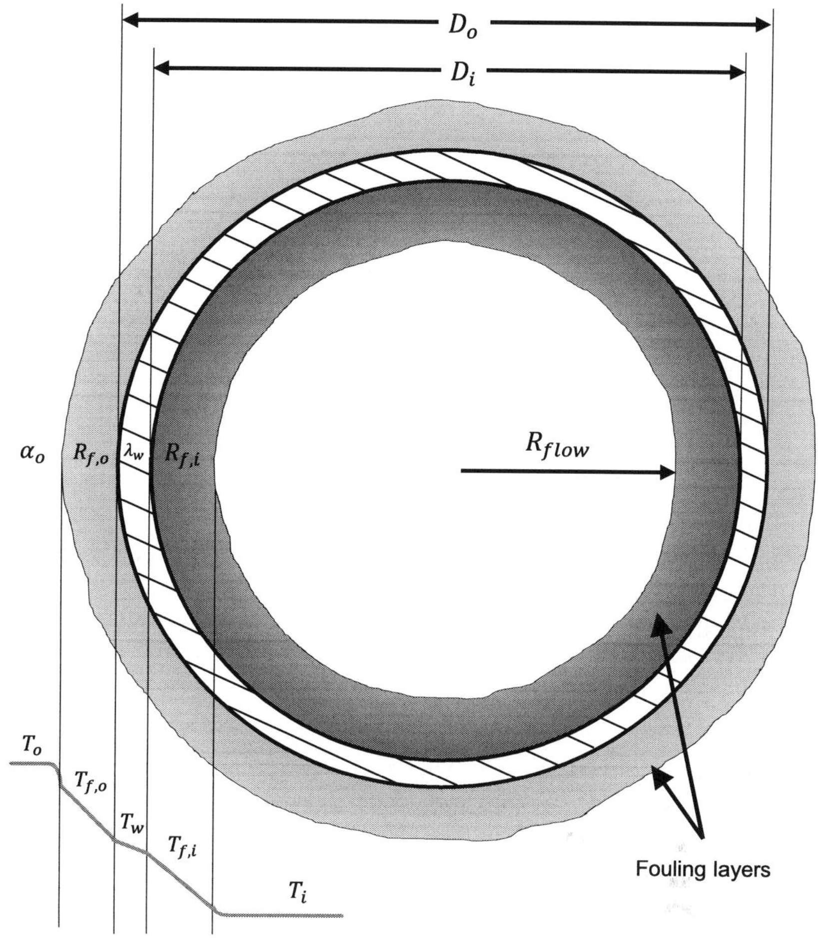

Figure 5.2 Temperature profiles on a heat transfer surface and nomenclature used in this section. *(Adapted from Hewitt et al., 1994).*

it has been argued that this is in fact not the case and the use of the TEMA tables attracted criticism over the years by several authors (Epstein, 1983; Chenoweth, 1988a; Hays, 1989; Rabas and Panchal, 2000; Bott, 2001; Bennett et al., 2007) for a number of reasons discussed in the next section.

5.1.1.1 Criticism of the Fouling Factor Design Approach

Chenoweth (1995) described the developments over the years of the first edition of the TEMA tables, questioning their very origin. He reported

Table 5.1 Typical fouling resistance of crude distillation unit streams (Bott, 1990)

Stream	$R_f \times 10^{-4}$ (K m^2 W^{-1})	Stream	$R_f \times 10^{-4}$ (K m^2 W^{-1})
Crude oil (<120 °C)	3.7–7.0	Kerosene	3.54–5.3
Crude oil (120–180 °C)	5.3–7.0	Light gasoil	3.5–5.3
Crude oil (180–230 °C)	7.0–9.0	Heavy gasoil	5.3–9.0
Crude oil (>230 °C)	9.0–10.5	Heavy fuel oil	9.0–12.3
Gasoline	3.5	Vacuum tower bottoms	17.6
Naphtha and light distillates	3.5–5.3	Atmospheric tower bottoms	12.3

that no written records of the original source could be located by TEMA itself, but he was able to track the origins back to a committee composed by eight experts. However, this version clashes with that by Kern, reported by Somerscales (1990), who proposed that they were originated by E. N. Seider in 1937.

Not only do the TEMA fouling factors have a controversial history, but there is also debate on their future. Chenoweth (1990) reported fundamental disagreement within the members of a joint committee comprising experts from TEMA and HTRI on how fouling should be incorporated in the design of heat exchangers. Although the majority supported the use of the fouling factors, some suggested that the use of 20–25% excess area is an easier and more appropriate way of dealing with fouling. With this approach, also supported by Bennett et al. (2007), the designer does not assume any allowance for fouling and specifies an additional heat transfer surface based on a percentage of the clean one. However, it has been shown (Chenoweth, 1988a) that using an excess area is equivalent to specifying a fixed fouling resistance.

In the case of crude oil fouling Chenoweth (1995) pointed out, quite appropriately, that at the time when the TEMA tables were originated, a 3 month run in a refinery was considered a great success. This time period is certainly not considered satisfactory anymore as refineries aim at operating over 4–5 year runs. Assuming that the design approach were acceptable at the time it is clear that the use of the same values of fouling resistances is certainly not acceptable in modern days.

Regardless of the historical and philosophical considerations on the TEMA tables, there is wide consensus among heat transfer specialists on the following limitations:

1. They barely recognize the variation of the fouling resistance with process variables (e.g., fluid velocity, bulk and surface temperature, crude composition). The TEMA document states that the specified fouling resistances are a *minimum* value to be used in the design without reporting the operating conditions under which this value should be used. Epstein (1983) highlighted that there is inadequate indication of which value they are representing: on one hand, these cannot be steady-state values as fouling resistance develops with time; on the other hand, if they reflect the time of operation this time is not stated. This information, he pointed out, is needed to make an adequate economic evaluation.
2. The dynamics of the fouling process is not taken into account. Fouling is considered as if, at start-up, the system instantaneously reached a steady state with a fixed value of R_f. As a direct consequence of the use of these fixed fouling factors, the designer is forced to select a bigger surface area (with consequential increases in cost) for a given heat duty than that required for a clean heat exchanger. This may have two effects, one on the performance of the installed exchanger and one on the design of heat exchangers for similar duties in the future; both these effects arise from the fact that fouling increases with reduced velocity and the consequential reduction in shear stress. The initial thermal overperformance (given by a larger surface area) produces higher initial temperatures and is usually adjusted by the use of a bypass, which in turn reduces the velocity and shear stress within the unit. As a consequence of higher temperatures and lower velocities fouling is actually exacerbated. As a result, the designer might think that the allowance for fouling is not large enough and will design an even bigger heat exchanger the next time, entering a "self-fulfilling prophecy" (Hays, 1989).

   Overdesign is so rooted in industry that very often extra surface area is built into the design without any discussion or input from the refinery (Jones and Bott, 2001a). As a demonstration of the ineffectiveness of this design philosophy, Yeap (2003) reports a study of refinery fouling data by Jones and Bott (1999) which showed that the thermal resistances at the beginning of the fouling period for all the exchangers were on average 2.5 times greater than the fouling factors recommended by TEMA and by a factor of three over an 8 month period.

3. Fouling is assumed to occur evenly throughout the heat exchanger. As noted by Fryer and Slater (1985), when the underlying mechanisms are based on chemical reaction, fouling does not occur evenly throughout the heat exchangers as it depends on local conditions. Figure 5.3 visually substantiates this statement showing that the extent of fouling in different passes (a) and within the same pass (b) is considerably different at shut down. Although flushing during shut down affects the final

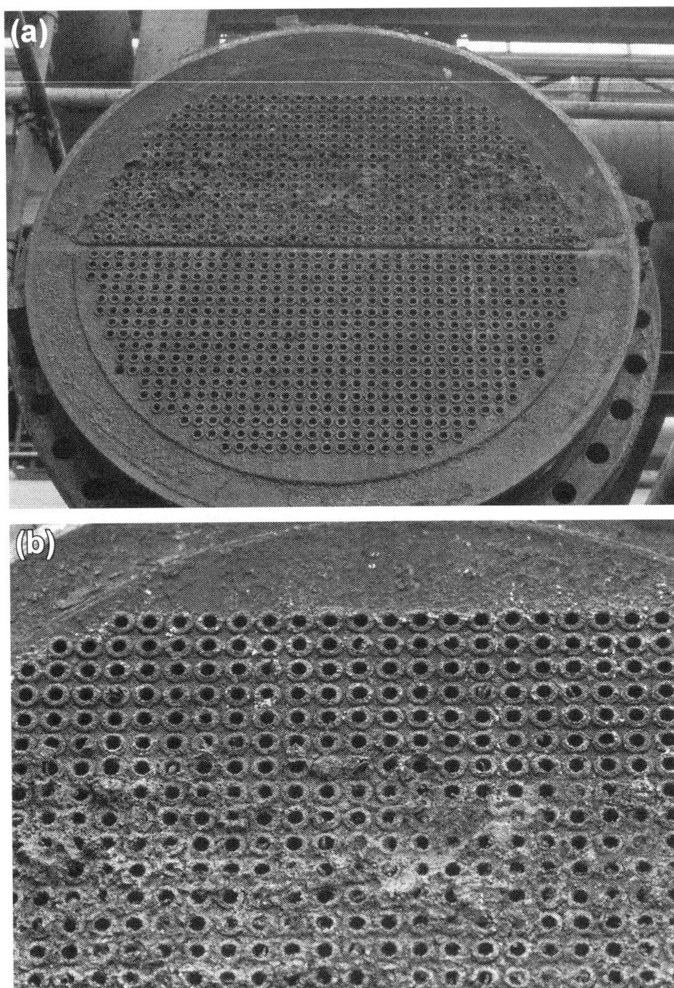

Figure 5.3 Difference in fouling behavior between two passes of the same heat exchanger (a) and effect of flow maldistribution within the same pass (b). Copyright B.D. Crittenden.

distribution of fouling seen once the heat exchanger is opened, the picture shows that spatial considerations are indeed important. The use of lumped fouling resistances fails to capture these localized effects. Thus even when the more accurate incremental method is used and the convective heat transfer coefficients are allowed to vary across the heat exchanger, the resistance produced by the deposits is considered constant and the effect that this has on heat transfer is neglected.

4. They lump together several uncertainties that have little to do with fouling. For example, errors in the estimation of physical properties used in the calculations of the heat transfer coefficient are shifted into the value chosen for R_f.

5.1.1.2 New Approaches to Heat Exchanger Design

In the 1990s a graphical approach was developed to visualize designs that meet heat duty and pressure drop requirements in one graph (Poddar and Polley, 1996; Butterworth, 1996). This methodology consists in plotting the tube count versus tube length in a two-dimensional graph, thus determining the region (the design space or envelope) of all exchanger designs that satisfies the constraints of maximum and minimum tube-side velocities, maximum allowable pressure drops, shell diameters, and tube lengths. Muralikrishna and Shenoy (2000) used the same approach, based on Kern's equations (Kern, 1988) but also included targets for minimum area and minimum cost designs. Serna-González et al. (2006) improved the accuracy of predictions implementing the Bell–Delaware method for the calculation of shell-side heat transfer coefficient and pressure drops as detailed in Taborek (2002a). The graphical representation provided by the design space methodology has the great advantage of supporting the designer with clear guidance regarding the influence of different design options and constraints (e.g., allowable pressure drop) on required geometry.

The design space method was also extended to account for crude oil fouling dependence on temperature and velocity (ESDU, 2000; Butterworth, 2000, 2002) by implementing the Ebert-Panchal fouling model (see Section 5.1.4.2) in software packages (Express Plus and DEVIZE). The estimation on fouling parameters is performed by fitting values of the fouling resistance calculated (via the LMTD method) from plant measurements. Polley et al. (2009a) used this approach to propose a design methodology to specifically include the identification of geometries that provide acceptable fouling levels throughout a predetermined operating period.

Shilling et al. (2009) proposed a risk-based design margin methodology aimed at quantitatively identifying uncertainties related to the design process. According to this new approach, area margins are added but the way these are determined takes into account errors in the heat transfer correlations, probability and severity of fouling, and operating uncertainties related to the specific process. Although this was an attempt to improve on the fouling factor approach, it still required a good level of engineering judgment and experience, for example, in determining the fouling propensity of a particular service.

Recently there has emerged some advanced methodologies for the design of heat exchangers based on both nondeterministic (e.g., genetic algorithm[3]) and deterministic (e.g., mixed integer nonlinear programing—MINLP) mathematical optimization techniques. Such optimization techniques offer the possibility of determining both continuous and noncontinuous design variables (e.g., number of tubes, pass, baffles) that minimize an objective function (e.g., heat exchange area) subject to specific constraints (e.g., maximum tube velocity). The main benefit of this approach is that the objective function specified can include both operating and capital costs; thus the design obtained does not merely take in isolation the heat exchanger for finding the minimum area required for a specific duty, rather it allows a truly economical design to be obtained. This framework can, in principle, be used also to incorporate fouling at the design stage if the underlying mathematical model is capable of capturing it. However, of all the papers reviewed in Table 5.2 only a few focus on incorporating fouling dynamics, albeit in a simplistic way (Pope et al., 1978a,b; Caputo et al., 2011).

5.1.2 Heat Exchanger Network Synthesis Methodologies

The aim of heat exchanger network (HEN) synthesis is to determine the optimal interconnection of heat exchanger units, together with the optimal design of these units, with the objective of maximizing energy integration between streams in a given process that need to be heated with those that need to be cooled. The HEN synthesis problem was introduced by Broeck (1944), although it was more rigorously defined only two decades later by

[3] Genetic algorithms were first proposed by John Holland in the early 1970s. These algorithms were inspired by the mechanism of natural selection in a competitive environment where stronger individuals are selected for a particular task. However, this approach, detailed in Tayal et al. (1998), does not guarantee global optimality.

Table 5.2 Summary of optimization approaches to heat exchanger design

Authors	Optimization algorithm	Objective function	Design variables	Fouling dynamics	Remarks
Fontein and Wassink (1978)	Steepest descend method	Depreciation and operating costs	D_s, NTU, P_x, d_o	x	
Pope et al. (1978a,b)	Proprietary	Total annual cost including W, maintenance, downtime	ΔT_{pp}, N_D, ΔP_t, ΔP_s	Linear model	Developed for geothermal systems
Sun et al. (1993)	Flexible tolerance	Total entropy generation	d_o, L, D_s, L_{bc}, B_c, N_b, N_t	x	
Reppich and Kohoutek (1994); Reppich and Zagermann (1995)	Second-order derivatives	Capital, operating and maintenance annual costs for HEX and pumps	N_t, L, D_s, B_c, N_b, L_{bc}	x	
Chaudhuri et al. (1997)	Simulated annealing	S or purchased cost index	N_s, N_p, d_o, L_{tp}, L, θ_{tp}	x	Coupled with HTRI design program
Rao et al. (1991, 1996)	Search method	Cost, weight and volume	D_s, d_o, θ_{tp}, L_{bc}	x	
Tayal et al. (1998)	GA	S or purchase cost	L, θ_{tp}, d_o, L_{tp}, N_s, baffle and shell type, shell angle, vibration	x	First GA application to design of shell-and-tube
Mizutani et al. (2003)	MINLP	Total annualized investment and operating costs	N_t, N_p, d_i, d_o, N_b, θ_{tp}, head type, fluid allocation	x	Introduced Bell–Delaware method
	MILNP	Investment cost	N_s, S, f_t, $LMTD$, ε	x	Multiple shell units

(Continued)

Table 5.2 Summary of optimization approaches to heat exchanger design—cont'd

Authors	Optimization algorithm	Objective function	Design variables	Fouling dynamics	Remarks
Ponce-Ortega et al. (2006)					
Selbaş et al. (2006)	GA	Capital, operating and maintenance costs	d_o, θ_{tp}, N_p, D_s, B_c, L_{bc}	x	
Ravagnani and Caballero (2007a)	MINLP	Costs including S, W	d_i, d_o, θ_{tp}, L_{tp}, L, N_p, N_t, D_s, D_{otl}, N_b, N_s, B_c, L_{bc}	x	Implemented TEMA standards
Costa and Queiroz (2008)	Search method	Minimum area	D_s, d_o, N_t, N_p, L, B_c, L_{bc}	x	
Fesanghary et al. (2009)	Harmony search	Capital and operating costs	D_s, d_o, N_p, θ_{tp}, P_x, L/D_s, B_c, L_{bc}/D_s, N_{ss}, materials of construction	x	Sensitivity analysis on geometric parameters
Ravagnani et al. (2009)	Particle Swarm	Cost for S and W or S	L, D_s, D_{otl}, d_i, d_o, L_{tp}, θ_{tp}, N_p, N_t, fluid allocation	x	Implemented TEMA standards
Caputo et al. (2011)	GA	Life cycle cost	d_i, d_o, θ_{tp}, L_{tp}, L, N_p, N_t, D_s, N_b, B_c, L_{bc}	Asymptotic	Includes maintenance considerations

Symbols are reported in nomenclature.

Masso and Rudd (1969). Since then, research in this area led to many substantial developments aimed at maximizing energy efficiency.

Extensive reviews of different approaches in HEN synthesis are available in the literature (Fryer, 1988a; Furman and Sahinidis, 2002). Given the large number of publications dealing with this subject, [4] here the discussion will be limited to the industry standard, the so-called "Pinch Technology" (Linnhoff and Hindmarsh, 1983), and to a brief overview of the new approaches based on advanced mathematical programing methodologies (Papoulias and Grossmann, 1983) with particular focus on the treatment of fouling.

5.1.2.1 Pinch Technology

According to Ebrahim and Kawari (2000), Pinch Technology (PT) was developed in the late 1970s by Ph.D. student Bodo Linnhoff from Imperial Chemical Industries (ICI) under the supervision of John Flower at the University of Leeds (Linnhoff, 1978) in response to the energy crisis of the 1970s and the need to reduce steam and fuel consumption in oil refineries and chemical plants by optimizing the design of heat exchanger networks (Linnhoff, 1993). Linnhoff further developed the methodology at the University of Manchester Institute of Technology and set up a consultancy firm known as Linnhoff March International Ltd. (later acquired by KBC Energy Services plc.).

In PT the overall thermal energy demand and availability profiles for a whole process are represented by composite curves for heating and cooling streams on a temperature–enthalpy graph. By means of this graph it is possible to determine the location of the process *pinch point* (of closest, preselected, temperature approach), and the minimum thermodynamic heating and cooling requirements or *energy targets*. This approach allows a systematic design of heat exchanger networks to achieve those energy targets with a minimum number of units. Hohmann (1971) in this Ph.D. thesis showed that the minimum number of units required in a network was one less than the total number of streams and laid some of the groundwork for the pinch method.

The benefits of using PT are clear and many successful plant implementations around the world exist. Worrell and Galitsky (2005) report that total-site pinch analysis (integrating heating and cooling demands

[4] Furman and Sahinidis (2002) identified over 460 relevant papers.

of various processes in the refinery) has been applied to refineries operated by major oil companies. Typical energy savings identified in these site-wide analyses were around 20–30%, although the economic potential was found to be limited to 10–15%. The PT approach has proven particularly useful also in retrofitting existing preheat trains (PHTs) to improve their energy efficiency (Markowski, 2000; Bulasara et al., 2009). However, the pinch point is highly affected by fouling (Szklo and Schaeffer, 2007) and pinch-based methodologies suffer from several limitations:

- They consider steady-state performance of exchangers (i.e., no fouling dynamics).
- They seek to match the hottest process streams with the crude at its highest temperature. This leads to wall temperatures being maximized. In turn, fouling, which is a function of temperature and velocity, is exacerbated.
- They rely on simple, lumped models that are limited by a number of assumptions on physical properties and temperature profiles inside the units.

The importance of including fouling at the individual heat exchanger level and accounting for the way in which several units interact in a network has been highlighted by Fryer (1988b). He considered the robustness of network design with regard to both the response to fouling and the controllability of the HEN when fouling occurs.

Brodowicz and Markowski (2003) proposed a modification of the classic PT to account for fouling based on the minimum sensitivity to the fouling effects by a single heat exchanger and the HEN. However, several aspects, such as the thermohydraulic interactions between fouling and pressure drops, remained to be addressed. Optimization approaches, reviewed in the next section, seem to be better suited to accommodate dynamic fouling models and generate optimal layouts of heat exchanger networks.

5.1.2.2 New Approaches to Network Synthesis

Optimization techniques offer a powerful way to automatically generating heat exchanger network configurations that satisfy specified targets and constraints, while minimizing total costs (i.e., including capital expenses and utility costs).

As noted in Section 5.1.1.2, optimization approaches in the synthesis of networks include the use of genetic algorithms (Ponce-Ortega et al., 2007), mathematical programing techniques based on the sequential solution of continuous and integer linear programs (Papoulias and Grossmann, 1983), and nonlinear optimization problems (Floudas et al., 1986).

The basic mathematical programing approach was expanded to account for pressure drops (Frausto-Hernndez et al., 2003; Serna-Gonzalez et al., 2004) and the need for flexibility (Floudas and Grossmann, 1987; Papalexandri and Pistikopoulos, 1994a,b). The approach was used for the retrofit of existing networks (Papalexandri and Pistikopoulos, 1993; Bjork and Nordman, 2005). Ravagnani and Caballero (2007b) proposed a network synthesis methodology for the simultaneous optimization of detailed equipment design and network synthesis.

Although extremely powerful, MINLP problems are computationally very intensive and in some cases many assumptions need to be made to solve a real-life scenario. To avoid the computational difficulties some recent work focused on removing nonlinearities and simplifying the formulation to a mixed integer linear programing (MILP) problem (Barbaro and Bagajewicz, 2005; Nguyen et al., 2010). Pan et al. (2013) developed new design methodology to maximize energy saving without any topology modifications of the heat exchanger network. Their approach, also based on a MILP optimization method, considered the implementation of both tube-side and shell-side intensification techniques simultaneously.

Table 5.3 summarizes the key contributions to the heat exchanger network synthesis problem identified by Furman and Sahinidis (2002).

Despite the simplifications introduced in solving complex MINLP problems, numerical difficulties still remain and incorporating fouling behavior in the synthesis problem is challenging. As noted by Yeap et al. (2005), if this is not done, the design of network layouts achieved might be optimal for energy recovery under clean conditions but not from a practical operating and, ultimately, economic point of view. This is also confirmed by practical refinery experience that highly optimized networks may not work due to fouling dynamics.

5.1.3 Fouling Monitoring Methodologies

In common refinery practice, monitoring of fouling consists of calculating the overall heat transfer coefficient from temperature and flow-rate measurements at a specific point in time and comparing this value with that calculated under clean conditions. The difference between the two is the fouling resistance. Monitoring of fouling is widely considered a pivotal activity in plant operations for identifying inefficiencies and avoid abrupt shutdown of operations.

Table 5.3 Summary of main contributions to the heat exchanger network synthesis problem (adapted from Furman and Sahinidis, 2002)

References	Remarks
Broeck (1944)	First HEN synthesis related paper
Westbrook (1961)	First use of mathematical programing in HEN synthesis
Hwa (1965)	First grassroots design and superstructure
Masso and Rudd (1969)	First formal definition of the HEN synthesis problem
Hohmann (1971)	Temperature—enthalpy composite curves
Umeda et al. (1978), Linnhoff and Flower (1978a,b)	Identification of the pinch point
Colbert (1982)	Dual temperature approach
Linnhoff and Hindmarsh (1983)	Pinch design method
Cerda et al. (1983)	Transportation model
Papoulias and Grossmann (1983)	Transhipment model
Floudas et al. (1986)	Automated optimization-based HEN synthesis
Li and Motard (1986)	Supertargeting
Yuan et al. (1989), Ciric and Floudas (1991), Yee and Grossmann (1990)	Fully simultaneous HEN synthesis

Calculations of fouling resistances and thus the conclusions that can be drawn from the monitoring activity are affected by the quality of the measurements available in industrial settings. Crittenden (1992) clearly showed how inaccuracies in plant measurements and heat duty calculations can readily propagate to give large errors in the values of the fouling resistances. They assumed an error of ±1% in the measured flow rates, ±1 °C in the temperatures, and assessed the error in the calculations of the fouling resistance on the order of ±20%.

The accuracy of the measurements is not the only factor affecting the quality of the estimates of the fouling resistance. In fact, the procedure and the underlying models used for the calculations have a significant impact as well. Gunness and Baker (1938) described a simple monitoring procedure based on the LMTD concept which is still in use by refineries worldwide. Over the years, fouling monitoring methodologies (Jones and Bott, 2001b; Glen et al., 2001) became more sophisticated, but they essentially remain based on the same simple overall heat balances and use several assumptions on fluid physical properties and heat transfer coefficients which introduce errors in the estimation of fouling resistance (Glen et al., 1997). Moreover, because thermohydraulic interactions between fluid flow and fouling buildup are not taken into account by these methodologies, errors are also introduced in estimates of the hydraulic performance of the heat exchangers. Values of velocity, pressure drops, and wall shear stress are likely to be underestimated by the use of the clean tube diameter in the calculations, which neglects the reduction in cross-sectional area caused by the very presence of the fouling layer. Merry and Polley (1981) discussed the difficulties in obtaining accurate estimation of the fouling resistances in four heat exchangers associated with a nitrogen compressor. They reported an error in the difference between heat duty on the tube-side and shell-side of around ±7% and suggested that calculations of fouling resistances have larger errors.

Another major limitation in using the LMTD approach, which seems not to have received much attention in the literature, is that it fails to provide a value for the fouling resistance if a temperature cross occurs in the unit being monitored. Figure 5.4 reports one such example for a typical heat exchanger in a refinery in the United Kingdom.

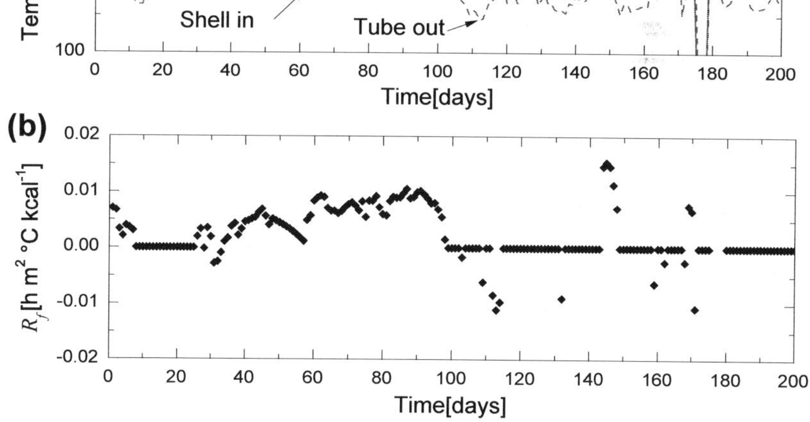

Figure 5.4 Fouling monitoring over 200 days based on LMTD methodology for a two shell unit (two tube pass per shell) from a UK refinery. Inlet and outlet temperatures are shown in (a). When process conditions are such that a temperature cross occurs, the fouling resistance cannot be calculated (b) and is automatically set to 0.

When the tube-side inlet and the shell-side outlet temperatures cross (Figure 5.4(a)), the argument of the logarithm at the denominator in Equation (5.4) becomes negative. As a result, no value can be calculated for R_f (reported as 0 in Figure 5.4). As seen for this particular unit, process conditions are such that temperature cross is an event occurring in 851 days of the 1819 monitored days, ca. 47% of data. Despite all the discussed limitations, most refineries still rely on the log mean temperature difference approach as a monitoring methodology (e.g., Akınç, 2013) for its simplicity to use. Howarth et al. (1997) discussed the importance of having appropriate tools for fouling monitoring and the barriers to the technology transfer of these tools to industry.

More recently, oil companies have been putting an effort into improving the way they monitor fouling especially in CDUs and have been using third party software for this purpose (Waters et al., 2009) or developed more advanced ones in-house (Liporace and de Oliveira, 2007; Heins et al., 2008).

Besides the accuracy of the calculations for the fouling resistances a key feature of monitoring tools is the ability to predict future trends of fouling as a function of oil composition and process conditions. If prediction of fouling trends is possible, the cleaning schedules can then be optimized so that maintenance is minimized and heat recovery of the network is maximized (Lavaja and Bagajewicz, 2004, 2005a,b; Rodriguez and Smith, 2007; Pogiatzis et al., 2012). Fouling models that enable this type of predictions are reviewed in the following section.

5.1.4 Fouling Models

Fouling models attempt to capture the dependency of fouling rate on fluid compositions, process conditions, and time. This section considers the development of fouling models with particular focus on chemical reaction models for crude oil fouling. Several reviews of fouling models for other mechanisms not covered here are available in the literature (Taborek, 1972; De Deus, 1980; Epstein, 1981a, 1987a; Crittenden et al., 1987b; Bott, 1995b; Wilson et al., 2005).

5.1.4.1 General Fouling Rate Models

Early attempts to model fouling in heat transfer equipment focused on developing relationships between fouling rates and process conditions. One classical and simple way of capturing fouling dynamics is based on the assumption that the fouling resistance, R_f, grows asymptotically with time (Kern and Seaton, 1958b):

$$R_f = R^* \left(1 - e^{-Bt}\right) \tag{5.8}$$

where R^* is the asymptotic value of the fouling thermal resistance and B is a parameter. Equation (5.8) is an idealized fouling model and cannot describe fouling in a given unit when it departs from asymptotic behavior. Moreover, it does not explain the variation of the fouling resistance as a function of process conditions.

Kern and Seaton (1958a,b) proposed that the net mass of foulant deposited, M_f, should be calculated as a function of two competing mechanisms, namely deposition and removal. The difference between the rates of deposition, $\dot{M}_d$, and removal, $\dot{M}_r$, gives the net fouling rate

$$\frac{dM_f}{dt} = \dot{M}_d - \dot{M}_r \tag{5.9}$$

which, in terms of fouling resistance, R_f, becomes

$$\frac{dR_f}{dt} = \phi_d - \phi_r \tag{5.10}$$

where the deposition, ϕ_d, and removal terms, ϕ_r, are in $m^2\ KW^{-1}\ s^{-1}$.

In his review of fouling models, Epstein (1981a) argued that there is a conceptual problem in defining the ϕ_r term in Equation (5.10) as the *removal* term due to reentrainment. His argument was that as a foulant particle approaches the surface, it would not be deposited in the first place because of the action of the fluid shear. Epstein noted that Kern, who first proposed the use of the term *removal*, redefined it later as a retardation process and called it *suppression* (Kern, 1966). In this process, aging is likely to play an important part as the structural change of the deposits to a harder material makes it difficult to remove. Wilson et al. (2005) noted that, in crude oil fouling, there is no evidence that removal actually happens thus the word *suppression* is indeed more appropriate.

Kern and Seaton (1959b) proposed that the thickness δ of the fouling layer could be calculated from

$$\frac{d\delta}{dt} = a_1 c_f \dot{M} - a_2 \tau \delta \tag{5.11}$$

where $\dot{M}$ denoted the mass flow rate and a_1 and a_2 are model constants. The deposition term is a function of the foulant concentration, c_f. The removal term assumes that deposits may be removed in chunks from the wall by effect of the shear action of the fluid, τ, and is proportional to the layer thickness. Bott (1995) showed how integration of Equation (5.11) allows the determination of a simple expression describing the thickness of the fouling layer:

$$\delta = \frac{a_1 c_f \dot{M}}{a_2 \tau}\left(1 - e^{-a_2 \tau t}\right) \tag{5.12}$$

which is substantially similar and leads to Equation (5.8) with the asymptotic fouling thickness δ^* calculated as

$$\delta^* = \frac{a_1 c_f \dot{M}}{a_2 \tau} \tag{5.13}$$

With their model, Kern and Seaton introduced the basic idea of capturing fouling as a function of process conditions and time. However, Equation (5.11) is a rather general fouling model that does not specifically take into account any particular fouling mechanisms. If the chemical reaction fouling is the underlying mechanism an appropriate model should take into account the kinetics of the reaction.

5.1.4.2 Rate Models for Chemical Reaction Fouling

Early attempts to describe chemical reaction fouling in crude oil systems were published by Nelson (1939a,b). Nelson considered the formation of coke in furnaces and described it as arising from chemical reaction in a deposited film and calculated the impact on heat transfer. He noted that the properties of the fouling layer would change with temperature and time (aging) and reported values of thermal conductivities of different carbonaceous materials for comparison. A following paper by Atkins (1962) focused on chemical reaction fouling in furnaces and the hottest exchangers in a preheat train. He proposed that the formation of the hard coke fouling layer occurred via two consecutive steps: the deposition of a tarry layer (or porous coke) which was then converted to a hard coke layer. The tarry layer was assumed to have the same thermal resistance as the values prescribed by TEMA whereas the resistance provided by the hard coke material was obtained from a graph as a function of time, temperature, and composition.

Watkinson and Epstein (1970) developed a transfer–adhesion–release model for fouling rate in gasoil and sand–water slurries systems that included a sticking probability, P:

$$\frac{dR_f}{dt} = a_1 P \dot{m} - a_2 \tau \delta \tag{5.14}$$

In this model, it was assumed that deposition was caused by mass transfer of suspended particles to the wall followed by adhesion and removal:

$$\frac{dR_f}{dt} = \frac{a_1 (c_b - c_w) \exp(-E_f / R_g T)}{u \sqrt{C_f}} - a_2 u^2 C_f \delta \tag{5.15}$$

where E_f denotes the activation energy of the fouling reaction, u the bulk velocity, C_f the friction factor and a_1 and a_2 are fixed coefficients. In Equation (5.15), the mass flux of foulant being deposited on the wall is assumed to be proportional to the difference between the particulate concentration in the bulk, c_b and at wall c_w. Watkinson and Epstein showed that Equation (5.15) predicts the dependency of initial fouling rate on temperature as found in experimental results. However, Crittenden et al. (1987b) suggested that Equation (5.15) does not predict the correct dependency of the asymptotic fouling on flow rate.

Taborek et al. (1972) summarized possible deposition terms in Equation (5.9). For chemical reaction fouling this was given by an equation in the form of a simple Arrhenius kinetics in order to account for the dependency of fouling on temperature:

$$\phi_d = a_1 A \exp\left(\frac{-E_f}{R_g T_w}\right) (c_r)^k \tag{5.16}$$

where A denotes the Arrhenius preexponential term, R_g, the ideal gas constant, and c_r, the concentration of the species reacting with reaction rate of kth order. In Equation (5.16), it is assumed that the chemical reaction producing fouling occurs at the wall by the use of the wall temperature, T_w. Taborek et al. assumed that the removal term was a result of a force balance between the fluid shear and the strength of the deposit bonds:

$$\phi_r = a_2 \frac{\tau}{R_b} \tag{5.17}$$

where R_b is defined as the "deposit bond resistance" and appears to be a difficult parameter to be estimated.

Fernandez-Baujin and Solomon (1976) developed a two-step model accounting for mass transfer and kinetics in the deposition of coke in steam cracking furnaces. They considered the mass transfer flux of the precursor to the reaction site, $\dot{m}_p$, as

$$\dot{m}_p = k_{m,p} (c_{p,b} - c_{p,i}) \tag{5.18}$$

where $k_{m,p}$ is the mass transfer coefficient and $c_{p,b}$, $c_{p,i}$ are the precursor concentrations in the bulk and at the fluid/deposit interface, respectively. The fouling rate was then given by

$$\frac{dR_f}{dt} = \frac{1}{\rho_l \lambda_l}\left(\frac{c_{p,b}}{1/k_{m,p} + 1/k}\right) \tag{5.19}$$

where ρ_l denotes the density of the foulant layer and λ_l its thermal conductivity.

A quasi-steady-state approach has been used by Sundaram and Froment (1979) for the description of gas phase coking in thermal cracking of propane. In this case, the authors assumed that since the coking rate is much less than the throughput of precursors the change in deposit thickness could be captured by updating the calculations periodically to account for the change in tube diameter:

$$\Delta\delta = \frac{A\exp\left(\frac{-E_f}{R_g T_w}\right) c_{p,b}\Delta t}{\rho_l} \tag{5.20}$$

The comparison of the model prediction with the plant data was considered to show an acceptable agreement.

Crittenden and Kolaczkowski (1979a) extended the Fernandez-Baujin and Solomon (1976) model to include convection of the foulant back into the fluid bulk, $\dot{m}_f$:

$$\dot{m}_f = k_{m,f}\left(c_{f,i} - c_{f,b}\right) \tag{5.21}$$

where $k_{m,f}$ is the foulant mass transfer coefficient. They calculated the thickness of the fouling layer, δ, as a function of the two mass fluxes:

$$\frac{d\delta}{dt} = \frac{1}{\rho_l}\left(\dot{m}_p - \dot{m}_f\right) \tag{5.22}$$

Assuming that the $c_{f,b}$ is very small they expressed the fouling resistance as

$$\frac{dR_f}{dt} = \frac{d(\delta/\lambda_l)}{dt} = \frac{1}{\rho_l \lambda_l}\left(\frac{c_{p,b}}{1/k_{m,p} + 1/k} - k_{m,f} c_{f,i}\right) \tag{5.23}$$

Writing the mass transfer coefficients in terms of flow rate and physical properties using the Chilton and Colburn (1934) analogy Equation (5.23) becomes

$$\frac{dR_f}{dt} = \frac{1}{\rho_l \lambda_l}\left[\frac{c_{p,b}}{\frac{\rho_l (D_i - 2\delta)^{1.8} Sc_p^{0.67}}{0.607\lambda_{lp}^{0.2}\dot{M}^{0.8}} + \frac{1}{A\exp\left(\frac{-E_f}{R_g T}\right)}}\right] - \left[\frac{0.607\lambda_{lp}^{0.2}\dot{M}^{0.8}}{\rho_l (A_i - 2\delta)^{1.8} Sc_f^{0.67}} c_{f,i}\right] \tag{5.24}$$

where Sc is the Schmidt number. Equation (5.24) also assumes that deposition is a function of a first-order reaction with an Arrhenius equation describing the rate dependency on temperature and the effects of mass flow rate on fouling are included. Crittenden and Kolaczkowski (1979a) noted that for low temperatures, small tube diameters, and large mass flow rates, the kinetics would be controlling in Equation (5.24) whereas for high temperatures, large tube diameters, and small mass flow rates, the precursor diffusion would be the controlling mechanism.

Crittenden and Kolaczkowski (1979b) further extended Equation (5.24) to consider the buildup of two layers (namely tarry and coke) on the tube wall conceptualized by Atkins (1962). This is, noticeably, one of the first attempts to formalize in a mathematical way the aging phenomenon (described in Section 2.3.5) by including its effect in Equation (5.10). They assumed that the increase in fouling resistance is governed by deposition of tarry layers but added two new terms to consider the interactions between the tarry and the coked layers:

$$\frac{dR_f}{dt} = \phi_d - \phi_{tr} - \phi_{cr} - \phi_r \tag{5.25}$$

where ϕ_{tr} and ϕ_{cr} are, respectively, the fouling rate decrease by mass transfer of tars to the bulk of the fluid and by conversion of tarry layer to coke. By considering mass transfer and reaction kinetics in the system, they proposed the following form of the terms in Equation (5.25):

$$\frac{dR_f}{dt} = \left[\frac{1}{\frac{a_1 (D_i - 2\delta)^{1.8}}{\dot{M}^{0.8}} + \frac{\exp\left(\frac{-E_{tr}}{R_g T_{tr}}\right)}{a_2}}\right] - \left[\frac{a_3 \dot{M}^{0.8}}{(D_i - 2\delta)^{1.8}}\right] - \left[\frac{a_4}{\exp\left(\frac{-E_c}{R_g T_c}\right)}\right] - \left[\frac{a_5 \dot{M}^{0.8} R_f}{(D_i - 2\delta)^{3.8}}\right] \tag{5.26}$$

where T_{tr} and T_c denote, respectively, the temperature in the tarry and in the coke layer, D_i denotes the tube inner diameter, and a_i denotes proportionality constants and other physical properties of the system. Crittenden et al. (1987b) used the equations above to model the experimental data for styrene deposition from kerosene by Crittenden et al. (1987a).

Epstein (1994) developed a model for the initial fouling rate in which attachment to the thermal surface is treated as a process in series with mass transfer and is assumed to vary with residence time of the fluid at the surface:

$$\frac{dR_f}{dt} = \frac{1}{\rho_l \lambda_l} \left[\frac{k_{m,f} c_{p,b}}{1 + a_1 u^2 (C_f/8)(\mu/\rho) K_p \exp\left(\frac{E_f}{R_g T_w}\right)} \right] \tag{5.27}$$

where ρ and μ denote the fluid density and dynamic viscosity, respectively. Equation (5.27) was fitted to the experimental data by Crittenden et al. (1987a) with an average absolute deviation of 14.2%. Details on the regression accuracy were provided in Vascaronák and Epstein (1996).

The review of fouling models so far has shown how the correlations have grown in complexity over the years with the improved understanding of the underlying fouling mechanisms. However, the increased complexity of the correlation used to describe fouling also meant that the number of unknown and difficult to measure parameters (e.g., concentration of foulants in the bulk and at the wall) has grown significantly, thus limiting the applicability of the models.

In the 1990s, models that did not require any information on composition began to emerge. These do include less physics but they are of more practical use. Crittenden et al. (1992) used a model similar to that of Equation (5.16) to fit refinery PHT data where they dropped the dependency on concentration:

$$\frac{dR_f}{dt} = A \exp\left(\frac{-E_f}{R_g T_w}\right) \tag{5.28}$$

The authors found a value for the activation energy $E_f = 33$ kJ mol^{-1} and suggested that a value below 40 kJ mol^{-1} indicates that both chemical and physical phenomena are important.

Ebert and Panchal (1995) proposed the following form for Equation (5.10) to fit data collected in a flow loop for crude oil fouling:

$$\frac{dR_f}{dt} = A\mathrm{Re}^{\beta} \exp\left(\frac{-E_f}{R_g T_f}\right) - \gamma\tau \tag{5.29}$$

where the deposition term is as a function of the film temperature, T_f (as opposed to T_w in Equation (5.27)) and the Reynolds number, Re. The suppression term depends on the wall shear stress, τ. Parameters $A = 8.3\ \mathrm{Km^2\,W^{-1}\,s^{-1}}$, $\beta = -0.88$, $E_f = 68\ \mathrm{kJ\,mol^{-1}}$, and $\gamma = 4.03 \times 10^{-11}\ \mathrm{m^4\,NK^{-1}\,s^{-1}}$ were determined via regression to experimental data and are expected to vary between different crudes.

The model was developed after experimental evidence of the existence of a threshold for given operating conditions below which fouling would not occur. Since the threshold is given by the balance of the rate of deposition with the suppression one, its location can be determined by setting the rate to zero, equating the two terms, and solving for T_f, τ, and Re. As seen in the following sections, the significance of this is that Equation (5.29) can be applied in a relatively simple way to the design of heat exchangers by determining a fouling threshold curve and changing design parameters so that operating conditions of the units are below it (Figure 5.5).

In a following work Panchal et al. (1997) determined a new set of constants using the experimental data for three different crudes and added the Prandtl number, Pr, to Equation (5.29), as this was seen to vary between 2.5 and 8.2. As for the previous model, in this new one the foulant is assumed to be removed from the boundary layer before it is deposited:

$$\frac{dR_f}{dt} = A\mathrm{Re}^{\beta}\mathrm{Pr}^{-0.33}\exp\left(\frac{-E_f}{R_g T_f}\right) - \gamma\tau \tag{5.30}$$

In this new version of the model the regression with a larger data set returned lower values of parameters $A = 0.0139\ \mathrm{Km^2\,W^{-1}\,s^{-1}}$, $\beta = -0.66$,

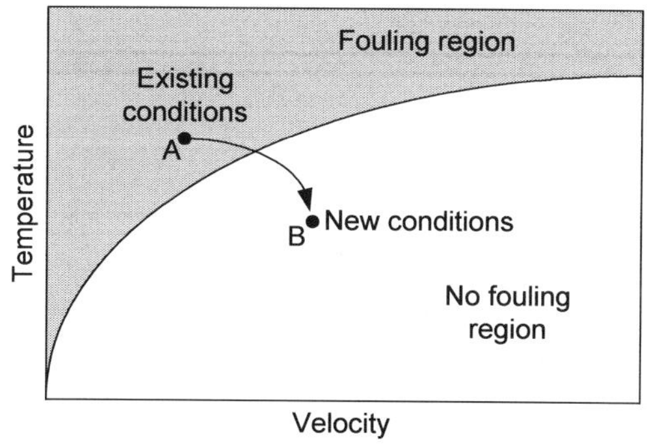

Figure 5.5 Fouling threshold concept in heat exchanger design (after Rodriguez and Smith, 2007). If operating conditions are changed from A to B, the unit moves from a fouling to a no-fouling region.

and $E_f = 48$ kJ mol^{-1} whereas γ did not change from the original value of 4.03×10^{-11} m^4 NKW^{-1} s^{-1}. It should be noted that only 60% of experimental data points followed the fouling trend predicted by Equation (5.30), which was accepted as a "quite encouraging" agreement by the authors of the paper given the difficulty of obtaining high-quality, reproducible data. However, Asomaning et al. (2000) demonstrated that the model predicts field data less successfully and suggested that this was due to the discrepancies existing between the laboratory data—obtained under controlled conditions—and the field data which tend to be situation specific.

Polley et al. (2002a) further modified Equation (5.30), assuming that the dependency of fouling on velocity is associated with transport phenomena and replaced the wall shear stress term, τ, with the Reynolds number, Re:

$$\frac{dR_f}{dt} = A\mathrm{Re}^{-0.8}\mathrm{Pr}^{-0.33}\exp\left(\frac{-E_f}{R_g T_w}\right) - \gamma \mathrm{Re}^{0.8} \tag{5.31}$$

The model with a new value for the exponent of the Reynolds number and the dependency on wall—rather than film—temperature showed a better agreement with the experimental results by Knudsen and Hays (1995) compared to Equation (5.30). The advantage of writing the suppression term as a function of Re number is that the need for the calculation of τ is removed. This enables the calculation of fouling on the shell-side where shear stress can only be quantified with a computational fluid dynamics (CFD) simulation.

Srinivasan and Watkinson (2005) fitted a simple correlation to experimental fouling data collected for three Canadian crudes in a recirculation fouling loop with an annulus test section:

$$\frac{dR_f}{dt} = a_1 u_a^{-0.35}\exp\left(\frac{-E_f}{R_g \widehat{T}_f}\right) \tag{5.32}$$

where u_a denotes the velocity in the annulus and $\widehat{T}_f$ is a modified film temperature weighted more heavily on the surface temperature:

$$\widehat{T}_f = 0.3T_b + 0.7T_w \tag{5.33}$$

The different definition of the film temperature allowed Srinivasan and Watkinson (2005) to better fit their rate data with Equation (5.32), achieving an accuracy of ±8%.

Saleh et al. (2005b) proposed a power-law equation to fit fouling data of a light Australian crude with a low asphaltene and sulfur contents obtained in a flow loop:

$$\frac{dR_f}{dt} = a_1 P^{a_2} u^{a_3} \exp\left(\frac{-E_f}{R_g T_f}\right) \tag{5.34}$$

where P denotes the pressure, a_1 denotes the power-law constant, and a_2 and a_3 are the exponents of, respectively, pressure and velocity.

A further modification of Equation (5.31) was proposed by Nasr and Givi (2006):

$$\frac{dR_f}{dt} = A\mathrm{Re}^{\beta} \exp\left(\frac{-E_f}{R_g T_w}\right) - \gamma \mathrm{Re}^{0.4} \tag{5.35}$$

in which the dependency on the Pr number is dropped. A better fit of fouling data by Saleh et al. (2005b) was shown by this form of the equation.

Polley (2010) proposed a different sticking probability model in which the effect of fluid shear was captured not via a suppression term but with a probability function:

$$P = 1 - \left(\frac{\tau - 2}{98}\right)^n \tag{5.36}$$

where n is a coefficient, arbitrarily set to 0.5. The fouling resistance is then given by

$$\frac{dR_f}{dt} = \frac{A}{\alpha} \exp\left(\frac{-E_f}{R_g T_f}\right) P \tag{5.37}$$

where α is the film heat transfer coefficient and A the preexponential parameter for the Arrhenius kinetics.

The success of the model by Ebert and Panchal (1995) reported in Equation (5.29) can be judged not only by the numerous variations that different authors have developed over the years but also by the fact that its simplicity allowed its use in designing and retrofitting of industrial heat exchangers (Butterworth, 2000; Polley et al., 2002b) and whole PHTs (Yeap et al., 2001a; Bories and Patureaux, 2003; Yeap et al., 2004; Polley et al., 2005; Coletti and Macchietto, 2010; Coletti et al., 2011), and in assisting the scheduling of cleaning, thus improving network operability and mitigating costs related to maintenance (Wilson and Vassiliadis, 1997; Smaïli et al., 2001; Ishiyama et al., 2009a; Sikos and Klemes, 2010; Coletti and

Macchietto, 2010). As noted above, the threshold concept has also been implemented in graphical tools (Butterworth, 2000, 2002; ESDU, 2000) and exploited in commercial software (Ishiyama et al., 2011; Hexxcell, 2014) and used to improve the design and monitoring of heat exchangers.

Although the different models reviewed here have been used more or less successfully in specific cases, the question remains on how much difference various formulations of the fouling rate really make. Yeap et al. (2004) compared the performance of three models in predicting fouling resistances calculated using 10 sets of data collected from experimental and plant measurements. The models considered were a modified version of the model by Epstein (1994), the model by Panchal et al. (1997), and the model of Polley et al. (2002a). Although the authors stated that the agreement was "only fair," comparison results show that roughly only half of the (few) data points considered stay within ±50% error for the models by Panchal et al. (1997); Polley et al. (2002a) achieved an overall better agreement by the modified Epstein (1994) model. This can be explained by the fact that fouling rate models (summarized in Table 5.4) aim at explaining with one simple correlation all the complex and interacting phenomena described in Chapter 2 (see, in particular, Figure 2.6). Unless a fouling rate model is integrated with a thermohydraulic description of the unit it would not be capable of capturing the complexity of the interactions existing in the system. Integrated heat exchange and fouling rate models are reviewed in Section 5.5.1.

5.1.4.3 Other Approaches to Modeling of Fouling in Heat Exchangers

To overcome the difficulties involved in modeling the complex fouling phenomena, several authors have attempted nondeterministic approaches to describe the fouling process.

Sheikh et al. (1996) presented an economic analysis based on a probabilistic generalization of the deterministic modeling strategy proposed by Casado (1990). Sheikh et al. developed a cost-based optimization model to calculate the optimum operation of a shell-and-tube heat exchanger in a PHT. Sheikh et al. (1996) further generalized Casado's analysis by incorporating the stochastic nature of the linear fouling growth law as discussed by Zubair et al. (1995, 1997a,b). They incorporated a power law, falling rate, and asymptotic random falling growth laws developed by Zubair and Sheikh (2000), Sheikh et al. (2000), and Zubair et al. (2001). However, these models consider fouling as being essentially random in nature and thus fail to explain in a satisfactory way the interactions between different phenomena involved in the deposition/removal process.

Table 5.4 Summary of rate models for chemical reaction fouling

References	Fluid	System	Mass transfer	Aging	Main assumptions	Remarks
Nelson (1939a)	Crude	Furnace tube	✗	✓	Fouling rate dependent on thickness of thermal boundary layer.	No removal term considered.
Kern and Seaton (1959b)	Water	HEX	✗	✗	Removal due to shearing action.	Conceptual model of fouling rate given by mass deposition–removal.
Atkins (1962)	Crude	Furnace tube	✗	✓	Deposit made up of two distinct layers.	No removal but considered aging.
Watkinson and Epstein (1970)	HC	HEX	✓	✗	Fouling controlled by mass transfer and adhesion of suspended particles.	Dependency of initial fouling rate on T predicted correctly but not of asymptotic fouling factor on flow rate.
Crittenden and Kolaczkowski (1979a)	HC	Heated tube	✓	✗	First-order kinetics and mass transfer control. Removal by back convection or shear.	Several parameters of difficult evaluation.
Crittenden and Kolaczkowski (1979b)	HC	Heated tube	✓	✓	R_f governed by deposition of tarry layers and their interactions with coked layers.	One of the first attempts to formalize in a mathematical aging.

(Continued)

Table 5.4 Summary of rate models for chemical reaction fouling—cont'd

References	Fluid	System	Mass transfer	Aging	Main assumptions	Remarks
Crittenden et al. (1987b)	HC	Heated tube	✗	✗	Initial fouling rate is first-order kinetics and mass transfer controlled.	Estimate model parameters from plant data.
Crittenden et al. (1992)	Crude	HEX	✗	✗	R_f controlled by reaction kinetics as a function of T_w.	Validated against refinery PHT data.
Epstein (1994)	Crude	Heated tube	✓	✗	Mass transfer controlled.	Good agreement with experimental data by Crittenden et al. (1987a).
Ebert and Panchal (1995)	Crude	Heated tube	✗	✗	Foulant formed in boundary layer by a one-step reaction. Suppression not function of film thickness.	Predicts the T_f at which initiation of fouling would occur at a given velocity.
Panchal et al. (1997)	Crude	HEX	✗	✗	Foulant is removed from the boundary layer before it is deposited.	Added the Pr number to previous E-P model. Only 60% of data followed the predicted fouling trend.

Polley et al. (2002a)	Crude	HEX	✗	✗	Modification of the E-P model assuming deposition as a function of T_w not T_f and suppression term being diffusion related, thus a function of Re, not τ.	Could be used to predict shell-side fouling without the use of CFD.
Srinivasan and Watkinson (2005)	Crude	Heated tube	✗	✗	R_f is a function of film temperature weighted more heavily on T_w.	Fitted on fouling data of three sour Canadian crude oils.
Saleh et al. (2005b)	Crude	Heated tube	✗	✗	R_f follows a power law.	Fitted on data of a light crude with low asphaltene and sulfur contents.
Nasr and Givi (2006)	Crude	HEX	✗	✗	No Pr number dependence.	Better fit than Polley et al. (2002a) on data by Saleh et al. (2005b).

HC = Hydrocarbons

Sheikh et al. (1997) developed a probabilistic approach with which they described fouling as a time random process where the fouling resistance is determined by

$$R_{f(t)} = f(a_1, a_2, t) \tag{5.38}$$

where a_1 and a_2 are parameters of the model that contain all the cumulative effects of the uncertainties related to the process.

Artificial neural network techniques have been successfully utilized to correlate the data during the entire fouling process in several types of fouling mechanisms (Malayeri and Müller-Steinhagen, 2001; Sheikh et al., 2001; Malayeri and Müller-Steinhagen, 2007; Radhakrishnan et al., 2007) and this approach has also been used for crude oil systems by Aminian and Shahhosseini (2008, 2009).

Li (2005) based his analysis on the fact that fouling data typically present fluctuations that are difficult to model. He proposed to separate the fouling rate into two parts: a mean component that exhibits a smooth and asymptotic behavior and a fluctuating component that exhibits an oscillatory behavior. However, this approach fails to explain the reason for the fluctuations and cannot be considered a reliable way of predicting fouling behavior.

In general, all the statistical methodologies reviewed above are essentially black box approaches that do not offer any physical understanding of the process. Another drawback is that this kind of model requires extensive data for "training" and appears to have limited extrapolation capabilities compared to either first principle or empirical models.

Other approaches to assess fouling behavior of industrial heat exchangers include the use of principal component analysis (Zabiri et al., 2006) and expert systems (Afgan and Carvalho, 1996).

5.1.5 Integrated Heat Exchange and Chemical Reaction Fouling Models

To successfully model fouling in a tubular heat exchanger, two main aspects need to be considered: the thermal-hydraulic model used to describe the heat exchanger under clean conditions and the fouling model used to describe the rates of deposition. It should be noted that these are not separate, but they closely interact.

The fouling rate models, reviewed in Section 5.1.4.2, are relatively simple correlations that provide a means to capture the fouling dependence on process conditions and time. Although these models are very easy to use, their accuracy in fitting the data is often not satisfactory and their applicability appears to be very limited. One reason for this is that fouling rate models do not include a thermal model (e.g., they do not allow calculation of outlet temperatures in the system considered) and thus they cannot be fitted to primary measurements (i.e., temperatures). Fouling resistance data, which are derived quantities (from experimental or plant measurements) that suffer from a propagation of errors, must be used instead. For its simplicity, developing and fitting rate models to fouling resistance values calculated from experimental (e.g., Ebert and Panchal, 1995; Polley et al., 2002a) or plant data (e.g., Costa et al., 2013; Ishiyama et al., 2013) are common practices. On the other hand, heat exchanger models that do not incorporate fouling provide erroneous predictions of heat exchanger performance.

Thermal models of heat exchangers that integrate fouling have the key feature of providing calculations of the fouling resistance as a function of process conditions within the system considered (single tube or whole heat exchangers). This approach allows, for example, the capturing not only of the thermal interactions between tube-side and shell-side but also the thermohydraulic effects generated by the presence of the fouling deposits, leading to more accurate calculations of the velocity, wall shear stress, and convective heat transfer coefficient.

Some models also include a detailed mass balance of precursors and fouling species together with a complex reaction scheme. This is particularly useful, for example, for milk fouling models where—unlike crude oil fouling—the reaction kinetics is better known.

5.1.5.1 Models of Fouling in Single Tubes

Takatsuka et al. (1989) proposed a distributed model for residual thermal cracking in a single furnace tube where the cracked products were assumed to come from an *n*-heptane-soluble component of the feedstock.

One of the most detailed models for organic fouling was developed by Panchal and Watkinson (1994). They included kinetics of the autoxidation of indene in kerosene and mass transport to the wall together with thermal exchange along the test section. The details of the reaction schemes hypothesized were reported in Section 2.2.6. The model was developed on the basis that the reaction mechanisms are known and includes several parameters which are difficult to estimate.

Watkinson and Zhang (2005) developed a model for organic fouling and simulated the material deposition from heavy hydrocarbon vapors in a single tube by solving a physical condensation rate–mass transfer model for the condensation of the vapors.

Coletti et al. (2010) developed a model for a single tube using a moving boundary approach that captured the growth of the fouling layer and the related thermohydraulic effects in crude oil fouling deposition. They extended the aging model developed in a previous work (Ishiyama et al., 2010) to account for localized aging effects across the axial and radial direction of the fouling layer. Aging was expressed via a change in the thermal conductivity of the fouling layer, between that of the freshly deposited material (the lower limit), λ^0, and a maximum value, λ^∞:

$$\lambda = \lambda^\infty + \left[\lambda^0 - \lambda^\infty\right] \times y \tag{5.39}$$

Values used for thermal conductivity (Watkinson, 1988) were $\lambda^0 = 0.2\ \mathrm{W\ m^{-1}\ K^{-1}}$ (initial gel-like deposit, λ similar to oil) and $\lambda^\infty = 1\ \mathrm{W\ m^{-1}\ K^{-1}}$ (coked deposit).

In Equation (5.39), y was defined as the "youth" variable and calculated assuming a first-order exponential decay dynamics function of the local deposit temperature, $T_l(z,r)$:

$$\frac{dy}{dt_{age}} = -A_a \exp\left(\frac{-E_a}{R_g T_l}\right) \times y \tag{5.40}$$

where E_a and A_a were, respectively, the activation energy (i.e., the temperature dependency) and preexponential constant characteristic of the aging process. The aging time, t_{age}, in Equation (5.40) referred to the lifetime of the deposit and was therefore a function of when it was formed. For an element formed at instant t_A,

$$t_{age} = t - t_A; \quad \forall t \geq t_A \tag{5.41}$$

Each element of deposit in a deposit layer featured a particular lifespan, between 0 and t, which needed to be tracked through time in order to evaluate the change in y. Coletti et al. (2010) presented a continuous formulation that captured the age of individual elements in the fouling layer using, in essence, a population balance which allowed Equation (5.40) to be written in the same time coordinate as the other differential equations used in the model:

$$\frac{dy}{dt} = (1 - \tilde{r})\left[-A_a \exp\left(\frac{-E_a}{R_g T_l}\right) \times y\right] \tag{5.42}$$

where $\tilde{r}$ was the nondimensional radial coordinate used to solve the moving boundary problem. It should be noted that when the deposition rate is not constant, the aging transformation (Equation (5.41)) is not exact. However, this was considered to be allowable, particularly since the underlying physics and associated parameters are not well known, and it was shown (Coletti and Macchietto, 2011) that the approximation was more than acceptable when used to describe industrial data.

The work by Coletti et al. (2010) also proposed two conceptual models to illustrate the effect that roughness dynamics has on the convective heat transfer coefficient as described in Section 2.4.4. The first was a simple linear model that assumed a growth of the surface roughness, e, as proportionally to the thickness of the fouling layer (calculated with Equation (5.30)) until a maximum roughness value was reached. In the second model proposed, the roughness at the interface between oil and deposit layer evolved asymptotically from the clean tube value, e^0, to a final value, e^∞, characteristic of the deposit type. In this case, the rate of change was a function of the foulant thickness and a dimensionless (adjustable) constant k_a:

$$e = e^0 + \left(e^\infty - e^0\right)\frac{\delta}{\delta + k_a} \tag{5.43}$$

A suitable choice of k_a made it possible to define faster or slower evolution of the roughness to its final value.

Sileri et al. (2009) went a step further by coupling not only heat and mass but also the momentum balances inside a square channel using CFD techniques. They described the two-phase flow of a hydrocarbon mixture in laminar flow. In their model, fouling was formed in the bulk of the fluid according to the reaction scheme by Panchal and Watkinson (1994) and transported to the wall where the deposit layer was formed. This was treated as a second liquid phase of changing viscosity as a function of process conditions and time. Yang et al. (2014) extended this model to turbulent flow and incorporated the aging model by Ishiyama et al. (2010) to account for the variation in thermal conductivity over time of the fouling layer. More details on the models by Sileri et al. (2009) and Yang et al. (2014) are reported in Section 5.3.

5.1.5.2 Models for Industrial Heat Exchangers Subject to Chemical Reaction Fouling

Although models for individual tubes or annuli are useful for explaining and interpreting fouling data from experimental rigs, the development of models for complete heat exchangers is crucial in enabling the necessary improvements in heat exchanger design methodologies, monitoring practice, and network configuration.

One of the first models for heat exchangers that incorporate chemical reaction fouling was proposed by Fryer and Slater (1985) who developed a dynamic and distributed model for a single pass heat exchanger undergoing milk fouling. Although the fluid treated was different from crude oil, the fouling mechanisms are very similar (i.e., chemical reaction fouling) and the fundamental thermal equations are relevant to the discussion. Fryer and Slater (1985) calculated fouling rates as a function of local temperature in terms of the Biot number for fouling, Bi_f:

$$Bi_f = \alpha_t^0 + \frac{\lambda_f}{\delta} \tag{5.44}$$

where α_t^0 is the tube-side convective heat transfer coefficient under clean conditions. The Biot number is a dimensionless fouling coefficient. Fryer and Slater (1985) captured the dependence of the Biot number as a function of both T_w and T_f:

$$\frac{dBi_f}{dt} = k_d \exp\left(-\frac{E_f}{R_g}\frac{1+\sigma+Bi_f}{T_w+(\sigma+Bi_f)T_f}\right) - k_r Bi_f \tag{5.45}$$

where k_d and k_r denote, respectively, the rate constants for deposition and removal, defined as a function of the wall shear stress, τ. In Equation (5.32) σ is

$$\sigma = \alpha_t^0\left(\frac{1}{\alpha_t^0} + \frac{\lambda_f}{\delta}\right) \tag{5.46}$$

The assumption underlying the model is that the convective heat transfer coefficient, α_t, does not change over time and is not a function of fouling layer thickness (i.e., $\alpha_t = \alpha_t^0 = \text{const}$).

For the specific system they examined, Fryer and Slater were able to determine a correlation between the rate constants and the shear stress from Arrhenius plots of initial fouling rates. The model has also been used to explore alternative control strategies (Fryer and Slater, 1986).

Georgiadis (1998) and Georgiadis et al. (1998a) developed a dynamic and distributed model for milk fouling in shell-and-tube heat exchangers that included radial velocity, temperature, and foulant concentration profiles capable of distinguishing among bulk, boundary layer, and surface reaction rates. The fouling resistance was expressed in terms of the Bi_f number:

$$\frac{dBi_f}{dt} = a_1 k_m c_f \tag{5.47}$$

In this case a detailed first principle, distributed mass balance was used, which includes the reaction of the foulant species to calculate c_f. The model was then used to find optimal design and operation of single units (Georgiadis et al., 1998b) and networks (Georgiadis et al., 2000).

Yeap et al. (2001b) presented a model for simulating crude oil fouling in shell-and-tube heat exchangers. They used the Panchal et al. (1997) fouling model coupled with a quasi-steady-state distributed thermal model. They divided the exchanger into segments of equal length along the unit where the effectiveness and the number of transfer units were calculated. In this relatively simple way they were able to determine a temperature profile within the unit and associated local fouling rates. Although more sophisticated thermal models (e.g., dynamic models in which the heat transfer coefficient is calculated as a function of process conditions) exist, this was the first attempt reported in the literature to develop a distributed model of a full heat exchanger undergoing crude oil fouling.

Yeap et al. (2004) coupled a lumped ε-NTU thermal model for a shell-and-tube heat exchanger with three different fouling rate models available in the literature (namely Panchal et al., 1997; Polley et al., 2002b; and Epstein, 1994). Although the model relies on a quasi-steady-state approximation (i.e., no time-dependent differential equations are solved for the heat balances), it accounted for time-dependent interactions of fouling growth with the hydraulic behavior of refinery heat exchangers and captured the effects of fouling on overall pressure drops. The model was then used by Ishiyama et al. (2008, 2009a,b) for network simulation as described in Section 5.1.5.3 and modified by Ishiyama et al. (2011, 2013, 2014) to account for aging of the fouling deposits. One limitation of the model is the thin slab approximation that restricts the rigorous applicability to cases in which the thickness of the fouling layer is less than c.10% the tube diameter.

Coletti and Macchietto (2011) developed a dynamic and distributed model of shell-and-tube heat exchangers undergoing crude oil fouling. In this model, differential heat balances were written in cylindrical coordinates to account for the effects of tube curvature on the heat flux, thus overcoming the limitations of the thin slab approximation used in previous models. Fouling was captured as a function of local process conditions and time across the heat exchanger length using the Ebert-Panchal model in a distributed (rather than lumped) way. As a result, Coletti and Macchietto (2011) could account for the effects of the variation in local fouling resistance on the driving force of the heat transfer between the two fluids. Numerically, they used a moving boundary approach to capture the growth of the fouling layer over time at any given point across the tube length, and the corresponding reduction in cross-sectional area. This allowed accurate calculations of the effect caused by flow variation due to fouling on wall shear stress and heat transfer coefficient. As a result, a comparison of model simulations with plant data from different refineries showed an agreement within 1–2% of outlet temperature measurements even when tested for prediction accuracy over extended periods of time (Coletti and Macchietto, 2010). Lastly, the aging model developed by Coletti et al. (2010) was implemented to describe the structural changes of the fouling deposit over time and the related effects on thermal conductivity. More details on the model by Coletti and Macchietto (2010) are given Section 5.4.

Table 5.5 summarizes the mathematical models for chemical reaction fouling in heat exchangers.

5.1.5.3 Models for Preheat Train Networks Subject to Crude Oil Fouling

Integrated models of heat exchange and chemical reaction fouling, reviewed in the previous section, have been used to study the effects that fouling behavior in individual heat exchangers has on preheat train networks (Yeap et al., 2001a, 2004, 2005; Wilson et al., 2002; Polley et al., 2005, 2009b; de Oliveira Filho et al., 2007; Ishiyama et al., 2008, 2013; Tavares et al., 2009; Coletti and Macchietto, 2010). Such models can capture how fouling dynamics affects the interactions between different units in the network and retrofits its configuration (Coletti et al., 2011) to minimize deposition and maximize performance over time.

Heat exchanger network models capable of predicting fouling behavior can also be put to a good use in assisting the scheduling of cleaning (Wilson and Vassiliadis, 1997; Smaïli et al., 2001; Ishiyama et al., 2009a, 2013; Sikos and Klemes, 2010). Models can be used to study the impact predetermined cleaning decisions have on the heat recovery. However, to ensure *optimal* cleaning actions, the predictive capabilities of the fouling trends are more effectively exploited in a mathematical optimization framework. Using an approach similar to that proposed by Wilson et al. (2001), Lavaja and Bagajewicz (2004, 2005a,b) introduced the use of a rigorous MILP model that included fouling dynamics in the planning of cleaning scheduling. However, to reduce the numerical complexity of the problem, they used only linear fouling models, thus eliminating the complications of nonlinearities in the definition of the problem and simplified the underlying heat exchanger model by using simple, lumped heat balances. Despite these

Table 5.5 Summary mathematical models for chemical reaction fouling in heat exchangers. (HC = Hydrocarbon)

References	Fluid	System	Side tube/shell	T/H/M model[a]	Dynamic[b]	$T(z)/\omega(T)/\omega(z)$ $h(T)/h(z)/h(\omega)$[c]		Aging	Remarks
Fryer and Slater (1985)	Milk	HEX	✓/✗	✓/✗/✓	✓	✗/✗	✗/✗/✗	✗	Single pass units only. HTC not function of fouling layer thickness.
Takatsuka et al. (1989)	Vacuum resid	Furnace tube	✓/✗	✓/✗/✓	✓	✗/✗	✗/✗/✗	✓	Fouling is related to sedimentation of precursors. Several parameters of difficult evaluation required.
Panchal and Watkinson (1994)	HC fluids	Tube	✓/✗	✓/✗/✓	✗	✗/✗	✗/✗/✗	✗	Compared model simulations with experimental data to determine which one of the three fouling mechanisms proposed may be occurring.
Georgiadis et al. (1998a)	Milk	HEX	✓/✗	✓/✓/✓	✓	✗/✗	✗/✗/✗	✗	Considered the radial dependency of velocity, temperature, and concentrations.
Clarke and Nicolas (2000)	Oil	HEX	✗/✓	✓/✓/✗	✓	✗/✗	✗/✗/✗	✗	Crude fouling on the shell-side.
Yeap et al. (2001b)	Oil	HEX	✓/✗	✓/✗/✗	✓	✗/✗	✗/✗/✗	✗	
Yeap et al. (2004)	Oil	HEN	✓/✗	✓/✓/✗	✗	✓/✗	✗/✗/✗	✗	

(Continued)

Table 5.5 Summary mathematical models for chemical reaction fouling in heat exchangers. (HC = Hydrocarbon)—cont'd

References	Fluid	System	Side tube/shell	T/H/M model[a]	Dynamic[b]	$T(z)/\omega(T)/\omega(z)$ $h(T)/h(z)/h(\omega)$[c]	Aging	Remarks
								Thin slab approximation.
Watkinson and Zhang (2005)	HC vapors	Furnace tube	✓/✗	✓/✗/✓	✓	✗/✗ ✗/✗/✗	✗	2D model for two-phase flow and deposition.
Ishiyama et al. (2011, 2014)	Oil	HEX/HEN	✓/✗	✓/✓/✗	✗	✗/✗ ✗/✗/✗	✓	Incorporates method for improving cleaning scheduling.
Coletti and Macchietto (2011)	Oil	HEX/HEN	✓/✗	✓/✓/✗	✓	✓/✓ ✓/✓/✓	✓	Tested with refinery data.

[a]T, thermal; H, hydraulic; M, mass transport model.
[b]Models that do not rely on quasi-steady-state approximation.
[c]$T(z)$: temperature dependence on the axial coordinate, $\omega(T)/\omega(z)$, generic physical property (i.e., ρ, λ, c_p) dependence on temperature and/or axial coordinate; $h(T)/h(z)/h(\omega)$, dependence of heat transfer coefficient on temperature, axial coordinate, and/or physical properties.

simplifications, Lavaja and Bagajewicz set up a sound framework to approach maintenance planning in a crude distillation unit.

When more complex, nonlinear fouling models (e.g., threshold models described in Section 5.1.1.2) are used, a MINLP problem must be solved. The biggest challenges faced when developing such a framework is the nonlinearity and the convexity of the resulting problems, which have limited so far effective implementation in refineries. To overcome these problems alternative algorithms have been proposed (Pogiatzis et al., 2012; Gonçalves et al., 2014). Although these methods allow incorporating more realistic fouling models, they do not guarantee a mathematical optimal solution.

5.1.5.4 Computational Fluid-Dynamic Approach

CFD is beginning to be applied to the design of heat exchanger equipment, and in particular for the shell-side flow in shell-and-tube exchangers (Andrews and Master, 2005; Vessakosol and Charoensuk, 2010). These studies have included assessments of the effects of baffle spacing (Mohammadi et al., 2009) and the benefits of helical baffles (Wang et al., 2009; Zhang et al., 2009).

Most works that use a CFD approach to study fouling behavior focus their attention on flat plate heat exchangers, commonly used in the food industry (Kho and Müller-Steinhagen, 1999; Zettler and Müller-Steinhagen, 2001; Jun, 2005).

Brahim et al. (2003a,b) integrated a model for the fouling layer deposition and suppression with a CFD analysis of the parallel flow across flat plates. The authors used a mechanistic model for fouling of calcium sulfate. Using a structured grid for the calculations of the flow between parallel flat plates, they considered both the velocity and the heat flux distribution as a function of the time-dependent total thickness of the layer. To avoid the need for a moving boundaries approach, they considered a "fictitious crystal growth model," This consists in varying the inlet velocity as a function of the thickness of the layer, which is calculated through the deposit mass (given by the deposition/suppression model) by its density.

Various other uses of CFD in fouling research include the method of Saghatoleslami et al. (2010) who used a commercial package to calculate the thickness of the fouling layer in a crystal, Al-Anizi and Al-Otaibi (2009) who used CFD to design enhanced impingement plates subject to fouling, Bergeles et al. (1997) who predicted particle deposition in lignite utility boilers, and Bouris et al. (2005) who studied the effects of a novel tube cross-sectional area on the fouling behavior of a gas.

Unlike other areas noted above, CFD has not often been applied to fouling analysis in heat exchangers processing crude oil. Clarke and Nicolas (2000) explored the effects of shell-side crude oil fouling in a 14 baffle unit. The shell volume was modeled as a porous medium and the tube bundle was represented by a reduction in porosity. This is a method commonly used to reduce the computational effort (Tierney, 1992). Impermeable surfaces were used for the baffles. The Ebert–Panchal model (the exact form of the equation used was not specified) was then used on the shell-side to take into account the formation of deposits on the outer surface of the tubes. To represent deposition, the model artificially increases the fluid viscosity, as the deposit thickness increases, in the cells around the tubes. This appears to be a very simplistic approach that suffers of a number of assumptions (e.g., the interactions between tube and shell-side are neglected by setting an arbitrary fixed linear temperature profile on the tube-side). However, it seems that attempts in the literature to simulate fouling on the shell-side of heat exchangers are limited to this one paper.

Yang et al. (2009b) showed that a CFD approach can be used to interpret crude oil fouling data in a batch stirred vessel. The approach was then used by Crittenden et al. (2013) to study the effects of heat transfer enhancement devices on wall shear stress.

The work by Sileri et al. (2009) and by Yang et al. (2014), which uses a CFD approach noted in the previous section, will be described in more details in Section 5.3.

5.1.6 Concluding Remarks

In this section several limitations of the traditional TEMA fouling factor approach to the design of heat exchangers and heat exchanger networks are highlighted. It has been discussed why the result of the use of a fixed fouling factor at the design stage does not solve, and in some cases exacerbates, the fouling problem for refinery operators. Two main factors have favored over the years the use of these design methodologies. The difficulties exist in capturing knowledge and translating it into a reliable model capable of accurately predicting fouling on one hand and the lack of an effective implementation of available models into user-friendly tools on the other.

To be able to improve design, retrofit, and monitor both heat exchangers and networks of heat exchangers, several key aspects must be addressed:

- Fouling, aging, surface roughness dynamics phenomena, and their effect on the thermohydraulic behavior of the system considered must be captured simultaneously as a function of process conditions.

- A suitable modern integration tool should be used to allow the solutions of a large number of equations in a user-friendly way. The implementation should be flexible enough to allow the analysis of important phenomena at several levels of investigation and serve a variety of purposes:
  - Analysis of laboratory data (typically obtained in tube sections). At this scale of investigation, it is important to include phenomena such as surface roughness dynamics.
  - Performance monitoring of individual industrial units and networks. This may include online and/or control applications and assessment of running costs. The correct estimation of model parameters is pivotal. A model-based parameter estimation approach using primary measurements (i.e., flow rates and temperatures) rather than derived quantities should be preferred.
  - Single unit design and retrofit. Here aging plays an important role and should be taken into accounted. Model validation with available plant data, although challenging, is important to give confidence in the accuracy of the predictions.
  - Heat exchanger network design and retrofit that take into account fouling dynamics. Assessment of the impact of selected network structures should be based not only on energy recovery under clean conditions, but on key performance indicators that inform not only design but also operational decisions.
  - It should be possible to easily replicate the single heat exchanger model in a flowsheeting environment to readily generate specific network topology. This would eventually allow the optimization of each single heat exchanger design together with the optimization of the network structure to minimize fouling and maximize economic performance over time.

The following sections will describe different modeling on the development of mathematical models capable of predicting fouling behavior at the microscale (Section 5.2), fundamental transfer processes (Section 5.3), and plant-scale (Section 5.4) levels that incorporate all the features above.

5.2 THERMODYNAMIC AND MOLECULAR MODELING

G. Jimenez-Serratos, A. J. Haslam, G. Jackson, E. A. Müller

Despite considerable efforts both from the scientific and the industrial community, most efforts at mitigating both asphaltene precipitation and fouling are based on an empirical and heuristic knowledge. As compared to other fields, such as bioindustry, fine chemicals, or polymers, the fundamental knowledge of the physico-chemical processes associated with oil and gas production is lagging far behind. Much of the fault lies in the current inability to fully characterize the working fluids and the simple impossibility of accounting for a full molecular description of a mixture of tens of thousands of distinct components with wide distributions of sizes and weights and whose complex interactions are difficult to understand. Molecular thermodynamics and statistical mechanics are the pillars of any modern description of the phase behavior of these complex systems. Special note is made of the contribution that computational simulations at different length scales can make to the understanding of the intricacies of this problem. The uniqueness of the problem and the lack of clarity in the methodologies that must be used have allowed the coinage of the distinct scientific discipline of "petroleomics" as the attempt to link molecular structure and the observed behavior (Breure et al., 2013).

5.2.1 Introduction

The complexity of a petrochemical mixture is such that a description based on the accounting of the individual distinct chemical species is simply impossible. Hence, diverse strategies have been suggested to reduce the unmanageable number of degrees of freedom by reducing the number of components considered in the model, by "lumping" groups of molecules with similar mass and chemical behavior into pseudocomponents, which are then treated classically. A more elegant approach is the consideration of a nonclassical treatment in terms of continuum thermodynamics (Cotterman et al., 1985; Cotterman and Prausnitz, 1985; Jaubert et al., 2000). Although this is an excellent conceptual approach, its implementation has not been widespread, presumably due to the reluctance of researchers and industry to modify the existing models to any large extent and particularly to make changes to software to account for a different approach to solving the phase equilibria. Most approaches to describing oil and gas mixtures make use of the idea of simplifying the mixture into pseudocomponents, the number of which depends on the application, but may vary from two, as in the black-oil–water models common in reservoir simulations to typically several dozen, as in equation-of-state modeling of light condensates. One should note, however, that studies with over 100 pseudocomponents have been reported (Avaullee et al., 1997). A common lumping criterion for classifying compound types instead of individual compounds is the SARA analysis method, in which four fractions are distinguished: saturates, aromatics, resins, and asphaltene. A key problem in the description is the characterization of asphaltenes, which are polyaromatic, heavy, and polar components considered the major precursors of sediments. In terms of solubility, asphaltenes are defined as soluble in aromatics like benzene and toluene but insoluble in alkanes like *n*-pentane and *n*-heptane (Ortega-Rodríguez et al., 2004; Pedersen and Christensen, 2006; Ancheyta et al., 2010; Artola et al., 2011; Breure et al., 2013; Ungerer et al., 2014).

Any attempt to gain a working knowledge in the field must inevitably consider the prediction of the behavior of the system as a function of temperature, pressure, and composition through modern thermodynamic tools involving the molecular description of the crude-oil components, the understanding and modeling of their intermolecular forces, and the ultimate integration of this information in the form of a (hopefully) closed-form equation of state (Aguilera-Mercado et al., 2006; Ancheyta et al., 2010; Artola et al., 2011; Ungerer et al., 2014). As simple as the previous statement might sound, the issue of constructing an equation of state has been a considerable challenge and is the subject of this section. Particularly, from a theoretical perspective, equations of state (EOS) are reliable tools for predicting the phase diagram and precipitation behavior. Although thermodynamic treatments of this problem date back as far as the late 1970s and early 1980s, the problem is as yet far from being resolved. The aim of this section is to give an overview of the platforms that thermodynamics and molecular modeling offer to contribute to the understanding of petroleum industry problems and particularly of the precipitation of asphaltenes and the consequent fouling.

5.2.1.1 Molecular Models of Asphaltenes

To be able to model appropriately systems containing fractions (such as asphaltenes) that may precipitate out of solution, it is vital to have an accurate description of the molecular structure. This in itself is a very challenging step as asphaltenes represent a solubility class of molecules, rather than a distinct molecule. As described earlier, an assumption must be made that there is a representative molecule or group of molecules that have chemical characteristics similar to those of asphaltenes and be can be used as a proxy for the description of a continuum of asphaltene molecular structures. It is not the intention of this review to give a detailed examination of the literature concerning asphaltene molecular structure. Two recent reviews give an overview of the available experimental data and the contentions regarding the size of the aromatic core and the corresponding molecular weight (Mullins, 2011; Herod et al., 2012).

As already noted in Chapter 2, in the case of asphaltenes and resins, there is no precise molecular formula or structure because by definition, a range of compounds falls into each of these categories and they invariably vary from crude to crude (Gray et al., 2011; Breure et al., 2013; Ungerer et al., 2014). Asphaltenes are usually defined operationally to be the (toluene soluble) fraction precipitated from oil on the addition of *n*-heptane; however, analogous precipitation from the oil can also be induced by the addition of other precipitants or, in particular, by changes in the thermodynamic conditions (temperature and/or pressure); such precipitates are also often referred to as asphaltenes. The detailed properties of these components are far from being well understood, and even their molecular weights are the subject of controversy (Mullins et al.; 2008; Herod et al., 2008). In the case of asphaltenes, polydispersity and solvent conditions affect the estimates giving numbers that realistically range from 700 to 3000 g mol^{-1} for asphaltenes (Aguilera-Mercado et al., 2006; Ancheyta et al., 2010). Some trends in the precipitation of asphaltenes have nevertheless been found in terms of general molecular properties, such as C/H ratio or the aforementioned molecular-weight range.

Molecular models for asphaltenes are sorted into three prototypical categories: continental, island, and archipelago. In the continental model, asphaltenes are represented with aromatic cores with some aliphatic branches; the island model has a similar structure and C/H ratio but a lower molecular weight (with a vague separation between both models); and the archipelago model consists of small aromatic regions bridged by alkanes (Ancheyta et al., 2010; Wiehe, 2012; Ungerer, 2014). Each model can stack differently, which can be useful for modeling specific behavior. It is likely that most crudes will have a combination of these structures, with possibly a predominance of a particular one. A vast fauna of supramolecular formations is reported in the literature: stacking of molecules into clusters, fractal aggregates, liquid-crystal domains, micelles, fibers, tubes, gels, and porous materials (Sirota, 2005; Gray et al., 2011; Breure et al., 2012). Experimental efforts using techniques such as small-angle neutron scattering, small-angle X-ray scattering, X-ray diffraction, nuclear magnetic resonance, and dynamic light scattering have helped to understand the structure and aggregation of asphaltenes (Pedersen and Christensen, 2006; Ancheyta, 2010), but the problem still resembles that of the Hindu fable retold by Saxe (1872) of the attempt of blind men to describe an elephant (Figure 5.6).

Figure 5.6 Classical attempt to describe asphaltenes by indirect measurements are not distinct from those described by the Hindu tale of the blind men and the elephant. In this story each blind man touches the elephant in a different part and they independently conclude that the elephant is like a snake, a wall, spear, tree, fan, or rope, depending on where they touch. Despite their stubborn discussions, they never resolve their conflict. Attempts to describe asphaltenes from what is inferred by indirect experimentation suffer from the same shortfalls.

Intermolecular interactions are a key factor in the description of asphaltene-containing systems. As described early on by Murgich (2002), the list includes acid–base interactions, hydrogen bonding, formation of charge-transfer complexes, apolar associations, and π–π stacking. The relative importance of the different types of interactions in typical heavy oils and aslphaltenes is far from clear (Gray et al., 2011), again despite the crucial importance of these interactions as drivers for precipitation. The use in experiments and at theoretical level of proxies or pseudofluids which might represent a fraction of crude oil with simpler or better defined molecules trying to resemble the intermolecular interactions or behavior of the mixture has been attempted (Sirota, 2005; Gonzalez et al., 2007; Artola, 2011; Greenfield, 2011; Breure et al., 2012). Examples of molecules that can be used as pseudocomponents to represent asphaltenes are polystyrene and hexa-*tert*-butyl hexa-peri hexabenzocoronene. These molecules share fundamental characteristics with asphaltenes: they contain aromatic and aliphatic structures and have similar solubility properties (Sirota, 2005; Artola, 2011; Breure et al., 2012).

5.2.1.2 Aggregation Models

One of the existing theories for asphaltene aggregation suggests that the dominant role in the aggregation process is the interplay between asphaltenes and resins. Polar regions of resins and asphaltenes would exert attractive forces by hydrogen bonds or dipole–dipole interactions (Ancheyta et al., 2010). Following this line of thought one would have to recognize resins as stability agents preventing asphaltene deposition. Under such a description, asphaltenes are considered as micelles or colloidal systems surrounded by resins that screen the interactions with the aliphatic compounds in the solvent (because resins are soluble in *n*-heptane, contrary to asphaltenes) (Pfeiffer and Saal, 1940; Pedersen and Christensen, 2006; Pina, 2006; Ancheyta et al., 2010). Presumably, if there is a lack of resins, the micelles attract each other forming bigger structures. Consequently, the resins/asphaltene ratio is found to be a decisive factor for the stability of petroleum (Pfeiffer and Saal, 1940; Ancheyta et al., 2010). However, other authors claim that a direct relation between the resin/asphaltene ratio cannot be established as there is a dependency on saturates content that must be considered (Buenrostro-González et al., 2001). Furthermore, other tests have revealed stability or instability effects of resins, depending on the nature of the crude (Pereira, 2007). An interesting overview of the contrasting descriptions is given by Ancheyta et al. (2010).

The micellar model can be traced to work of the 1920–1940s (Nellensteyn, 1923; Pfeiffer and Saal, 1940) and its historic persistence over the intervening years (Fahim et al., 2001) is a testament of the capacity of the model to explain complex observations. However, from a conceptual and fundamental point of view it is inaccurate as asphaltenic systems do not exhibit a critical micellar concentration or the expected bimodal behavior of the distributions of aggregate sizes (namely, one peak related to actual micelles and the other one to individual amphiphiles, in the case of surfactant systems) (Sirota, 2005; Aguilera–Mercado et al., 2006; Pedersen and Christensen, 2006). Well-known variations of these micellar/colloidal models are the Yen (renamed Yen-Mullins) model (Buenrostro-González et al., 2001; Mullins, 2010). Under such conception, asphaltenes are assumed to form nanoaggregates from the stacking of aromatic sheets and then those aggregates form larger clusters defining a hierarchical organization (Mullins et al., 2007; Mullins, 2008; Wiehe, 2012). Again, some statements disproving the conceptual basis of colloidal models can be found in the literature. As an example, Sirota (2005) argues that even the interpretation of scattering experiments assuming asphaltenes as particles can lead to a misunderstanding of phase separation scattering data. Another interesting point is the precipitation exhibited in asphaltenic systems at high temperatures. In the colloidal model resins are expected to respond to the temperature increase by mixing in the solvent destabilizing the asphaltenes, but according to Sirota (2005) this explanation is inconsistent because the difference in molecular weight between asphaltenes and resins to reproduce such behavior should be larger than the experimental estimates.

A more-modern line of thought aims at describing the aggregation as a liquid–liquid phase separation. Even in the case of "athermal" systems, the interactions play a secondary role and it is the asymmetry in sizes that leads to the driving force in precipitation processes (Frenkel and Louis, 1992). This in itself is not sufficient to describe all the features of the aggregation process, but is definitively an important driving force. The description of the problem may be made in the context of solution theory of molecules, where asphaltenes are not assumed to be suspended but dissolved in the oil and the precipitation is seen as a phase separation process (Sirota, 2005; Pina et al., 2006). Solubility models of this type are typically based on the Flory–Huggins theory for polymer solutions (Flory, 1941; Huggins, 1941); at this level, resins are not typically considered an explicit component, having a less-relevant function than in colloidal models (Hirschberg et al., 1984; de Boer et al., 1995; Wu et al., 1998). A glass transition phenomenon as a possible model has also been suggested (Sirota, 2005).

5.2.2 Thermodynamic Approaches

From the abridged discussion presented in the previous sections, it is obvious that there is a fundamental lack of physical insight into the asphaltene deposition problem and the modeling lags behind because of this. This is not to say that there have not been a large number of attempts to develop a consistent and workable theory. The unique nature of the problem requires, however, a unique tailor-made methodology. Instead, most current attempts to model these mixtures are based on the adaptation of current techniques used to model surfactants, polymers, colloids, and/or the phase equilibria of simple mixtures. The resulting strategies fail mainly because of this lack of correspondence between the real problem modeled and the forced idealization of theoreticians.

Experimental research constitutes the central mechanism for understanding the behavior of crude oil and conditions for asphaltene precipitation. Improvements in the technology over the last decades have provided enhanced experimental data under extreme conditions and with higher sensitivity (Mullins et al., 2007). However, complications related to data analysis, operational cost, control of experimental conditions, and in general, the ultimate aspiration of establishing predictive connections between the characteristics of the mixture and the phase behavior require the use of theoretical and computational approaches. The petroleum industry is an innate and demanding consumer of thermodynamic tools for the description of the phase behavior of crude oil and related systems (Kontogeorgis and

Folas, 2010) and this has been instrumental in the developments of recent years.

Semiempirical methods, correlations, and engineering-type correlative procedures represent a basic attempt to characterize pressure–temperature–composition behavior of pure compounds or mixtures of industrial interest. However, the complexity of the system of interest leads to numerous empirical adjustable parameters and correlations that do not possess a clear physical meaning (Ortega-Rodríguez et al., 2004; Aguilera-Mercado et al., 2006; Pedersen and Christensen, 2006). Empirical cubic equations of state, molecular-based thermodynamic theories, and their combination with computer simulations represent useful alternatives for characterizing the system from a physical perspective. In particular, a statistical thermodynamic approach gives a further connection between the structural features and the corresponding macroscopic properties (Pedersen and Christensen, 2006; Mullins et al., 2007). Herein some of the most common and successful of such thermodynamically based approaches will be discussed.

5.2.2.1 Polymer Solution Models

As noted previously, Flory–Huggins (FH) polymer solution theory has been recurrently used to model precipitation of asphaltenes in combination with the Hildebrand regular solution theory (RST) (Hildebrand, 1929; Hildebrand et al., 1970). In the framework proposed by Hirschberg et al. (1984), a mixture of asphaltenes, maltenes, and liquid can be represented as a simple binary system wherein the asphaltenes are one of the components and the maltenes and liquid the second component. The maximum volume fraction of asphaltenes soluble in the oil phase assumes the form (Hirschberg et al., 1984; Pedersen and Christensen, 2006):

$$\phi_{max} = \exp\left(\frac{V_A}{V_L}\left[1 - \frac{V_L}{V_A} - \frac{V_L}{R_g T}(\delta_A - \delta_L)^2\right]\right) \tag{5.48}$$

where R_g is the universal gas constant, T is the temperature, V_L and V_A are the molar volumes of the asphaltenes in the liquid and asphaltene phases, respectively, and δ_L and δ_A are the average solubility parameters of the nanoaggregates or asphaltenes in the corresponding component of the binary mixture. These authors note that the main weakness of their model is that it excludes the effect of resins. An implicit way to incorporate this is by modifying the asphaltene solubility to higher values in the liquid phase than that of pure asphaltene. In the same work, the vapor–pressure equilibrium is solved by using the Soave–Redlich–Kwong (SRK) cubic EOS, but there is a thermodynamic inconsistency as the solution is not unique (Pedersen and Christensen, 2006).

As can be seen from Equation (5.48), the amount of asphaltenes in solution strongly depends on the average solubility parameter. Further modifications by other authors include better estimates of the average solubility parameter of the oil, e.g., the study of Wang and Buckley (2001), where a correlation with the refractive index is used.

Alboudwarej et al. (2003) have obtained a qualitative description of precipitation in Athabasca bitumen using a more-sophisticated model. The author included *n*-paraffin, saturates, aromatics, resins, and asphaltenes as pseudocomponents of the mixture; the asphaltenes (considered with a molecular-weight distribution) and resins populate the asphaltene-rich phase.

Other notable studies employing lattice models are those of Burke et al. (1990), Kawanaka et al. (1991), Kokal et al. (2003), and de Boer et al. (1995). For example, Kawanaka et al. (1991) used an extension of the Flory–Huggins polymer solution theory, developed by Scott and Magat to incorporate the effect of polydispersity through heterogeneous distributions of polymers (Scott and Magat, 1945; Scott, 1945).

Considering the complexity of asphaltenes and crude oils, it is remarkable that simple descriptions like the FH-RST can offer a suitable qualitative description of the precipitation phenomenon. Importantly, these models are not predictive and the parameters must be obtained from experimental data (Kontogeorgis and Folas, 2010). The deficiencies of these theories come from deep features of the system that require more complete characterizations: the association phenomena of asphaltenes with resins, the description of the asphaltene–oil equilibria (is it a liquid–liquid, solid–liquid, colloidal instability?), the extensive association in the components of the mixture, etc.

For the interested reader, a rich discussion on the Flory–Huggins solution model applied to bituminous systems is made by Wiehe (2012) and a recent and detailed review is presented by Forte and Taylor (2014). The foundations of these approaches and their applications to other systems can also be gleaned from Kontogeorgis and Folas (2010).

5.2.2.2 Cubic Equations of State

Cubic equations of state are classical models for systems at pressures high enough so that the intermolecular interactions have a significant effect on the vapor–liquid equilibrium (VLE), i.e., nonideal conditions (Kontogeorgis and Folas, 2010). Bernoulli at the end of the eighteenth century proposed the inclusion of the volume of the molecules as a repulsive term, but it was van der Waals (vdW) and works derived from his doctoral thesis in 1873 that resulted in the EOS that accounts for such a repulsive contribution and a pressure correction related to attractive forces (van der Waals, 1873; Valderrama, 2003). Although the vdW EOS is not quantitatively accurate for most applications, the prediction of a critical point and the description of vapor–liquid coexistence below the critical point represented a giant leap in the representation PVT behavior. The vdW equation and the countless modifications can be written as a generic cubic equation (Valderrama, 2003):

$$P = \frac{R_g T}{V - b} - \frac{a}{V(V + d) + c(V - d)} \tag{5.49}$$

where a, b, c, and d are constants or functions of temperature, acentric factor, critical compressibility, etc. The third degree in the volume V gives a simple mathematical form and the extension to mixtures is relatively direct.

Modifications to vdW EOS that keep the cubic dependence in volume usually change the attractive pressure term of Equation (5.49). The original work of Redlich-Kwong (1949) proposed an attractive term which is temperature dependent through a function $\alpha(T)$ and opened up the field for a myriad of different empirical versions of this attractive term; it was rapidly understood that it was this second term in Equation (5.49) that carried all the important weight into the parameterization of fluid phase equilibria. Wilson (1964) introduced Pitzer's acentric factor, $\alpha(T_R,\omega)$ as a third parameter in the parameterization and Soave (1972) improved the functional definition of the $\alpha(T_R,\omega)$ function. The Peng-Robinson (1976) modification consists of improvements to Soave's version by changing the form of the attractive term and recalculating the $\alpha(T_R,\omega)$ function (Kontogeorgis and Folas, 2010). Further versions usually follow three approaches (Valderrama, 2003): a modification of $\alpha(T_R,\omega)$; a modification of the volume dependence of the attractive term (or both by applying volume translations, Peneloux et al., 1982); and the use of a third substance-dependent parameter. Numerous variations and the efforts to improve mixing and combining rules result in a wide tailor-made suite of cubic EOS to describe diverse systems. Kontogeorgis and Folas (2010) and Valderrama (2003) present well-organized information and tables that summarize the wide variety of cubic EOS in use today.

In the field of asphaltenes, two major approaches have been used. The first is to group the mixture into a reduced number of pseudocomponents and describe each with specific critical properties. The second is to describe properties through continuous distributions to represent the reservoir mixture (Haynes and Mathews, 1991; Vakili-Nezhaad et al., 2001).

Yarborough used the PR-EOS (1979) to predict the fluid-phase behavior of crude oils and gas condensates. In his study, a procedure to characterize heptanes and heavier components was developed and comparisons of calculated and experimental VLE for ternary systems were made. Contemporary contributions to that of Yarborough are those of Vogel et al. (1983), Willman and Teja (1987), Firoozabadi (1988), and Skjold-Jørgensen (1984).

The Peng-Robinson EOS was used by Szewczyk and Béhar (1999) to study flocculation, considering it as the appearance of a second liquid phase. The mixture has three components: C10–, C10–20, and C20+ fractions. For pressures above the bubble pressure, flocculation appears as a liquid–liquid separation, while under the bubble pressure, there is a liquid–liquid–vapor equilibrium. The main problem in this study is (as is usual in modeling approaches) the large number of parameters required and the difficulty of obtaining them, which is an obstacle to implementation at an industrial level (Pina et al., 2006).

In the work of Nghiem et al. (1993) the PR-EOS was used to model the gas and liquid phases, whereas the asphaltene phase was considered to be a pure solid, the fugacity of which is given as function of temperature, pressure, and the asphaltene onset pressure. In a related work, Kohse et al. (2000) model the mixture with 12 pseudocomponents, where the asphaltene is again modeled as a pure solid. The solid model is also the basis of the work of Jamaluddin et al. (2000) and Castellanos Díaz et al. (2011) in which the complexity of the representation is increased, including 17 components in the first case and 16 to represent the boiling-point curve in the latter.

Examples of applications of EOS to asphaltene precipitation where a large number of compounds or pseudocompounds are included in the models are those of Sabbagh et al. (2006) and Behar et al. (2003) that use of the order of 30 to model bitumen, using the PR-EOS (Forte and Taylor, 2014).

In the context of gas-injection modeling, a comparative study by Danesh et al. (1991) analyzed and compared the phase behavior and volumetric properties of injection gas-reservoir oil systems obtained through cubic EOS. They concluded that the modified Patel-Teja (VPT) (1982) and the Zudkevitch-Joffe-Redlich-Kwong (ZJRK) EOS (Zudkevitch and Joffe,1970) were generally superior to other approaches, that the inclusion of the volume-translation concept improves the liquid density prediction, and that all of the tested EOS needed pertinent experimental information to be reliably used close to miscibility conditions.

Victorov and Firoozabadi (1996) proposed a description under the micellar model; the petroleum was assumed to be a dilute solution and the PR-EOS was applied to calculate the fugacity of monomeric asphaltene in the bulk of the petroleum fluid. The model captures the change in precipitation for different precipitants and it is sensitive to the quantity of resins in the mixture. Following the micellar model other works include that of Pan and Firoozabadi (1998), Fahim et al. (2001), and Mahdavi and Kharrat (2011), among others (Forte and Taylor, 2014).

The success of cubic EOS in the petroleum industry is based on their simplicity, which is suitable for fast calculations of relevant mixtures. However, even when they can provide accurate correlations, their predictive power can be restricted to some intervals of phase space (Artola et al., 2011). As the mixtures and corresponding interactions increase in complexity, cubic EOS present severe limitations. Polarity, associative features, and the large asymmetry in the molecular sizes involved are some of the aspects that cannot be included in an appropriate manner with a cubic EOS (Kontogeorgis and Folas, 2010; Artola et al., 2011). Increasing the predictive power of cubic EOS for multicomponent and multiphase mixtures containing hydrocarbons and polar/associating chemicals was the practical motivation of the development of the Cubic-Plus-Association EOS (Forte and Taylor, 2010; Kontogeorgis and Folas, 2010).

5.2.2.3 The Cubic-Plus-Association Equation of State

The Cubic-Plus-Association EOS (Kontogeorgis et al., 1996) is a combination of the SRK equation (although other cubic EOS can be used) with an association term, based on the Wertheim first-order perturbation theory (1984a, 1984b, 1986a,b) of the type used in the statistical associating fluid theory (SAFT), described later in Section 5.2.2.4. As presented by Kontogeorgis et al. (1996, 1999), the associative effect is introduced as

$$P = \frac{R_g T}{V-b} - \frac{\alpha(T)}{V(V+b)} - \frac{1}{2}\frac{R_g T}{V}\left(1+\rho\frac{\partial lng}{\partial \rho}\right)\sum_i x_i \sum_{A_i}(1-X_{A_i}) \tag{5.50}$$

where ρ is the molar density ($=1/V$) and X_{A_i} is the fraction of nonassociated molecules, which depends on the association strength between two sites belonging to two different molecules, given in terms of the radial distribution function g. A complete exposition of the general theory, details of the combination and mixing rules, and industrial applications has been given by Kontogeorgis and Folas (2010).

In the oil and gas industry, the introduction of associative effects finds a natural market. Applications of CPA-EOS to asphaltene precipitation can be found in the recent work of Li and Firoozabadi (2010a,b). In the first case the authors study asphaltene precipitation in model solutions (asphaltene + toluene) and in heavy oils and bitumens. The system is characterized through three pseudocomponents including saturates, aromatics/resins, and asphaltenes. The PR-EOS is the basis of the method and the associations between asphaltenes and those between asphaltenes and heavy components are contemplated with the Wertheim association term. The model describes the saturation and onset pressures, the amount of asphaltene precipitated, and the gas–liquid–asphaltene behavior for several live oils. In the second study the precipitation is modeled again as liquid–liquid or gas–liquid–liquid separation. The pseudocomponents of the mixture are: pure components, pseudohydrocarbon components, and the hydrocarbon residue, which is further divided into "heavy" (alkanes, heavy aromatics, and resins) and asphaltene. A successful description of the effect of pressure and temperature on the bubble point is reported, as well as the phase separation behavior. Shirani et al. (2012) have used PR and SRK EOS to describe the physical part of the interactions, and the association contribution is attributed to chemical interactions and hydrogen bonding. The model is based on the Whitson's splitting and lumping method (Ahmed, 1989), and the asphaltenes are considered as molecules with two association sites. The SRK description is found to be superior to that of PR-EOS when compared with experimental data for live oils.

Including molecular information in the thermodynamic description enables the connection between the structure (microscopic) and the function (macroscopic), contrary to pure macroscopic approaches (Ortega-Rodríguez, 2004; Prausnitz and Tavares, 2004). Although the equations of state discussed are useful tools for the modeling of asphaltene-deposition behavior, their parameters do not permit the aforementioned link to be established and follow the trail of the intermolecular potentials. On the other hand, models based on statistical thermodynamic theories, such as statistical associating fluid theory (SAFT, described in the next section), include parameters with a well-defined physical meaning (Müller and Gubbins, 2001; Kontogeorgis and Folas, 2010).

5.2.2.4 Statistical Associating Fluid Theory Equation of State

Based on statistical thermodynamics, Wertheim (1984a,b, 1986a,b) developed a first-order perturbation theory for the free energy of fluids with a repulsive core and strong directional short-range attractive sites. In the SAFT approach, as presented by Chapman et al. (1988, 1989, 1990), molecules are modeled as spheres (monomers) or chains of fused spheres with associating sites. The Helmholtz free energy is given as separate terms: the ideal term; the contribution from monomeric repulsive or dispersive interactions; the contribution due to formation of chains; and the contribution due to association. In general, when considering crude oils, the association term is dropped and the equation related only to terms linked with the underlying "monomers" and the chain length, m.

The original SAFT EOS has been adapted to describe a complex system with particular features. This flexibility gives SAFT key advantages over common cubic EOS. Some of the well-accepted versions of SAFT (Kontogeorgis and Folas, 2010) are the following: CK-SAFT (Huang and Radosz, 1990, 1991), Simplified SAFT (Fu and Sandler,1995), LJ-SAFT (Kraska and Gubbins, 1996a,b), SAFT-VR (Gil-Villegas et al., 1997; McCabe et al., 1999), Soft SAFT (Blas and Vega, 1998), PC-SAFT (Gross and Sadowski, 2001), Simplified PC-SAFT (von Solms et al., 2003; Tihic et al.,2006), and the recent and versatile SAFT-VR Mie (Lafitte et al., 2013) and SAFT-γ (Papaioannou et al., 2014). Reviews of the SAFT methodology and more details about the versions can be found in Müller and Gubbins (2001), Economou (2002), Kontogeorgis and Folas (2010), and McCabe and Galindo (2010).

The use SAFT in modeling alkane and oil-like systems is now well established in the oil industry (Leekumjorn and Krejbjerg, 2013). In particular, SAFT has been applied with some success in modeling the onset of asphaltene deposition from crude oil (Wu et al., 2000; Tan et al., 2008; Punnapala, 2013; Sedghi, 2014). Special mention is due to the efforts of the Chapman group at Rice who have a particularly exhaustive set of papers on the different aspects of modeling asphaltene systems with the SAFT EOS (Ting et al., 2003; Gonzalez et al., 2005, 2007, 2008; Vargas et al., 2009a,b; Panuganti et al., 2013). By capturing experimental onset and bubble-point data, these studies have demonstrated the broad applicability of SAFT for the study of crude-oil systems.

Among the biggest challenges present in applying SAFT is the description of the asphaltene fraction; i.e., what is the molecular description of an asphaltene? The nature of the SAFT EOS allows only for a description in terms of distinct energies and molecular sizes, so there is very little justification for the use of a large number of adjustable parameters. The extent of the capabilities of a SAFT calculation in which most of the different components of the crude oil share the same parameters (distinguishing themselves by the different average chain lengths for each pseudocomponent fraction) is shown in Figure 5.7. The asphaltene fraction is treated as a single component and a single binary interaction parameter is employed to describe the experimental data (Artola et al., 2011).

The general trends of the experimental data are clearly reproduced. The bubble curve is seen to be the boundary delimiting a three-phase region of VLLE, while the location of the precipitation boundary is strongly related to the LLE boundary, clearly supporting the interpretation that the onset of precipitation of the asphaltene can be represented as a liquid–liquid instability. In addition, one can understand how the precipitation boundary would be affected by changes in the nature of the crude oil. Of course, it is important to bear in mind that in reality, the precipitation boundary would be very sensitive to the precise distribution of asphaltenic species

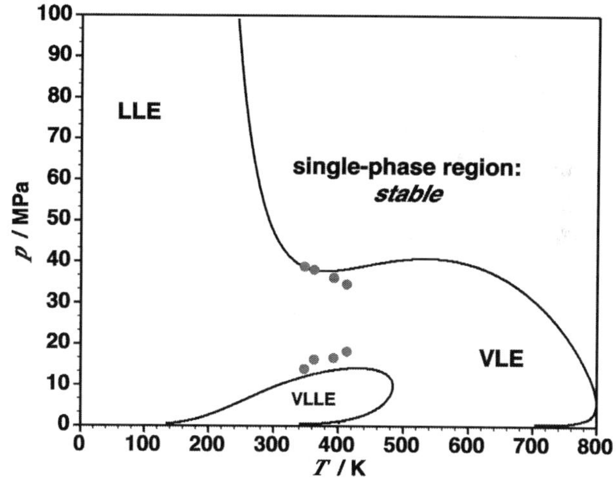

Figure 5.7 Calculated pressure–temperature fluid phase diagram of a simple model of an oil based on the SAFT EOS showing the phase boundaries and including both the liquid–liquid equilibria (LLE) and the vapor–liquid (VLE) and vapor–liquid–liquid (VLLE) regions. In this model the same potential parameters are used in representing three of the five components used to model the asphaltene-containing oil, with just a single adjustable parameter invoked to tune the alkane–asphaltene unlike interaction. *Symbols are experimental data from a Mexican crude oil taken from Buenrostro-Gonzalez et al. (2004). Adapted from Artola et al. (2011).*

and not just the average molecular weight. For example, a sample with a long-tail distribution of larger molecules would be more prone to be unstable than a system with an equivalent average molecular weight, but a sharper distribution; such sensitivity is well known in polymer–solvent phase equilibria.

The crude-oil bubble curve is essentially that of an asphaltene-free liquid phase; i.e., in terms of the light (or oil) components, the composition of the two liquids inside the two-phase liquid–liquid (LLE) region is essentially the same. By contrast, if one considers instead the composition of the asphaltene, a marked difference in the phases is found; at low pressure two phases are present, one in which the mole fraction of asphaltene is close to unity, and one (a gaseous phase) in which there is, essentially, no asphaltene. At c. 3.5 MPa, a third phase appears in which there is still very little asphaltene. In this three-phase region, the content of lights in the liquid phases increases with pressure, while that of the heavier components decreases accordingly. At higher pressure, once again only two phases are observed and although asphaltene is present in both phases, the composition of asphaltene in the asphaltene-rich phase remains at least an order of magnitude higher than that in the other phase. This analysis of the two phases, one of them largely free of any asphaltene, provides a physically relevant picture of the phase behavior.

5.2.3 Molecular Modeling as Applied to Crude Oils

The options available to describe the behavior of matter in silico resemble a ladder (Figure 5.8), where each of the steps from the bottom up can be thought as representing different scales of ever-decreasing physical detail.

At the bottom of the ladder is the full electronic quantum mechanical description of matter: essentially exact numerical solution of the Schrodinger equation for the many-body wavefunction, using, for example, configuration interaction or renormalization group methods. This is only feasible for a handful of atoms at most because the complexity of the many-body wavefunction increases exponentially with the number of particles. If one is willing to accept a few simplifications, density functional theory (DFT) offers a very successful combination of accuracy and relatively low computational cost, and makes it possible to run simulations on hundreds of atoms as a matter of routine, and thousands of atoms on large supercomputers (Burke, 2012). Though DFT has become a widely accepted tool it is burdened with two important consequences of the inherent simplifications: the use of more complicated wave functions do not necessarily provide more accuracy and

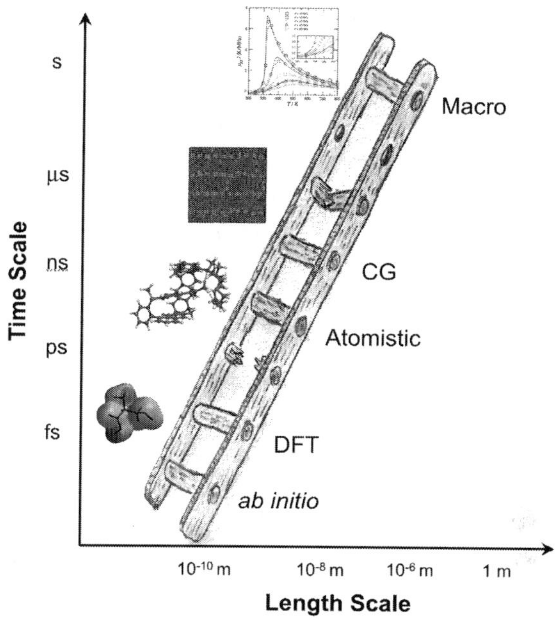

Figure 5.8 The computer simulation ladder. Information should ideally transfer up and down the ladder with ease. Note that there are two areas where the ladder is broken, thus presenting challenges for truly multiscale modeling. The most crucial missing aspect for crude oil description is the link between the continuum description and the coarse grained modeling. *Adapted from Muller and Jackson (2014).*

there is no rigorous way of accounting for dispersion forces within a DFT treatment, which for an accurate description of fluid-phase equilibria is a crucial requirement. Larger scales must be sought. Up a rung of the ladder is a description of atoms or tightly bound chemically distinct groups, such as the CH_3 and CH_2 groups of *n*-alkanes, using the simpler force fields of classical mechanics, which are commonly represented with spherical symmetrical pair potentials with Coulombic charged sites. At this level of abstraction one therefore considers the interaction sites to be made up of a collection of atoms and atom groups or even small molecules, e.g., carbon dioxide or benzene. Such a methodology is commonly referred to a treatment at the coarse grained (CG) level. The intermolecular interactions at this CG level are typically described with semiempirical analytical functions of the interseparation taken to represent the pairwise potential energy, $u(r)$. It is common practice to estimate the parameters of the force fields by attempting to reproduce target structural (radial distribution function, structure factor, etc.), thermodynamic (second virial coefficient, liquid density, vapor pressure, internal energy, enthalpy, etc.), or transport (viscosity, diffusivity, etc.) properties of the fluid or a combination of these properties. In an effort to resolve the dilemma of the degeneracy of macroscopic parameters for the intermolecular potential, it is then tempting to employ a more detailed, higher-resolution atomistic model (a step down the ladder) as the basis for determination of the parameters for CG models by means of an appropriate integration of the degrees of freedom which are to be removed. A review of current practice including a discussion of the key approaches to the simultaneous modeling of several time and/or length scales (*multiscale modeling)* is the subject of a book (Voth, 2009) and themed journal issues (Faller, 2009; Visscher et al., 2011). Also reviews on the topic have been published by Klein and Shinoda (2008), McCullagh et al. (2008), and Brini et al. (2013) to name just a salient few. It is at this level of abstraction that one may start envisaging the modeling of crude oils as the molecular level of representation must be coarsened with the complexity in order to allow for a tractable description of the molecular behavior. Models such as those employed to describe microphase separation, e.g., dissipative particle dynamics techniques (Groot and Warren, 1997), are representative examples of this next step up the ladder. These high-level CG models exhibit some of the consequences of the selective removal of degrees of freedom, echoed in issues of robustness, transferability, and representability. Further abstraction into more coarse grained models commonly removes all type of molecular descriptors and moves one into the realm of continuum models. The formal connection between these steps higher up our ladder is ever more sparse and very few models effectively bridge these gaps. This is an area of very dynamic and topical research.

5.2.4 The Combination of Equations of State with Molecular Modeling

While not the subject of this section, one could, in principle, attempt to obtain information and ultimately model the behavior of crude oils and gases through direct molecular modeling. A recent review by Greenfield (2011) summarizes the state of the art in terms of molecular modeling of asphaltene-containing crudes; unfortunately space limitations preclude any further discussion here. Molecular modeling of these systems is in itself is an immense challenge, as the difficulties present in equation-of-state modeling simply become enhanced, most importantly due to the finite size of the simulation sampling cell that must be used. However, very recently, a paradigm change has been proposed: the simultaneous use of an equation of state in conjunction with molecular modeling of fluids at the coarse-grained level (Müller and Jackson, 2014; Avendaño et al., 2011, 2013; Lafitte et al., 2012; Mejía et al., 2014).

Another important consequence of the simplicity of the SAFT approach noted in the previous section is that all of the component species are treated using models defined by integer values of the parameter m, which denotes the number of segments in the chain representing the molecule. Together with the low number of components in the model, this will allow the same system to be studied using molecular-simulation techniques, such as molecular dynamics. By contrast, had the crude-oil model comprised many components or (as would be usual in SAFT-type studies) any components would have been characterized by noninteger m (i.e., the model molecules consisted of a noninteger number of segments); such a complementary study would not be feasible.

Not all versions of SAFT have a direct link between the Hamiltonian and the final closed form EOS. A recent version of the equation denoted the SAFT–VR–Mie model (Lafitte et al. 2013) provides such a link in a tractable way. The direct molecular simulation of the SAFT molecular models provides a reliable route to properties that are not accessible from the equation of state such as the structure, the self-assembly into mesophases, the interfacial tension, the adsorption on surfaces, and the transport properties, all key to understanding the properties of asphaltenic systems. The group-contribution nature of the methodology (Papaioannou et al., 2014) allows for the description of the behavior of these complex fluids in a consistent and quantitative way. Furthermore, the one-to-one correspondence of the force fields to the equation of state provides a unique avenue to the multiscale modeling of crude oils.

5.2.5 Conclusions and Perspectives

The lack of a deep understanding of the physical phenomena leading to the instability of the fluid phase, aggregation, and posterior precipitation of asphaltenes is an open problem that is particularly important in crude-oil production, transport, and refining, because the economic cost associated with reservoir, pipeline, or heat exchanger fouling due to asphaltenes is substantial. Clearly, any insight into the precipitation phenomenon that could be provided by an improved understanding of the phase behavior of crude oil would be of great value, and this can be perceived as one of the principal goals of thermodynamic and molecular modeling.

Partly because of a paucity of experimental data, but also because the asphaltene fractions of crude oils are far from being well characterized, the problem of modeling crude-oil behavior from a thermodynamic perspective is as yet unresolved; it is this lack of a precise characterization that has represented the major barrier to improving our understanding in this area. Nevertheless, significant progress has been made; although a deep understanding remains some way off, studies such as those highlighted in this chapter have certainly provided useful and valuable insights. Within industry and academia there remains a substantial effort in continuing research both at the level of experimental characterization and in development of modeling techniques; together these promise continued improvement in our understanding of the important, yet perplexing, phenomenon of asphaltene precipitation.

5.3 FUNDAMENTAL TRANSPORT MODELING

J. Yang, D. Sileri, O. K. Matar

Current models for asphaltene fouling (reviewed in Section 5.1) are based mainly on the use of empirical or semiempirical correlations (Epstein, 1994; Ebert and Panchal, 1995; Panchal et al., 1999; Polley et al., 2002; Nasr and Givi 2006; Yeap et al., 2004). These models, which are of the form

$$\text{Fouling rate} = \text{Deposition rate} - \text{Removal rate} \tag{5.51}$$

rely on the fitting of laboratory- or plant-scale data rather than on achieving a fundamental understanding of the transport processes underlying fouling, which include aggregation, diffusion, deposition, aging, removal, and their dependence on temperature, pressure, and flow conditions. Such an understanding is crucial to the development of *predictive* models, based on fundamental conservation principles, which could be used for effective thermal and fouling management, and mitigation. The differences between the various models are due to the assumptions underlying the removal and deposition terms. The former is modeled using expressions related to shear stress and mass transfer; the latter are described using reaction models or a combination of reaction, phase separation, and transport. Although models such as those given by Equation (5.51) are used widely throughout industry due to their ease of implementation, they are of limited applicability, and often the quality of their predictions is poor.

Recently, numerical models relying on the use of computational fluid dynamics (CFD) have been developed to provide more reliable predictions of the fouling process. Examples of this include the work of Sileri et al. (2009), which focused on the laminar regime, and accounted for deposition through a chemical equilibria model, removal through hydrodynamic interactions, and aging (which impacted the deposit rheology). Bayat et al. (2012) used a two-dimensional CFD model to simulate fouling in a shell-and-tube heat exchanger. This model, which uses lumped surface chemical reactions, overpredicted the fouling rates during the induction period; this may be because the model over- and underestimated the preexponential factor and the activation energy of the fouling reactions, respectively, and simply represented the precipitation and aging mechanisms by a species diffusion model with constant diffusion coefficient. Yang et al. (2013) also developed a CFD model that comprised two-step, multiphase, heterogeneous reactions to describe the chemical reactions, in addition to a chemical equilibrium model to describe asphaltene precipitation; their model also accounted for rheological aging of the deposit. The turbulence present in the flowing crude-oil phase was modeled using the large eddy simulation (LES) approach. Yang et al. (2013) used their numerical model to simulate fouling in a single tube of a heat exchanger in a typical refinery preheat train, and found that the chemical reaction fouling rate is an order of magnitude faster than asphaltene precipitation, with the fouling and removal rates increasing with surface temperature and bulk velocity, respectively. CFD models of the type developed by Yang et al. (2014) are promising and can yield valuable input into plant-scale models, through appropriate upscaling algorithms, which can improve their predictions and enable their use in the design of effective fouling mitigation strategies.

In the following sections, a concise review of the concepts relevant to the class of flows to which fouling belongs, nonisothermal, multicomponent, and multiphase flows, is provided. Some of the techniques used to solve the equations that govern such systems are also reviewed and the results of a parametric study presented for situations in which the flow is in either the laminar or the turbulent regime. Finally, perspectives for future work are reported in the conclusions.

5.3.1 Numerical Method for Multiphase Flows

In order to describe the behavior of a system comprising a flowing crude-oil mixture and an asphaltenic deposit, an initially separated, multiphase flow model is chosen. In this section, the mathematical formulation and associated assumptions are reviewed, paying particular attention to the conditions that can lead to deposit entrainment, and hence removal. In such a model, the interface that separates the phases might exhibit waviness during its motion. Depending on the flow conditions and geometry of the problem, the interfacial dynamics can become increasingly complex: increasing the relative significance of inertia leads to the formation of large-amplitude waves, which develop a three-dimensional (3D) structure. These 3D waves can lead to the development of elongated ligaments that break up to give rise to droplet detachment (Villermaux, 2007). This mechanism contributes to the removal of the asphaltenic deposit from the walls that bound the flow, and therefore merits special attention.

CFD allows one to solve the governing equations of the considered model, by discretizing them in time and space, and solving iteratively the equations for each element of the domain. Direct numerical simulation (DNS) involves the solution of the governing equations with minimal assumptions, and the resolution of all relevant scales of motion; in particular, no models are required for the small-scale flow physics. Thus, for turbulent flows, the computational grid must be sufficiently fine in order to resolve the flow down to the Kolmogorov scale; these are the smallest scales in turbulent flows, and depend on the average rate of energy dissipation per unit mass and the kinematic viscosity. The computational cost of DNS scales like Re^3 where Re denotes the Reynolds number, making it prohibitively expensive for Re ranges of typical industrial applications ($Re > 10^4$). It is possible to use high-fidelity DNS to construct subgrid-scale models for use in LES, which resolve the largest scales of motion in turbulent flows.

Predicting multiphase flow dynamics is complicated by the multiscale nature of the flow, and the tremendously complex interfacial topology. The position of the interface must also be determined as part of the solution, which adds to the simulation complexity. Interface tracking uses a number of marker points along the interface. Its variants include the surface-marker approach in which the marker points form a separate surface grid, which is moved with the kinematic condition in a fixed grid; this approach does not conserve mass inherently, and cannot handle topological changes, a key feature of complex interfacial flows. Of the alternative interface-capturing methods, the volume-of-fluid (VOF) method conserves mass but suffers from excessive numerical diffusion unless attention is paid to this issue; level-set methods do not conserve mass inherently but improvements are possible through adaptive mesh refinement, by either coupling them to the VOF method or via appropriate reinitialization of the level-set function. DNS with interface-tracking or interface-capturing capabilities can, in principle, resolve all the relevant length and time scales in the flow; for highly turbulent flows (Re $\sim 10^5$), this is computationally highly expensive with current methods even for single-phase flows.

Unlike single-phase DNS, additional length scales must be resolved, which may be similar to the Kolmogorov length scale, related to, for instance, the smallest droplets sheared from a film or a jet, events controlled by capillary forces. Including these forces is crucial and they are appended to the momentum conservation equation, confined to act in the interfacial region; the methods of discretizing them vary in the literature. Also, DNS-based approaches require additional relations to account for rupture of films separating drops or bubbles prior to their coalescence, for instance, or for contact line motion, or filament breakup when its radius approaches zero; this is due to the difficulty in resolving such small-scale flow features. RANS-based approaches employ filtering of all eddies and are generally applied to determine time-averaged quantities in steady or quasi-steady flows, appropriate for dispersed flows but not for stratified or annular flows as the averaging procedures filter out interfacial waves, an essential "ingredient" of these flows. They also cannot predict the small-scale features of "slug" and "churn" flows.

LES-based methods with interface tracking represent an excellent compromise, capable, in principle, of simulating all the flow regimes and phenomena in complex multiphase flows. They have a cutoff turbulent eddy size that can be resolved, set by capillary forces. The explicit filtering process applied leads to information loss: smearing of discontinuities demarcating interfacial boundaries, and loss of subgrid droplets and/or bubbles and/or interfacial waves. In the case of subgrid waves, associated breaking, bubble-capturing, or droplet entrainment events may also be lost (depending on the numerical resolution), as could information associated with deposition of subgrid bubbles and/or droplets. The predictive capabilities of these models are therefore limited by their numerical resolution and subgrid models.

5.3.2 Numerical Models for Crude Oil Fouling

5.3.2.1 Asphaltene Precipitation Model

Heavy compounds (e.g., asphaltene) that are dissolved in the crude oil can separate out via the precipitation fouling route. In this work, a chemical equilibrium model based on the Gibbs free energy that was originally developed to describe wax deposition (Svendsen et al., 1993) was adapted to predict asphaltene precipitation. This model accounts for the phase equilibrium, phase transition, thermodynamics, and the fluid's mechanical effects.

In addition, it takes advantage of the treatment of the crude oil as a multicomponent mixture of hydrocarbons to quantify the amount of mass phase separation in the oil, which leads to diffusion and the adherence of heavy compounds to the walls of the cylinder (Sileri et al., 2009).

The phase equilibrium constant for the oil mixture of n hydrocarbons can be evaluated using the equation

$$K_i = \frac{d_i}{x_i} = K_{i0}exp\left[-\frac{\Delta H_{Pi}}{R_g T}\left(1-\frac{T}{T_{Pi}}\right)\right] \quad i = 1, 2, \ \ldots \ , n \tag{5.52}$$

where d_i and x_i are the mole fraction of the foulant and oil phases for each component, respectively; T denotes the temperature of the oil phase; ΔH_{Pi} and T_{Pi} are the precipitation enthalpy and precipitation temperature for each component, respectively; K_{i0} is the equilibrium coefficient when $T = T_{Pi}$; and R_g is the universal gas constant.

It is assumed that the mass flux, J, from the oil phase to fouling phase, follows Fick's law of diffusion:

$$J = D_d\left(\frac{\rho\eta}{T}\frac{\partial T}{\partial \boldsymbol{n}}\right) \tag{5.53}$$

where D_d is the diffusion coefficient of the diffusing foulant species, ρ is the total density, and $\boldsymbol{n}$ is the normal vector. The normal derivative term indicates the fouling phase diffusing to the wall. Equation (5.53) shows that the mass flux J can be expressed as a function of dimensionless functions, η_i, with the following form:

$$\eta = \sum_{i=1}^{n} \eta_i, \quad and \quad \eta_i = \chi_i \frac{\left[T(1+\theta)^2 \frac{\partial L_x}{\partial T} + \theta\frac{\Delta H_{Pi}}{R_g T}\right] K_i}{(1+\ \theta K_i)^2} \tag{5.54}$$

Here, χ_i is the weight fraction of each component of the mixture and θ is a parameter that depends on the number of moles L_x present for each phase. The formulations for χ_i, θ, and L_x are given and discussed elsewhere in the literature.

To model asphaltene precipitation, individual components of crude oil were lumped into two groups: asphaltene and nonasphaltene components. The oil phase is assumed to consist of two (asphaltene/nonasphaltene) components with characteristics reflective of those of practical crude oil (see Table 5.6). The fouling phase is assumed to be a single component "mixture" consisting exclusively of asphaltene.

Table 5.6 Precipitation properties for each component

Property	ΔH_P	χ	T_p
Nonasphaltene	34.9 kJ mol^{-1}	0.58	600 K
Asphaltene	107.8 kJ mol^{-1}	0.42	423 K

5.3.2.2 Chemical Reaction Fouling Model

Chemical reaction fouling is typically attributed to hydrocarbon autoxidation and has been identified as the main cause of heat exchange fouling in many cases (Watkinson et al., 1997). This fouling route involves a very complicated chemical kinetic scheme in which very large numbers of intermediate species are formed via a similarly large number of elementary reactions. It would therefore be impractical to develop a detailed chemical reaction fouling model. although such models could be created in principal, current computing facilities would take an extremely long time to solve the very large number of chemical rate equations that would be involved, which are groups of nonlinearly coupled extremely stiff ordinary differential equations.

Current chemical reaction fouling models (reviewed in Section 5.1) are based on the assumption that the key chemical process can be described as a two-step reaction involving the initial generation of a soluble precursor followed by the formation of an insoluble foulant (Panchal and Watkinson, 1993), as shown below:

$$\begin{matrix} Reactants \\ (soluble) \end{matrix} \overset{r_1}{\rightarrow} \begin{matrix} Precursor \\ (sparingly\ soluble) \end{matrix} \overset{r_2}{\rightarrow} \begin{matrix} Foulant \\ (insoluble) \end{matrix} \tag{5.55}$$

Here, r_1 and r_2 are the reaction rates for the first and second reactions:

$$r_1 = k_1 c_{r,1}^m c_{r,2}^n \tag{5.56}$$

$$r_2 = k_2 c_p^l \tag{5.57}$$

In the above expressions, $c_{r,i}$ and c_p are the concentration of reactants and precursor, respectively; m, n, and l are the orders of the reactants and precursors, respectively. k_1 and k_2 are reaction rate constants that have the form specified by the Arrhenius equation:

$$k = A\ exp\left(-\frac{E}{R_g T}\right) \tag{5.58}$$

where A is the preexponential factor, E is the activation energy, R_g is the gas constant, and T is the temperature.

Table 5.7 Reaction constants for the two-step chemical reaction fouling model

A_1, L mol^{-1} s^{-1}	E_1, kJ mol^{-1}	A_2, L mol^{-1} s^{-1}	E_2, kJ mol^{-1}	m	n	l
1.565×10^{10}	96.3	1.1×10^{8}	96.3	1.5	0.5	0.6

Note that the activation energy E_1 was estimated by using oxidation data for styrene, which has a reaction mechanism similar to that for indene; E_2 was assumed to be the same as E_1.

Due to the complex nature of crude oil, it is difficult to decouple the different fouling routes and measure the reaction rate of chemical reaction fouling in isolation. However, many experimental studies have been conducted to investigate the chemical fouling rate for simple hydrocarbons (e.g., Jet A, JP-7, kerosene, and indene) (Watkinson and Wilson, 1997). Panchal and Watkinson (1993) analyzed batch-reaction data gathered by Russell (1956) and derived reaction rates for the two-step process of indene autoxidation; their suggested reaction rates are listed in Table 5.7.

In Section 5.3.2.1, a two-component model of crude oil consisting of asphaltene and nonasphaltene in a volumetric ratio of 58:42 was proposed to describe asphaltene precipitation. This model was adapted for the description of chemical reaction fouling by treating the dissolved asphaltene component (in the oil phase) as a soluble precursor that goes on to form the insoluble foulant (the foulant phase) via a heterogeneous process that has the same reaction constants as the second reaction of the two-step fouling process (see Table 5.7).

5.3.2.3 Aging Model

As noted previously, the exposure of the fouling layer to distinct temperature and shear stresses over extended periods of time causes its molecular structure and liquid properties to change via a series of physico-chemical processes. The gel-like foulant layer eventually degrades to a hard material (Ishiyama et al., 2010). During the aging process, changes in the liquid properties of the foulant layer such as its viscosity (Coussot, 2007) and thermal conductivity (Ishiyama et al., 2010) alter the adhesion forces that bind it to the wall and hold it together as well as its heat transfer properties. This in turn affects the removal processes.

To account for changes in the viscosity of the foulant layer, a rheological model for thixotropic substances developed by Coussot (2007) was used to characterize the aging of the deposited material. The apparent viscosity μ is the ratio of the shear stress τ and shear rate $\dot{\gamma}$:

$$\mu = \frac{\tau}{\dot{\gamma}} = \mu_0(1 + \xi^n) \tag{5.59}$$

where μ_0 is the viscosity of fresh foulant material and μ is a fluid parameter. ξ is a structure parameter that describes the local degree of interconnection of the material's microstructure and depends on the flow history according to the following constitutive relation:

$$\frac{d\xi}{dt} = \frac{1}{\theta} - a\xi\dot{\gamma} \tag{5.60}$$

The above equation indicates that the rate of change of the structure parameter is the result of a competition between a "structuration" term (the first term on right-hand side) and a "destructuration" term (the second term on the right-hand side) (Sileri et al., 2009). The structuration term depends on θ, which is the characteristic time for the formation of polyaromatic and graphitic structures in the fouling layer (Fan and Watkinson, 2007). Conversely, the destructuration term depends on the local shear rate $\dot{\gamma}$, which is responsible for the breakdown of these microstructures, and the material-specific constant a.

The combination of Equations (5.59) and (5.60) yields

$$\mu = \mu_0 \left[1 + \left(\frac{\frac{1}{\Theta} - e^{-t-C}}{\alpha|\dot{\gamma}|} \right)^n \right] \tag{5.61}$$

which describes the aging process of the foulant layer in terms of the variation in its liquid viscosity. The values of the dimensionless constants α and Θ are reported to be 0.01 and 0.1 for the asphaltene aging process (Sileri, 2011). The constant n is postulated to be an integer and assigned a value of 1. By calculating the initial viscosity value at time $t = 0$, the C value can be determined as 0.032. The local shear rate is calculated by CFD modeling.

The impact of aging on the foulant's thermal conductivity was modeled using a simple chemical principle that incorporates first-order kinetics to generate quantitative comparisons of different aging rates (Ishiyama et al., 2010). This aging model assumes that the foulant's thermal conductivity varies between the minimum value for the fresh foulant and a maximum value for a coke-like deposit. The thermal conductivity of the local foulant at time t can be estimated using the equation

$$\lambda = \lambda^{\infty} + \left[\lambda^0 - \lambda^{\infty}\right] \times y_0 \times exp\left(-k^t \times t\right) \tag{5.62}$$

where λ^0 is the thermal conductivity for freshly deposited material, λ^{∞} is the maximum value of 2 W m^{-1} K^{-1}, which is the same as the value for a coke-like deposit (Ishiyama et al., 2008), y_0 is the initial youth variable and is equal to 1, and t is the lifetime of the deposit (which depends on when it was formed). k^t is an aging rate constant whose temperature dependence is described by an Arrhenius-type equation:

$$k^t = A\ exp\left(-\frac{E_a}{R_g T}\right) \tag{5.63}$$

where E_a and A are the activation energy and preexponential factor of the aging process, taking values of 10 kJ mol^{-1} and 1.157 s^{-1} (Ishiyama et al., 2010), respectively. T is the local fouling layer temperature.

5.3.3 Laminar Flow Regime

In the laminar regime, the work of Sileri et al. (2009) represents one of the first attempts at integrating explicit simulations of the hydrodynamics with mechanisms for deposition and aging. The aim of this two-dimensional model is to capture the essential features of the fouling problem. Sileri et al. (2009) consider the dynamics of a two-fluid, initially stratified, pressure-driven flow in a horizontal channel in which only the crude-oil fluid flows from the inlet; no deposit is present initially. A phase separation mechanism (asphaltene precipitation) in the crude-oil fluid leads to the formation of the deposit, which diffuses to toward the walls to form wall layers. The aging of the wall layers is assumed to impact their viscosity through rheological models that capture the increase in viscosity due to the building of the internal microstructure of the deposits. The governing equations are those of mass, momentum, and energy conservation, which are coupled by the Cahn-Hilliard equation (Jacqmin, 2000; Ding and Spelt, 2007a,b; Ding et al., 2007) for the volume fraction of the deposit. These equations are complemented by a chemical equilibria model (see Section 5.3.2.2), which requires knowledge of the thermodynamic properties of each component of the crude-oil mixture, as well as the temperature field in the computational domain in order to furnish information about the mass flux from the crude-oil bulk to the deposit; the latter is a source term in the Cahn–Hilliard equation at the channel wall and along the interface separating the cruder-oil mixture and the wall layer. The deposited layer is also assumed to possess a time-dependent viscosity in order to model "aging," where a structure parameter is responsible for its change. An evolution equation comprising "structuration" and "destructuration" terms governs the change in time of the viscosity using a rheological model (see Section 5.3.2.3). Simulations were carried out for a range of system parameters, which demonstrated the rich dynamics exhibited by the flow phenomena associated with fouling: interfacial waves, ligament formation, droplet pinchoff and entrainment, and entrapment of crude-oil droplets within the fouling deposit.

The DNS calculations performed by Sileri et al. (2009) are based on DIM for incompressible two-phase flows. The Coussot-like rheological relation governs the aging process of the deposit, where its viscosity increases with a strengthening of its internal microstructure. The Navier–Stokes equations for the conservation of mass, momentum, and energy are complemented with the Cahn–Hilliard convective equation for the volume fraction that is part of the diffuse interface method chosen for this model. Simulations have been performed varying the parameters governing the strength of the deposition and aging and the Reynolds number. Also in this case, it is possible to find connections between the time-varying deposit rheology and the flow dynamics.

A two-dimensional model that captures the dynamics of a two-fluid, stratified pressure-driven channel flow is studied here. In this model only the core fluid flows from the inlet, whereas the fouling layer results from a phase separation in the core fluid that results in the formation a distinct phase as a wall layer. The flow of two immiscible, incompressible fluids of equal density ρ and high viscosity ratio m is considered in a horizontal, rectangular two-dimensional channel of length L and height H. A Cartesian coordinate system $(x, 0, z)$ is used in which x and z correspond to the horizontal and vertical directions, respectively. The channel walls are rigid and impermeable, and located at $z = 0$ and $z = \mathrm{H}$, respectively; the more viscous layer is present at the walls with an initial thickness h. The case with an initially clean-wall configuration is also considered. The inlet and outlet of the domain are located at $x = 0$ and $x = \mathrm{L}$, respectively. Hereinafter, the less viscous fluid will be referred to as the "bulk" and the more viscous fluid as the "deposit."

The Navier–Stokes (N–S) and mass conservation equations are used to model the flow dynamics, in addition to a convective-diffusive Cahn–Hilliard (C–H) equation for the concentration of the more viscous fluid. The temperature is governed by an energy transport equation and, in the cases studied here, the temperature at the channel walls is higher than the temperature of the flow at the inlet. The calculated mass flux, J, will be used as a source term in the C–H equation at the wall and at the interface between the two fluids to feature the phase change. To account for phase change and aging phenomena, the original formulation of the diffusive interface method was modified and a clean-wall and an initial film wall thickness configurations have been adopted. The conservation of mass, momentum, C–H, and energy equations in their nondimensional form are then formulated as follows:

$$\nabla \cdot \boldsymbol{u} = 0 \tag{5.64}$$

$$\boldsymbol{\rho}\left[\frac{\partial \boldsymbol{u}}{\partial t} + \boldsymbol{u}\cdot\nabla\boldsymbol{u}\right] = -\nabla \boldsymbol{p} + \frac{1}{\boldsymbol{Re}}\nabla\cdot\left[\boldsymbol{\mu}\left(\nabla\boldsymbol{u} + \nabla\boldsymbol{u}^{T}\right)\right] + \frac{\phi\nabla C}{\boldsymbol{ReCa}}[1-\boldsymbol{\beta}(\boldsymbol{T}-1)] \tag{5.65}$$

$$\frac{\partial \boldsymbol{C}}{\partial t} + \boldsymbol{u}\cdot\nabla C = \frac{1}{Pe}\nabla\cdot(MW\nabla\phi) + \frac{\eta}{D_s}\frac{|\boldsymbol{n}\cdot\nabla T|}{T} \tag{5.66}$$

$$\frac{\partial \boldsymbol{T}}{\partial t} + \boldsymbol{u}\cdot\nabla T = \frac{1}{RePr}\nabla\cdot(k\nabla T) \tag{5.67}$$

where Re is the Reynolds number, Pe, the Peclet number, Pr, the Prandtl number, Ca, the Capillary number, and D_s an equivalent of a Peclet number related to the diffusive rate of the component. Their expressions are given in Table 5.8. The procedure of nondimensionalization on the above equations was given in Sileri (2011).

The above nondimensionalized governing equations were solved numerically on a staggered grid to discretize Equations (5.64)–(5.67) using

Table 5.8 Typical parameter values for the DIM simulations

Nondimensional parameters	**Expression**	**Value**
Re	$\frac{\rho_0 VH}{\mu_0}$	500
Pr	$\frac{C\rho_0\mu_0}{k_0}$	5.435
Ca	$\frac{\mu_0 V}{\sigma_0}$	0.00865
Pe	$\frac{LV}{M_c\phi_c}$	1/EPS
D_s	$\frac{Hu}{D}$	10^9
r_d	$\frac{\rho_d}{\rho_0}$	1.5
r_k	$\frac{k_d}{k_0}$	10
r_ν	$\frac{\mu_d}{\mu_0}$	10
β	$\frac{\partial\sigma_0}{\partial T}/\sigma_0$	0.3

the finite-volume method. The scalar variables pressure and volume fraction are defined at the center of each cell, whereas the velocity components are evaluated at the cell faces. The interfacial dynamics are strongly dependent on the velocity field, thus the C–H and N–S equations are strongly coupled. First, the C–H equation is solved using the velocity field at the time step n and the volume fraction updated to time step $n+1$. Then, the velocity field is also updated to time step $n+1$, solving the N–S equation (Equation (5.65)) together with the continuity equation (Equation (5.64)). The surface tension and other body forces are evaluated in an intermediate step.

In order to achieve convergence on mesh refinement, several simulations were performed using different grid sizes (1200 × 80, 3000 × 200, 2400 × 160, and 3000 × 200 grid points in x- and z-directions, respectively). It was found that the grid with a resolution of 2400 × 160 yields a converged fouling mass within a reasonable computational time; see Figure 5.9. This grid size was adopted for the rest of the simulations presented herein.

5.3.3.1 Varying Deposition without Aging

Here, the results are presented for three cases in which the strength of the deposition is varied and for which no aging effects are taken into account. The typical values used for these simulations are summarized in Table 5.8. For the other parameters the following are chosen: $M_1 = 1500$ kg kmol^{-1}, $M_2 = 215$ kg kmol^{-1}, $K_{10} = K_{20} = 1$, $\beta_1 = 0.042$, $\beta_2 = 0.13$, $z_1 = z_2 = 0.5$, $T_{p1} = 423$ K, $T_{p2} = 600$ K, and $T_{in} = 473$ K; the latter

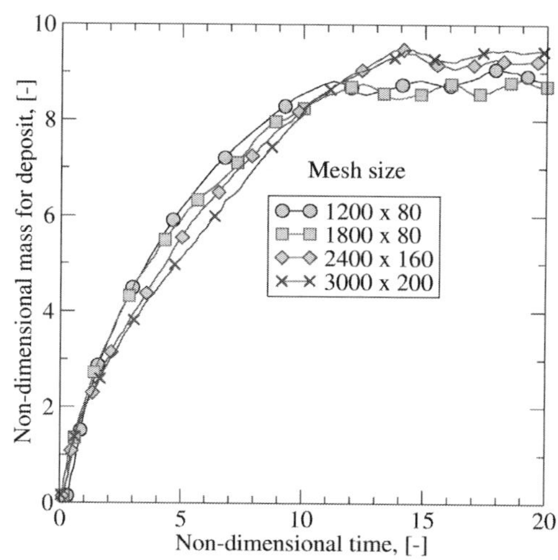

Figure 5.9 Deposit mass versus time for four different grid sizes, using the same set of parameters as the other simulations described later.

temperatures lead to $\Omega_1 = 0.895$ and $\Omega_2 = 0.705$. For each case, a different value of D_s has been used, in order to vary the amount of material that phase separates from the bulk to form a deposit. The start is from a deposit-free or "clean" wall configuration and the deposition induced by phase change is localized both at the clean wall surface and at the interface between the two fluids. Three values of D_s are chosen to vary the deposition rate: $D_s = 5 \times 10^9$, 1×10^9, and 5×10^8. As expected, the deposition rate has been found to be inversely proportional to the value of D_s, as shown in Figure 5.10 in which is displayed a plot of the volume occupied by the deposit divided by the total volume of the domain versus time for each of the three cases. Because the deposition term $J = \frac{\eta}{D_s} \frac{\mathbf{n} \cdot \nabla T}{T}$ is linked to the chemical equilibrium model by Equations (5.52)–(5.54), the formation of deposit is localized where the absolute value of temperature and its gradient are large enough to allow the phase change; in this case, this coincides with the clean wall at the interface.

As can also be seen in Figure 5.10 for the smaller values of D_s studied, the deposit volume fraction appears to asymptote to an essentially constant value for sufficiently large times. It is speculated at this stage that the "saturation" effect depicted in Figure 5.10 is due to a dynamic balance between deposition and removal. It is believed that the viscosity contrast between the two fluids is responsible for the formation and growth of interfacial instabilities.

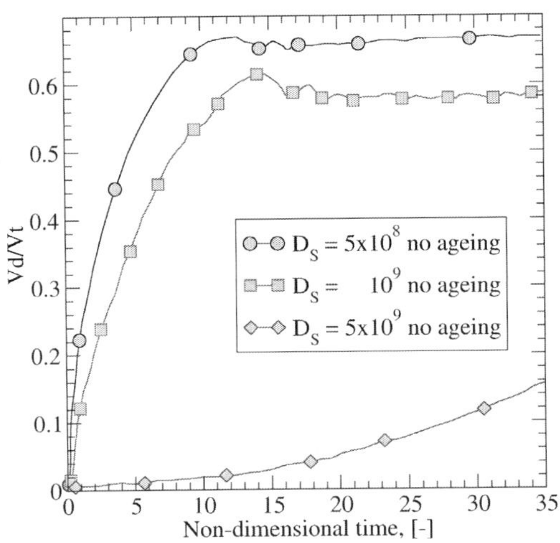

Figure 5.10 Deposit volume fraction versus time for three cases without aging for different values of the D_s parameter. The rest of the parameters values are listed in Table 5.8.

Such phenomena lead to the displacement of the deposit downstream of the inlet and its ultimate removal from the channel; this is in competition with the formation of fresh deposit due to phase separation and deposition. As discussed below, in some cases formation of ligaments and entrainment of deposit into the main flow in the form of droplets are also observed. Ligament formation is also responsible for some fluid from the main flow being trapped inside the deposit layer at the walls of the domain. At this point, it is necessary to make a distinction and clarify the difference between the deposition rate, and the rate of mass deposited at the wall. The deposition rate is dependent on D_s, T, and other parameters and it is evaluated using the chemical equilibria model; there is no direct influence of the fluid motion on the deposition rate. The rate of mass deposited at the wall, however, depends on the net balance between the deposition and the removal rate, the latter caused by fluid interaction and wave formation.

In Figures 5.11–5.13 there are three columns showing the spatio-temporal evolution of the volume fraction, temperature, and viscosity associated with $D_s = 5 \times 10^9$, 1×10^9, and 5×10^8, respectively, where the original nondimensional length of the channel is 15 and being unit in height. Note that all the spatio-temporal plots in this section have been expanded by a factor of four in the vertical direction in order to facilitate readability.

The volume fraction plot in Figure 5.11 shows "nucleation" of the deposit at the walls in the entrance region, gradually expanding downstream and displaced by interfacial waves. Downstream of the deposit, it is possible to observe a region of high recirculation, typical of regions in the wake of bluff bodies, characterized by vortex shedding. These regions are present in the temperature and viscosity plots of Figure 5.11 and are also due to the formation of von Karman vortex-shedding phenomena behind a bluff body at moderate to high Reynolds numbers (Ding et al., 2003; Almeida et al., 1993; Uchida and Ohya, 2001). The volume fraction and viscosity plots for smaller values of D_s (Figure 5.13) allow one to observe the nucleation of deposit along the wall for the entire length of the channel. Also, the absence of vortices and highly recirculating regions for these values of D_s may indicate that the nucleation of deposit downstream of the inlet region for high values of $D_s = 5 \times 10^9$ in Figure 5.11 may be inhibited by the recirculating fluid disrupting deposit formation and diffusing it back in the main flow. Another reason to support this point is given by the vortices observed in the temperature plot; see Figure 5.11. These vortices appear to be much more noticeable than in the volume fraction plot and, because the deposit

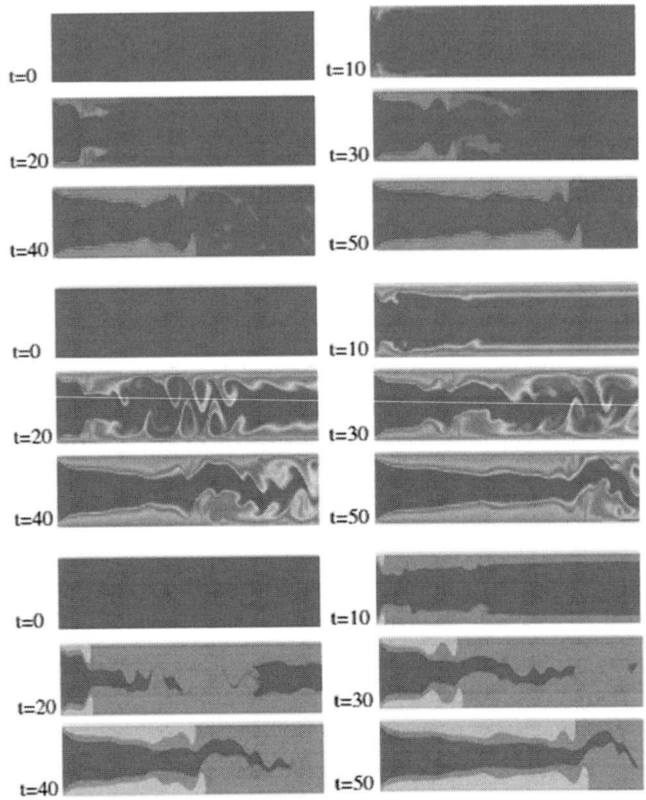

Figure 5.11 Spatio-temporal evolution of the volume fraction (top), temperature (center), and viscosity (bottom) for $D_s = 5 \times 10^9$ and $t = 0$–50. The rest of the parameters, unless stated otherwise, are the same as those in Table 5.8.

has a thermal conductivity much higher than that of the bulk fluid. It appears that the hot, newly generated deposit at the wall behind the deposit layer recirculates at the center of the channel forming counterrotating vortices (see, for example, in Figure 5.11 at time $t = 20$–30).

Figure 5.12 reports the volume fraction, temperature, and viscosity for $D_s = 10^9$. Contrary to the previous case, and as noted above, the nucleation of deposit is more uniform along the wall than in the $D_s = 5 \times 10^9$ case and, at close scrutiny, the formation of trapped droplets of bulk fluid within the deposit layer can be noted. Those trapped droplets vary in number and size and are probably linked to the formation and coalescence of micro/macroligaments, depending on the droplet size. Again, it is possible to observe a certain degree of diffusivity in the region presenting high

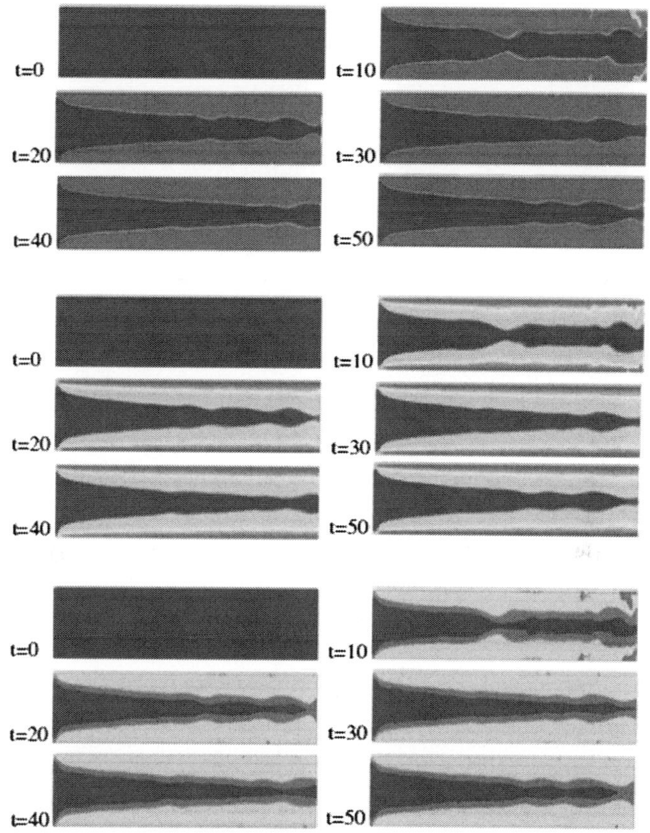

Figure 5.12 Spatio-temporal evolution of the volume fraction (top), temperature (center), and viscosity (bottom) for $D_s = 1 \times 10^9$ and $t = 0$–50. The rest of the parameters, unless stated otherwise, are the same as those in Table 5.8.

recirculation, although to a smaller extent than in the $D_s = 5 \times 10^9$ case. In Figure 5.13, the last case without aging for $D_s = 5 \times 10^8$ is reported. The relatively large value of the deposition term associated with the smaller value of D_s allows the rapid development of a uniform thick coating of deposit along the walls without the presence of recirculating regions and/or the presence of droplets inside the deposit.

These results indicate the validity of the previous assumption regarding the mechanisms responsible for deposit formation: the fast deposition rate does not allow the interaction between bulk and deposit to interfere with droplet nucleation at the wall, thus, leading to the formation of an essentially uniform wall layer.

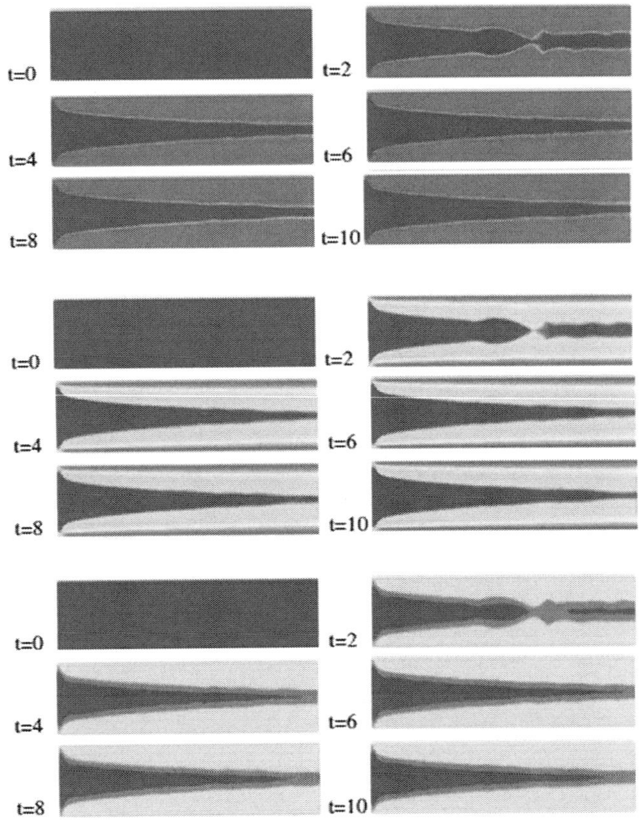

Figure 5.13 Spatio-temporal evolution of the volume fraction (top), temperature (center), and viscosity (bottom) for $D_s = 5 \times 10^8$ and $t = 0$–10. The rest of the parameters, unless stated otherwise, are the same as those in Table 5.8.

5.3.3.2 Varying Deposition in the Presence of Aging

This section shows and discusses three cases, with the same parameters as those in Section 5.3.3.1, but characterized by finite a and Θ values, which control the degree of aging in terms of viscosity property. The parameters governing the change of viscosity with time are the same for all three cases: $\Theta = 10$ and $a = 0.01$. The parameter Θ is linked with the characteristic time of deposit microstructure formation, leading to an increase of viscosity dictated. On the other hand, a is present in a term where the local shear rate decreases the value of viscosity. Hence, the total change of deposit viscosity depends on the overall balance of the LHS of Equation (5.61). A low value of Θ will lead to rapid development of internal structure within the deposit, increasing the viscosity; a high value of a, on the other hand, will lead to a

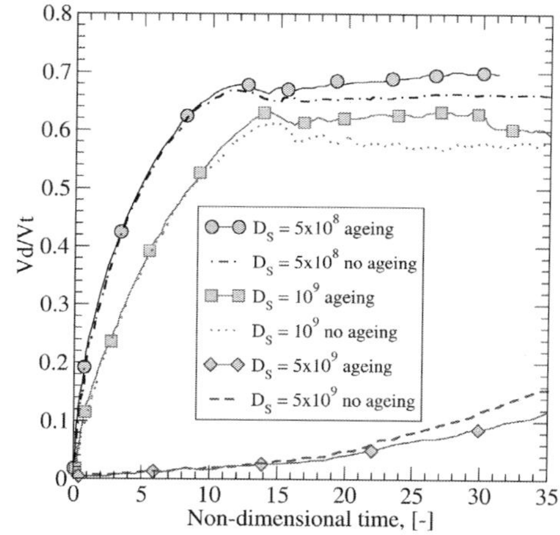

Figure 5.14 Deposit volume fraction V_d/V_t versus time for three cases in presence of aging (solid line) and absence of aging (dashed line) for different values of the D_s parameter; here V_d and V_t correspond to the volume of the deposit and channel, respectively. The rest of the parameters are listed in Table 5.8.

disruption of the microstructure due to shear forces, decreasing the local value of the viscosity. The observed deposit viscosity will then have a local value at each point of the domain depending on time, shear rate, and flow history. In Figure 5.14, the deposit mass plots for the three cases without aging is superimposed on the three cases in the presence of aging. It is evident from the figure that the total mass of deposit in the presence of the aging is higher for the two cases with high values of D_s. This behavior is in line with what is expected: the formation and growth of interfacial waves and displacement of deposit downstream are more difficult due to the increased viscosity. The mass plots for the aging case start to diverge from the case without aging (Figure 5.14) when they reach the asymptotic level and, as discussed below, the displacing effect of instabilities becomes more important. An increase of mass deposit up to a maximum of 13% and up to 4% is observed for $D_s = 10^9$ and $D_s = 5 \times 10^8$, respectively.

For the case with a low deposition rate ($D_s = 5 \times 10^9$), the mass deposited without aging is higher than that in the presence of aging; see Figures 5.15 and 5.16.The explanation for this behavior may be related to the interaction between vortices in the highly mixed zones and the deposit nucleation sites on the wall. The case without aging presents a more elongated deposit layer than that without aging, and therefore a higher interfacial area between the two fluids where deposition can take place. Since the

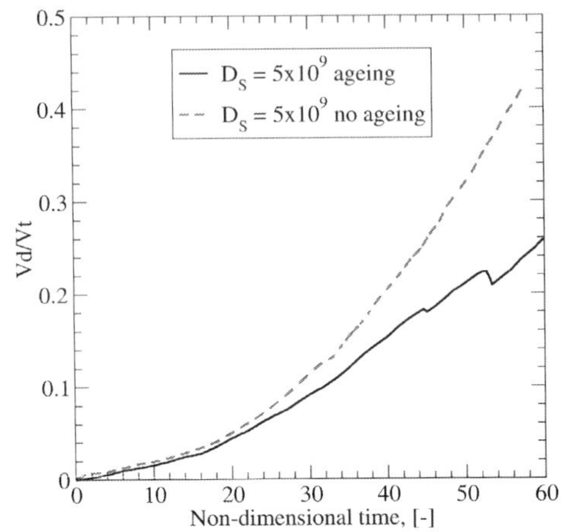

Figure 5.15 Deposit volume fraction versus time for $D_s = 5 \times 10^9$ for the same parameters as those in Figure 5.14.

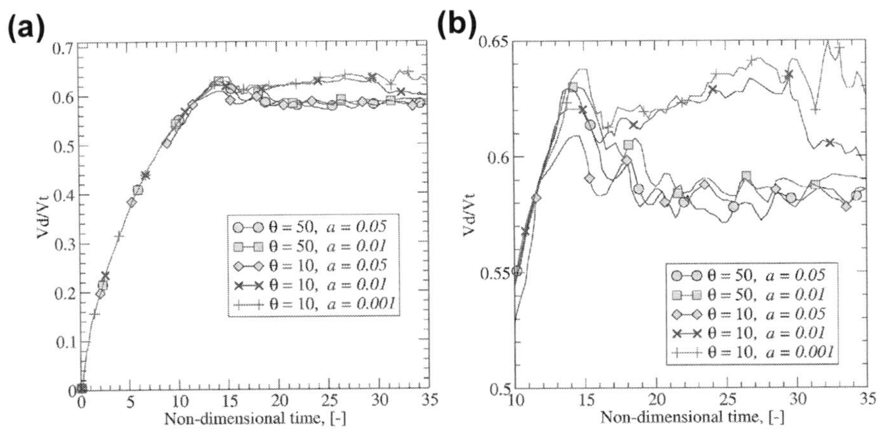

Figure 5.16 Deposit volume fraction versus time for different values of Θ and a parameters (a) and enlarged version of the deposit volume fraction versus time (b). The rest of the parameters remain unchanged with respect to previous cases.

interfacial region and the wall are the only regions where the temperature profile allows deposition to take place and since vortices and mixing inhibit nucleation at the wall, the higher interfacial area may allow a higher amount of new deposit to stay in place instead of being diffused back to the bulk. This will be discussed below, following the presentation of the spatio-temporal results in Figures 5.17–5.19, the spatio-temporal evolution of

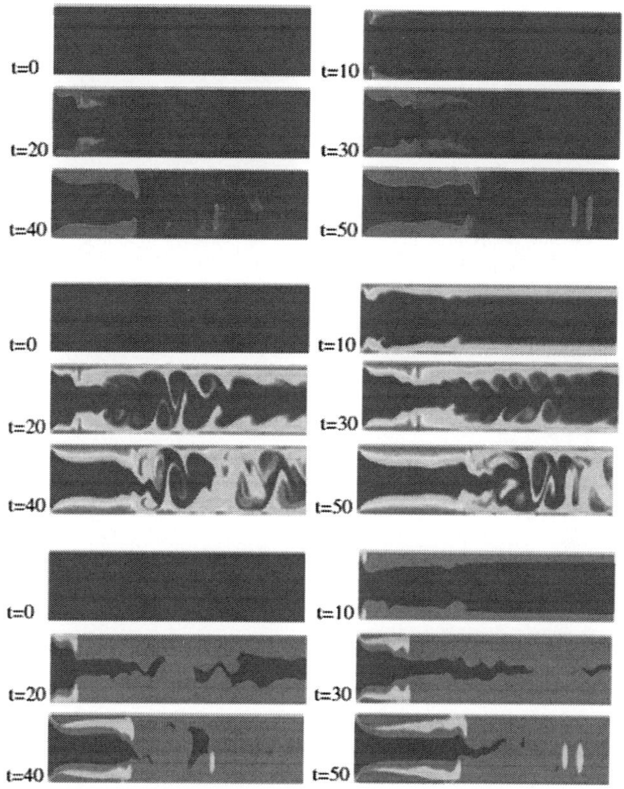

Figure 5.17 Spatio-temporal evolution of the volume fraction (top), temperature (center), and viscosity (bottom) for $D_s = 5 \times 10^9$, $\Theta = 10$ and $a = 0.01$ and $t = 0$–50. The rest of the parameters, unless stated otherwise, are the same as those in Table 5.8.

the volume fraction, temperature, and viscosity for the three different values of D_s are shown. For case $D_s = 5 \times 10^9$, a substantial difference with respect to the previous situation without aging is observed: as expected, the length of the deposit layer is shorter than before due to the increase in viscosity in the presence of aging. Moreover, the viscosity plot can be examined to determine the cause of droplet formation: the viscosity of the bottom part of the deposit layer greatly increases in relation to that of the upper part of the layer; this viscosity contrast drives instabilities increasing the growth rate of waves and leading to ligament formation. The interaction of the bulk to the deposit through the interface further destabilizes the deposit upper layer and increasing its kinetic energy leading to pinch-off of droplets (Ding et al., 2007, 2010; Scardovelli and Zaleski, 1999; Cao et al., 2004,

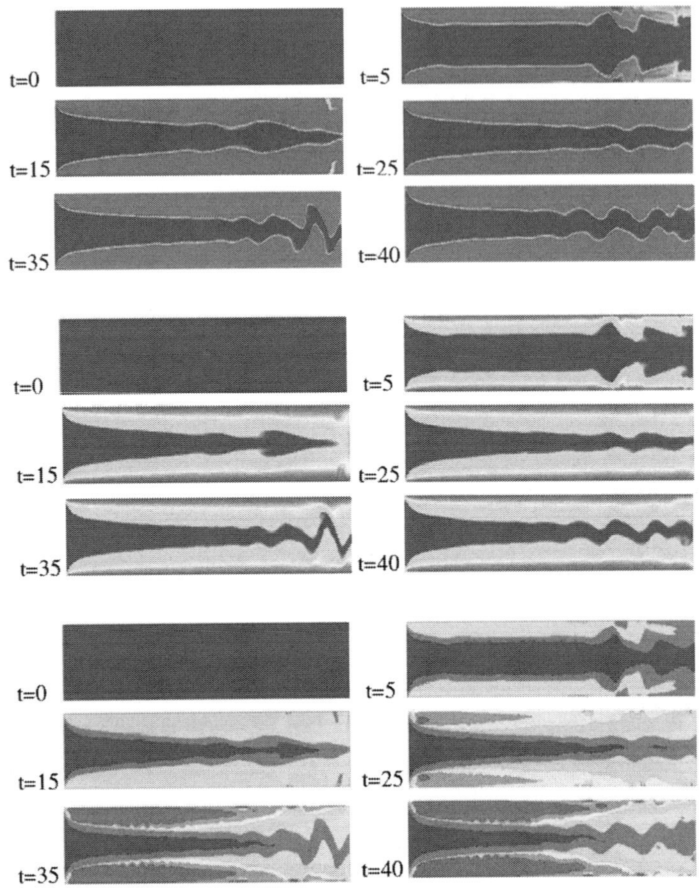

Figure 5.18 Spatio-temporal evolution of the volume fraction (top), temperature (center), and viscosity (bottom) for $D_s = 1 \times 10^9$, $\Theta = 10$ and $a = 0.01$ and $t = 0$–40. The rest of the parameters, unless stated otherwise, are the same as those in Table 5.8.

2006). Note that the plots have been expanded by a factor of four in the vertical direction making the droplets appear elliptic instead of circular. After pinch-off the droplets flow outside the domain pushed by the main flow, which is reflected by the oscillations in the mass plot in Figure 5.17 between time $t = 44$–53. By the inspection of the spatio-temporal evolution shown in Figures 5.12 and 5.18, no substantial changes in the flow features are observed until $t = 25$; at this point the formation and growth of instabilities in form of waves are more pronounced in Figure 5.18 than in Figure 5.12. The waves formed in Figure 5.18 seem to be increasing in amplitude after a point where the underlying highly viscous bottom part of the deposit close

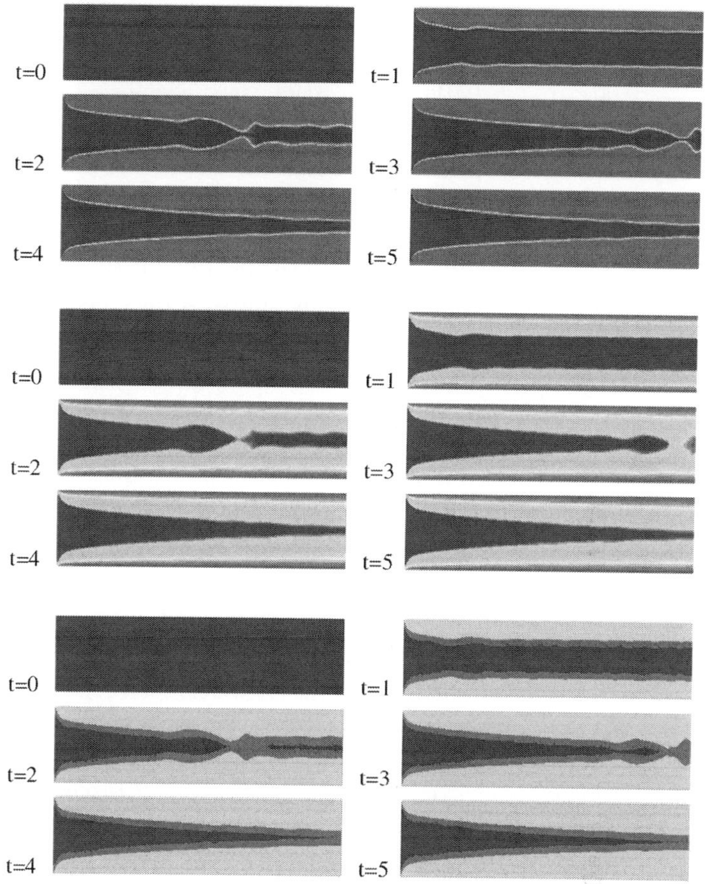

Figure 5.19 Spatio-temporal evolution of the volume fraction (top), temperature (center), and viscosity (bottom) for $D_s = 5 \times 10^8$, $\Theta = 10$ and $a = 0.01$ and $t = 0$–5. The rest of the parameters, unless stated otherwise, are the same as those in Table 5.8.

to the wall starts to thin after an initial increase in thickness. For cases with high deposition rates, shown in Figures 5.13 and 5.19, the plots are remarkably similar, given that the two cases are compared at an early stage and aging effects did not yet substantially modify the highly viscous bottom part of the deposit, shown in Figure 5.18.

5.3.3.3 *Varying the Significance of Aging in the Presence of Deposition*

In this section, a comparison of five different cases is made where the aging parameters ($\Theta = 10$, 50 and $a = 0.001$, 0.01, and 0.05) are changed but the

Table 5.9 Predicted maximum and averaged increases of deposit viscosity

	$\Theta = 10$			$\Theta = 50$	
a	0.001	0.01	0.05	0.01	0.05
$\Delta\mu_{max}$	95.9%	70.6%	48.5%	13.8%	11.8%
$\Delta\mu_{ave}$	81%	39%	12.3%	11%	10.12%

deposition rate is kept constant. The rest of the parameters are unvaried with respect to the previous cases. Figure 5.16(a) shows a mass plot for the five different sets of parameters used and Figure 5.16(b) is an enlarged version of the same plot for time $10< t <35$.

The effect of the different sets of parameters used becomes evident when the system reaches an asymptotic or pseudo-equilibrium point at around $t = 12$. Here the amount of deposit present in the channel is different for each case. As expected, the most severe aging case ($\Theta = 10$, $a = 0.001$) presents the highest amount of deposit and the maximum viscosities. The maximum and averaged values of deposit viscosity for all parameters studied are presented in Table 5.9.

A useful parameter for evaluating the importance of Θ and a on the flow corresponds to the viscosity of the deposit layer averaged over the volume occupied by the deposit, presented in Table 5.9 The effect of changing selectively Θ or a: for example, a change in a from 0.01 to 0.001 increases $\Delta\mu_{max}$ of around 36%, although the increase in $\Delta\mu_{ave}$ is 112%. This large difference is most likely caused by the effect of the smaller destructuring term in Equation (5.61), not allowing the breakdown of the molecular microstructure of the fluid in the presence of high shear stresses and consequently leveling the local viscosity value in the whole deposit. The opposite effect can be seen by comparing cases ($\Theta = 10$ and 50 with $a = 0.05$): $\Delta\mu_{max}$ in case of $\Theta = 10$ is more than three times than in the cases of $\Theta = 50$; however, $\Delta\mu_{ave}$ is comparable for both cases. Again, this increase of Θ is directly responsible for a faster formation of microstructures with time but it has a smaller effect on deposit viscosity as layers of deposit undergone aging for a different length in time are compared.

The spatio-temporal evolutions for different aging parameters are depicted in Figures 5.20–5.23. It is shown that increasing aging progressively. One possible reason for the increased waviness is the reduced cross-sectional area of the channel available for the bulk fluid to flow downstream: in this case, the higher velocity of the flowing bulk fluid exerts higher shear stresses at the interface promoting wave formation.

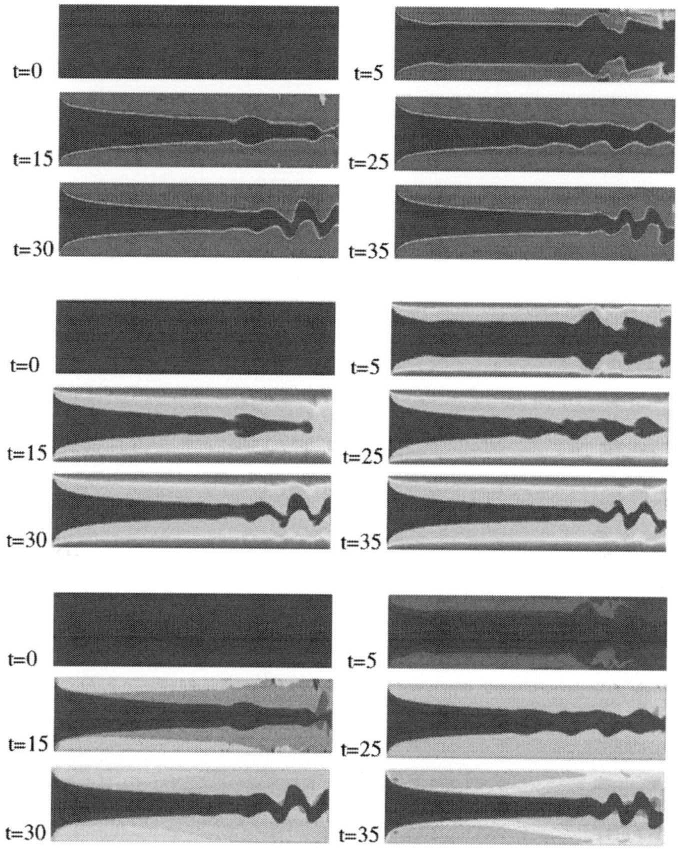

Figure 5.20 Spatio-temporal evolution of the volume fraction (top), temperature (center), and viscosity (bottom) for $D_s = 1 \times 10^9$, $\Theta = 10$ and $a = 0.001$ and $t = 0$–35. The rest of the parameters, unless stated otherwise, are the same as those in Table 5.8.

5.3.3.4 Varying the Reynolds Number

The impact of Reynolds number is also investigated herein. Two cases for $\mathrm{Re} = 750,\ 1000$ are presented together with the previous case ($\mathrm{Re} = 500$) for comparison. In Figure 5.24, it can be observed that higher Reynolds numbers result in a lower amount of deposit present for $t > 7$. The increased Reynolds number of the flow results in a larger number of interfacial waves forming in the domain and displacing a larger amount of deposit downstream. For the two cases analyzed in this section it can be observed how the waves tend to bridge at the center of the channel similar to what observed in the previous chapter. Analyzing the viscosity plots in Figures 5.25 and 5.26, it can be ascertained that high Re numbers are

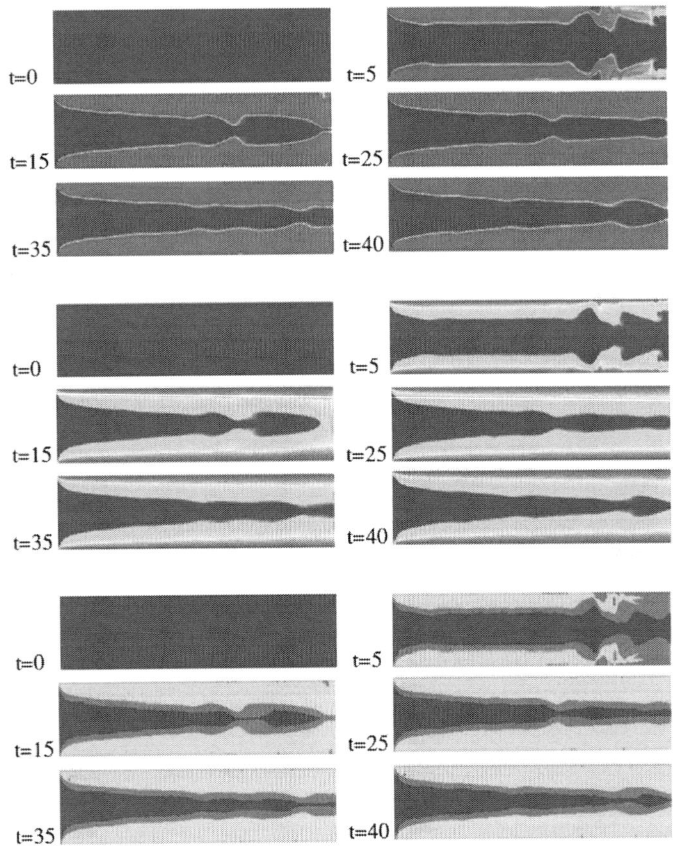

Figure 5.21 Spatio-temporal evolution of the volume fraction (top), temperature (center), and viscosity (bottom) for $D_s = 1 \times 10^9$, $\Theta = 10$ and $a = 0.05$ and $t = 0$–40. The rest of the parameters, unless stated otherwise, are the same as those in Table 5.8.

responsible also for a decrease in the thickness of the highly aged deposit sublayer, promoting the removal of the "fresh" upper layer. Comparing maximum increase values of viscosities in Table 5.10, a nonmonotonic trend that does not completely match with what is expected can be observed. On the other hand, the higher shear stresses corresponding to high Re regimes present lower values of average viscosity see in Table 5.10, and the viscous sublayer appears to be thinner with increasing Reynolds as in Figure 5.26.

5.3.4 Turbulent Regime

In this section, the focus is on the effect of turbulence on the fouling dynamics. The geometries and operating conditions of industrial heat

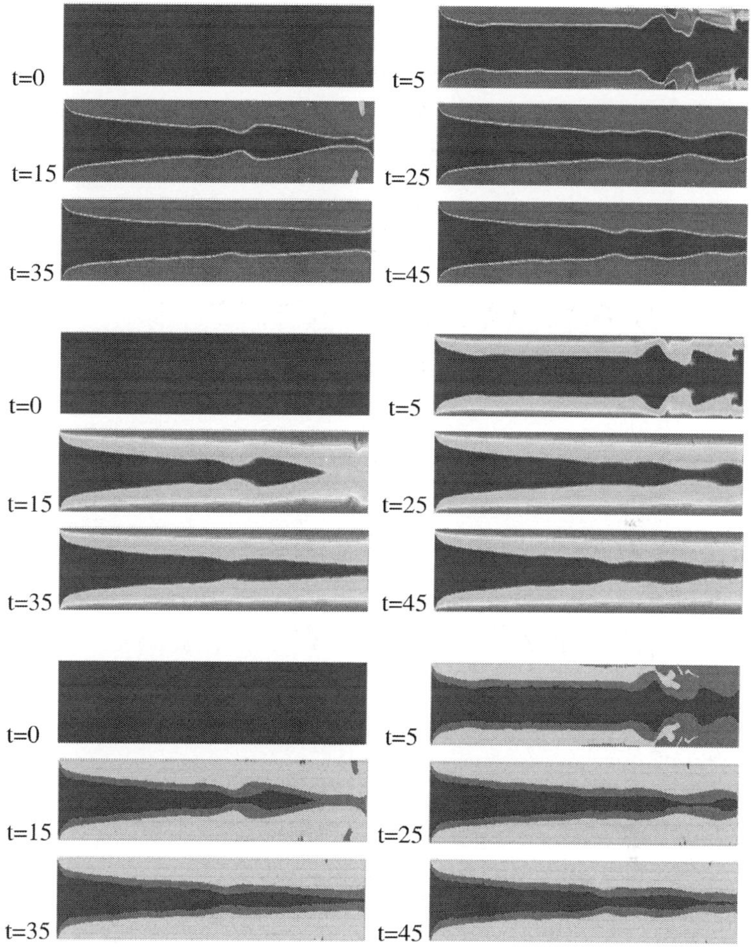

Figure 5.22 Spatio-temporal evolution of the volume fraction (top), temperature (center), and viscosity (bottom) for $D_s = 1 \times 10^9$, $\Theta = 50$ and $a = 0.01$ and $t = 0$–45. The rest of the parameters, unless stated otherwise, are the same as those in Table 5.8.

exchangers vary significantly depending on their service, crude blend processed, and preferences of the design engineer. Some fairly typical geometries and operating characteristics for a heat exchanger in a refinery preheat train (Coletti et al., 2011) are given in Table 5.11.

For the purposes of this work, a single tube 0.5 m long with the properties listed in Table 5.11 was modeled using the CFD tool. Such a short geometry renders the observations of the fouling behavior computationally feasible. The major disadvantage is that the pressure gradient along the

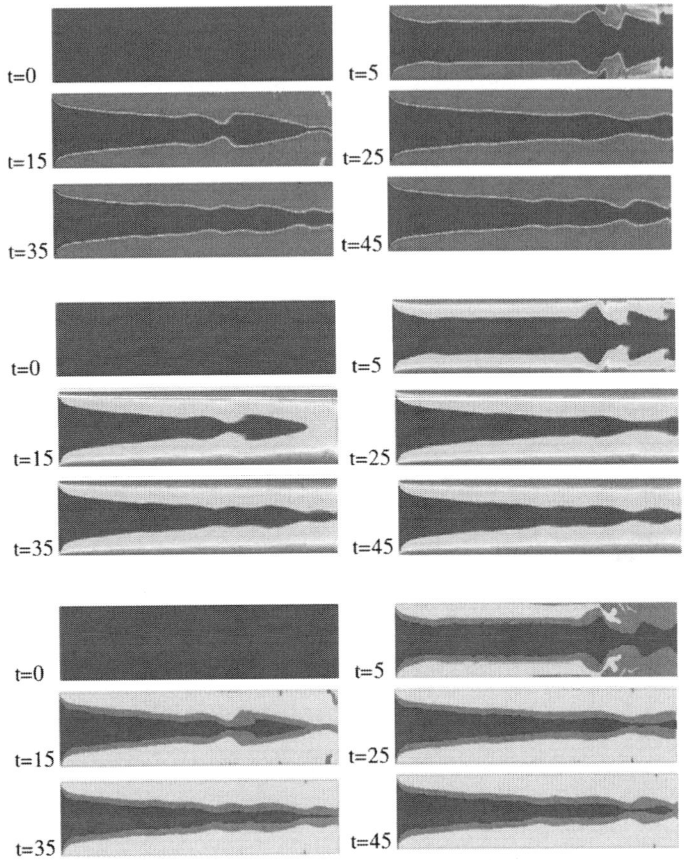

Figure 5.23 Spatio-temporal evolution of the volume fraction (top), temperature (center), and viscosity (bottom) for $D_s = 1 \times 10^9$, $\Theta = 50$ and $a = 0.05$ and $t = 0$–45. The rest of the parameters, unless stated otherwise, are the same as those in Table 5.8.

tube could not be predicted accurately compared to the industrial heat exchanger usually 5–10 m long, or even longer. In the formulation above, the crude oil and fouling deposits were treated as two immiscible, Newtonian, incompressible fluids in a horizontal cylindrical tube. The crude oil is a low-viscosity bicomponent fluid that flows in the turbulent regime (e.g., Reynolds number 30,000 based on a bulk mean velocity of 0.5 m s^{-1}) and undergoes fouling as it passes through the tube. The fouling deposit is a high-viscosity single component fluid that adheres to the laminar region of the wall. The insoluble foulant is formed via chemical reactions and asphaltene precipitation in the bulk fluids that are transported to the wall surface and accumulate into large droplets, which merge to form a thin

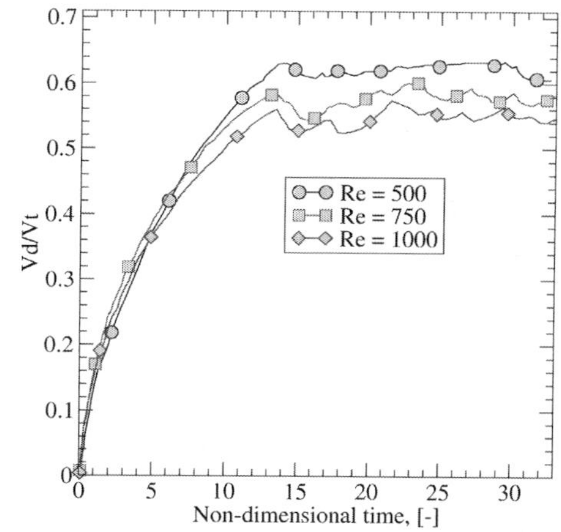

Figure 5.24 Deposit volume fraction versus time for different values of the Reynolds number.

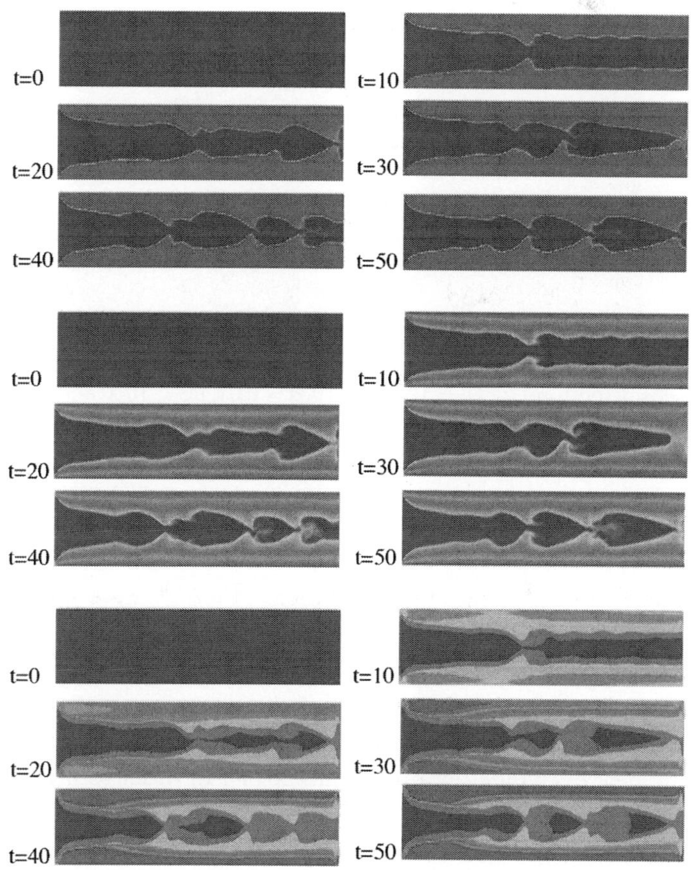

Figure 5.25 Spatio-temporal evolution of the volume fraction (top), temperature (center), and viscosity (bottom) for $D_s = 10^9$, $\Theta = 10$, $a = 0.01$, Re $= 750$ and $t = 0$–50. The rest of the parameters, unless stated otherwise, are the same as those in Table 5.8.

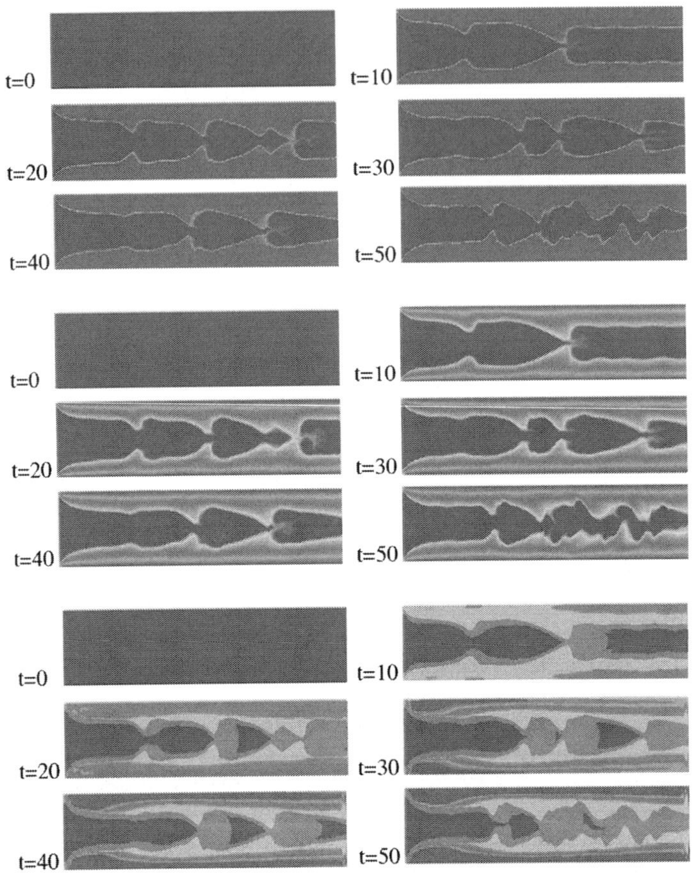

Figure 5.26 Spatio-temporal evolution of the volume fraction (top), temperature (center) and viscosity (bottom) for $D_s = 10^9$, $\Theta = 10$, $a = 0.01$, Re $= 1000$ and $t = 0$–50. The rest of the parameters, unless stated otherwise, are the same as those in Table 5.8.

Table 5.10 Predicted maximum and averaged increases of deposit viscosity

	Re = 500	**Re = 750**	**Re = 1000**
$\Delta\mu_{max}$	70.5%	73%	68%
$\Delta\mu_{ave}$	38%	32%	30.5%

fouling layer. It usually takes several 100 hours for a visible fouling layer to form in the heat exchanger (Young et al., 2011). In order to investigate the fouling behavior (e.g., interfacial evolution) without requiring an excessive amount of computing time, the cylinder wall was assumed to be coated with

Table 5.11 Typical geometries and operating characteristics for a heat exchanger in a refinery PHT

Inner diameter (m)	Outer diameter (m)	Reynolds number (–)	Averaged flow velocity (m s^{-1})	Inlet temperature (K)	Outlet temperature (K)	Operating pressure (bar)
0.02	0.025	60,000	1.0	463	483	20

a 2-mm-thick fouling layer from the start of the simulation in a baseline case. This fouling layer was considered to be subject to an aging process that converts it to a hard material similar to coke, and to interfacial evolution that causes it to break up into small droplets due to shear stresses and pressure at the interface between the two fluids.

To investigate the impact of mean bulk velocity ($\overline{V}_{bulk}$) and wall surface temperature (T_w) on the fouling formation rate, the cases with an initially clean-wall configuration will also been considered under the range of $\overline{V}_{bulk}$ 0.25–1.5 m s^{-1}, corresponding to Re 15,000–90,000, and T_w 525–625 K, respectively.

The computational domain inside the tube is discretized using grid points of 500 × 80 × 160 in the axial (x–), radial (r–) and tangential (θ–) directions, respectively, in cylindrical coordinates. To ensure that the first grid point is sufficiently close to the wall (y + <5), a grid with non-uniform spacing in the radial direction was generated; see Figure 5.27(a). In the axial direction, the grid spacing is uniform; see Figure 5.27(b). The whole domain consists of c. 6.5 million hexahedral cells that are sufficiently fine to provide useful levels of detail near the wall region but comparatively coarse in the core region of the tube. The main boundary conditions are shown in

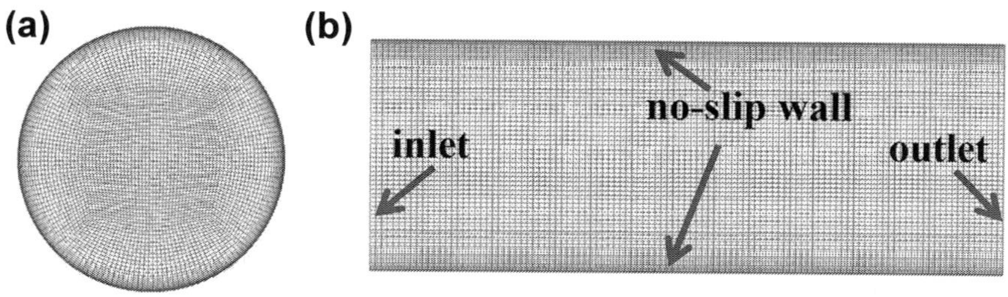

Figure 5.27 Left (a) and front view (b) of the computational domain and boundary conditions.

Figure 5.27(b). A power-law empirical velocity profile coupled with a uniform temperature and composition profile is imposed at tube inlet. The pressure outlet was constrained to have a zero gauge pressure with a constant temperature. A uniform temperature was assumed across the tube wall.

The submodels noted in Section 5.3.2 including the crude oil surrogate model, aging model, and fouling model have been incorporated into the commercial CFD solver ANSYS Fluent ver. 14.5 (ANSYS, 2014) via a user defined function interface. The code takes the advantages of the solver's finite volume discretization feature and the VOF method (Hirt and Nichols, 1981) for interfacial tracking to solve the two-phase Navier–Stokes system. The aging model is implemented through user-defined material properties. The fouling model was implemented via the phase interface definition: the asphaltene precipitation model defines the mass transfer rate of insoluble foulant from the oil phase to the foulant phase, and the chemical reaction fouling model is imposed using a user-defined reaction rate function for a heterogeneous reaction in which insoluble foulant is formed via a one-step reaction as described in Section 5.3.2.3.

In the VOF calculations, the continuum surface force model proposed by Brakbill et al. (1992) was employed to predict the surface tension force, which was then added as a source term of the momentum equation. The surface tension coefficient for crude oil/asphaltene, σ, reported as $0.043\ \mathrm{N\,m^{-1}}$ by Sileri (2011) was adopted herein. The turbulence in a sheared oil-foulant flow with a deformable interface is another crucial factor that determines the interfacial motions, e.g., wave/ligament formation and breaking, which in turn affects the turbulence in the near-interface region. To accurately explore the turbulent field in the current system, the LES approach with the Smagorinsky expression for subgrid scale stress was employed. In brief, LES resolves large turbulent eddies directly and accounts for smaller eddies using Smagorinsky–Lilly's subgrid scale model (Smagorinsky, 1963).

The concentration gradients of the different compounds in the fluid and the associated diffusion processes are the main driving forces of species transport within the system. In the CFD calculations, the mass fraction of each species is determined by solving a set of partial differential convection–diffusion equations in which the diffusion term is associated with the species' diffusivity. The diffusion coefficient of asphaltene in toluene has been reported to be in the range of 1×10^{-10} to $1 \times 10^{-11}\ \mathrm{m^2\,s^{-1}}$ under ambient conditions (Ostlund et al., 2004). However, there are very little data available concerning asphaltene diffusion under high-temperature conditions.

A value of $1 \times 10^{-7}\ m^2\ s^{-1}$ for the diffusion coefficient of asphaltene through petroleum at high temperatures (c. 540 K) has been suggested and validated using CFD simulations (Bayat et al., 2012). This value was therefore used to describe the molecular diffusion of asphaltene species in the current system.

The SIMPLE algorithm was used for pressure–velocity coupling and the second-order upwind scheme was used to discretize the convection terms in the energy, momentum, species transport, and volume fraction equations. More details concerning Fluent and the turbulence model can be obtained from ANSYS (2014). The simulations were conducted on an HP-Z820 24-core workstation. They were initiated with a velocity profile for a fully developed straight pipe flow onto which small random perturbations were superimposed. Each simulation was run for 30,000 time steps with a fixed time step size of 0.0001 s that limits the global Courant number to 0.25 and ensures the high accuracy of transient simulation. Each complete run took approximately 2300 CPU-hours.

A baseline model including all of the submodels was used to simulate the fouling process in a 3D tube in which it was assumed that a thin fouling layer had already formed across part of the tube wall. The time history of total foulant mass accumulation in the tube is plotted in Figure 5.28. The total foulant mass increases dramatically after the crude oil enters the pipe, rising rapidly toward its maximum value of c. 0.038 kg. During this period, the interface wave formed and propagated across the fouling layer. However, the degree of interfacial deformation was still weak at this point and no interfacial wave breaking occurred. At this stage, chemical reaction fouling and asphaltene precipitation driven by the high temperature started to produce

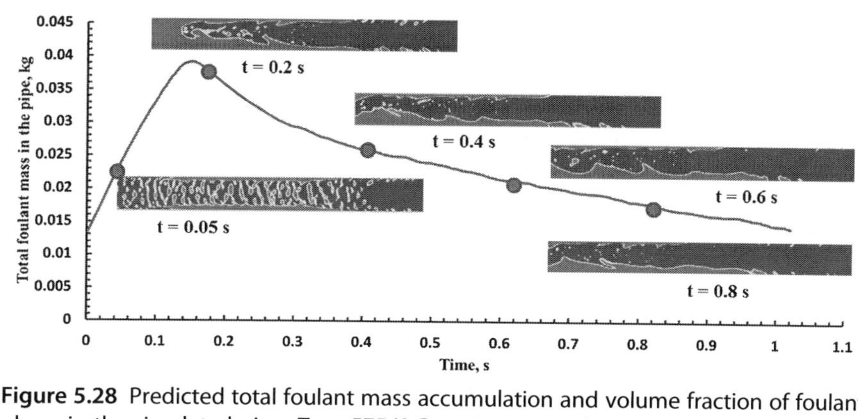

Figure 5.28 Predicted total foulant mass accumulation and volume fraction of foulant phase in the simulated pipe, $T_w = 575$ K, Re = 30,000, with aging effect.

insoluble foulant in the bulk fluid and near the wall region, as shown by the subcontour at 0.05 s in Figure 5.28. By around 0.16 s, a large mass of insoluble foulant had accumulated in the pipe and the total mass of foulant was approaching its maximum value. Beyond this point, the interfacial wave became steeper and eventually ruptured at the tip due to shear stress. Droplets formed from this interfacial wave disrupted the entrainment of the bulk fluid and began to exit the pipe via the outlet. During this period, the total mass of accumulated foulant in the pipe began to decline gradually. The baseline calculation clearly demonstrates the competition between foulant formation and removal process over time. In addition, it was also clear that the insoluble foulant was deposited near the wall and grew to form large droplets that eventually connected to the fouling layer, as shown by the sequence of subcontours for time points between 0.2 and 0.8 s in Figure 5.28.

By numerically switching each submodel on and off, the impact of the bulk velocity, wall temperature, and aging on the each fouling formation route can be investigated; this will be discussed in the following section.

5.3.4.1 Asphaltene Precipitation

The total foulant mass in the domain via the asphaltene precipitation route was predicted under different operating conditions; see Figure 5.29. As noted previously, the wall surface temperature is a key factor in fouling formation. The impact of varying the surface temperature was examined by comparing the total mass of the asphaltene phase at different temperatures (525, 575, and 625 K, represented by black, red, and blue curves, respectively). As shown in Figure 5.29, increasing the wall temperature increases the amount of asphaltene phase. Since the rate of deposition mainly depends on the temperature gradient in the direction normal to the wall (see Equation (5.53)), higher wall temperatures would be expected to accelerate asphaltene precipitation. Interestingly, the increase in asphaltene mass in going from 575 to 625 K was substantially more pronounced than that observed on increasing the temperature from 525 to 575 K at $Re = 15{,}000$. This pronounced trend gradually weakens from $Re = 30{,}000$–$60{,}000$, eventually reversing at $Re = 90{,}000$.

The impact of mean bulk velocity on the asphaltene precipitation is also depicted in Figure 5.29. The total asphaltene phase in the domain decreases with increasing the Re number from 15,000 to 90,000. This is because the relatively cold crude oil has longer residual time to be heated by the hot wall when the flow goes through the domain relatively slowly at the lower Re

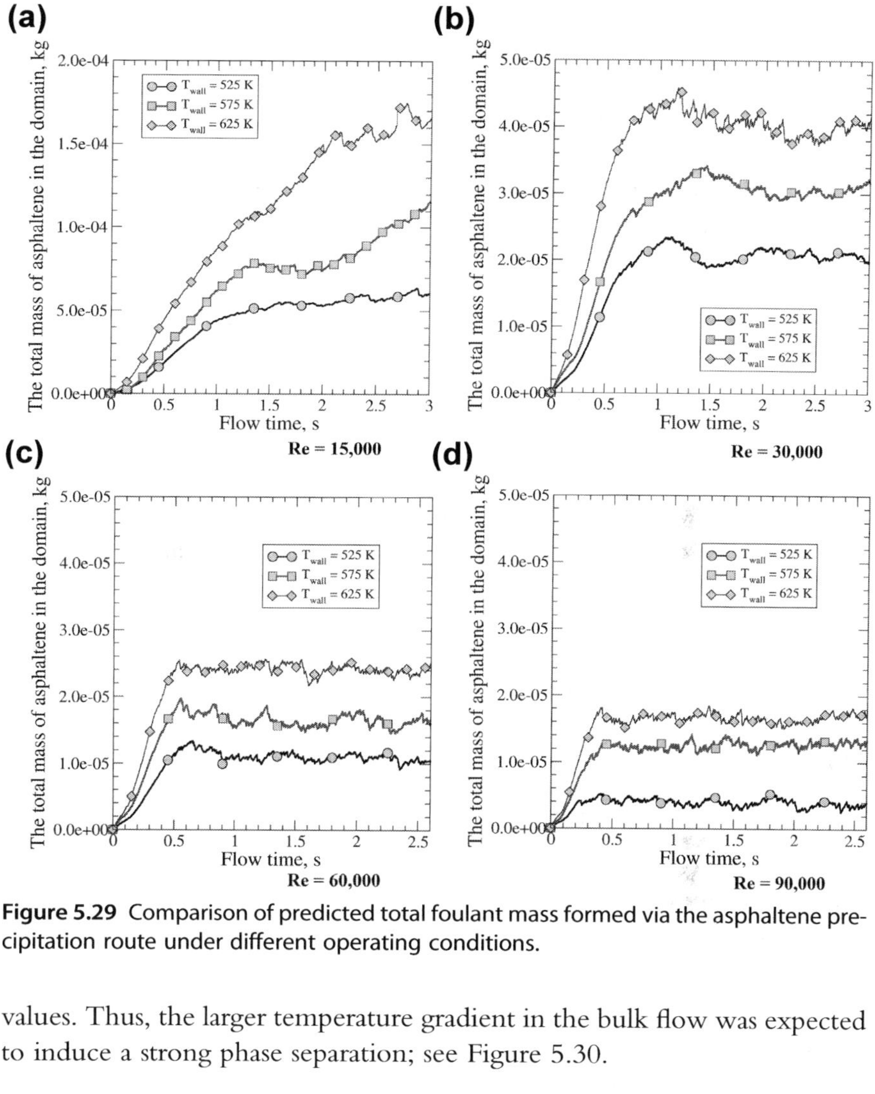

Figure 5.29 Comparison of predicted total foulant mass formed via the asphaltene precipitation route under different operating conditions.

values. Thus, the larger temperature gradient in the bulk flow was expected to induce a strong phase separation; see Figure 5.30.

5.3.4.2 Chemical Reaction Fouling

The impact of mean bulk velocity and wall temperature on the chemical reaction fouling route has also been studied (see Figure 5.31). In general, the total foulant mass increased with increasing wall surface temperature and decreased with increasing Re. Comparing the scale of the mass plot, it can be seen that the foulant formation rate via chemical reaction route is an order of magnitude higher than that via asphaltene precipitation for the full temperature and Re range. The contour of volume of fraction of

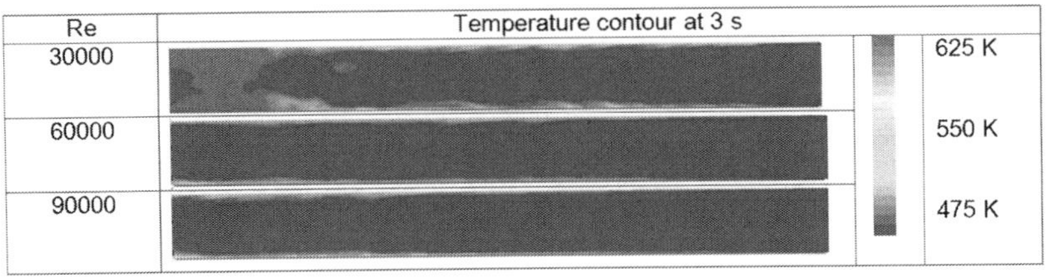

Figure 5.30 Predicted temperature contour inside the tube at different inlet velocity conditions.

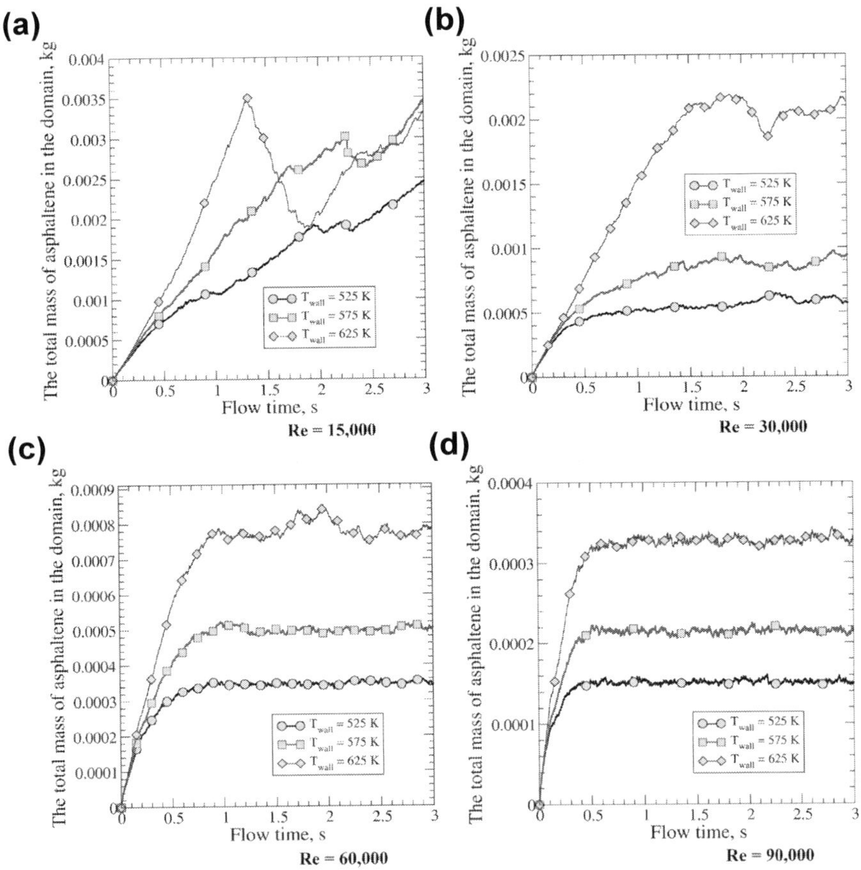

Figure 5.31 Comparison of predicted total foulant mass formed via chemical reaction fouling route under different operating conditions.

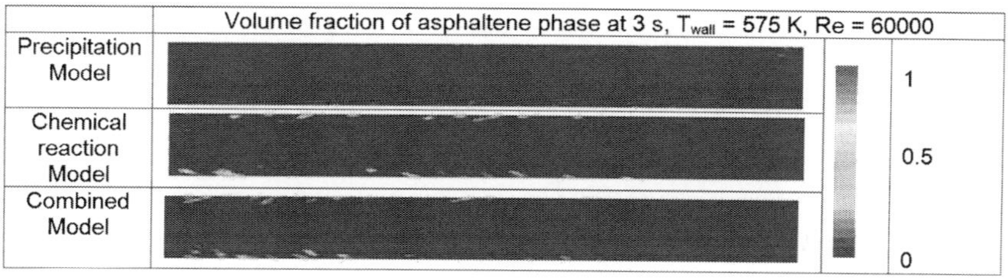

Figure 5.32 Comparison of predicted volume of fraction of asphaltene phase formed via different route.

asphaltene phase in Figure 5.32(a) and Figure 5.32(b) also indicates that the chemical reaction fouling route dominates the whole fouling process. The contour also indicates that the most foulant was formed and deposited at the region far from the inlet.

In addition, the increase in asphaltene mass in going from 575 to 625 K was more pronounced than that observed on increasing the temperature from 525 to 575 K at full Re range. An exception to this was found for low Re < 15,000 and high wall surface temperature (625 K), as shown in Figure 5.31(a), in which a large amount of asphaltene generated through the chemical reaction fouling route was expected between 0 and 1.5 s, then the asphaltene mass plot fell drastically (1.5–2.0 s). This is because large amounts of the asphaltene phase flowed out of the domain. The mass then increased to a large value due to the strong fouling formation rate.

5.3.4.3 Combined Fouling Model

To examine the interaction between asphaltene precipitation route and chemical reaction fouling route, the cases including both mechanisms, named the "combined model," were performed under the same conditions as in Sections 5.3.4.1 and 5.3.4.2. The predicted total asphaltene mass via individual route and the combined route are shown in Figure 5.33.

The model predicts the formation of smaller amounts of asphaltenes than in the case of purely chemical reaction-driven fouling under the wide operating conditions. In the combined model, the asphaltene phase could be formed via both precipitation route and chemical reaction route. By comparing the mass transfer rate contributed via the individual model (Figure 5.34), it can be seen that the contribution of the precipitation route was strengthened in the combined mode; this becomes more pronounced at low Re and high temperatures. In contrast, the contribution of the chemical reaction route is weakened in the combined model compared to the pure

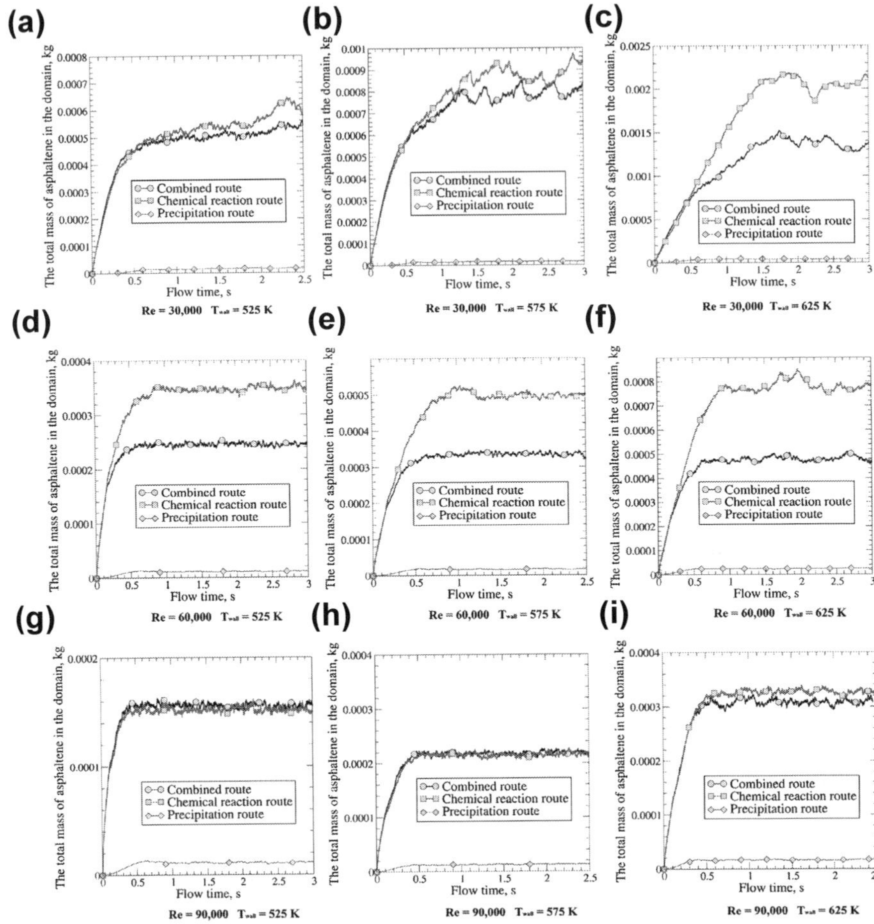

Figure 5.33 Comparison of predicted total foulant mass formed via combined model under different operating conditions.

model; this also becomes pronounced at high temperatures and low Re numbers. This effect outweighs that associated with enhanced precipitation-driven fouling, leading to an overall reduction in the fouling rate when both chemical and precipitation-driven fouling are accounted for in the model. The mechanism underlying this observation requires elucidation, which will be part of future work.

The volume of fraction of asphaltene phase under different temperature and Re conditions is depicted in Figure 5.35. As shown in this figure, the large amount of asphaltene is formed in tube at high-temperature low Re conditions and vice versa.

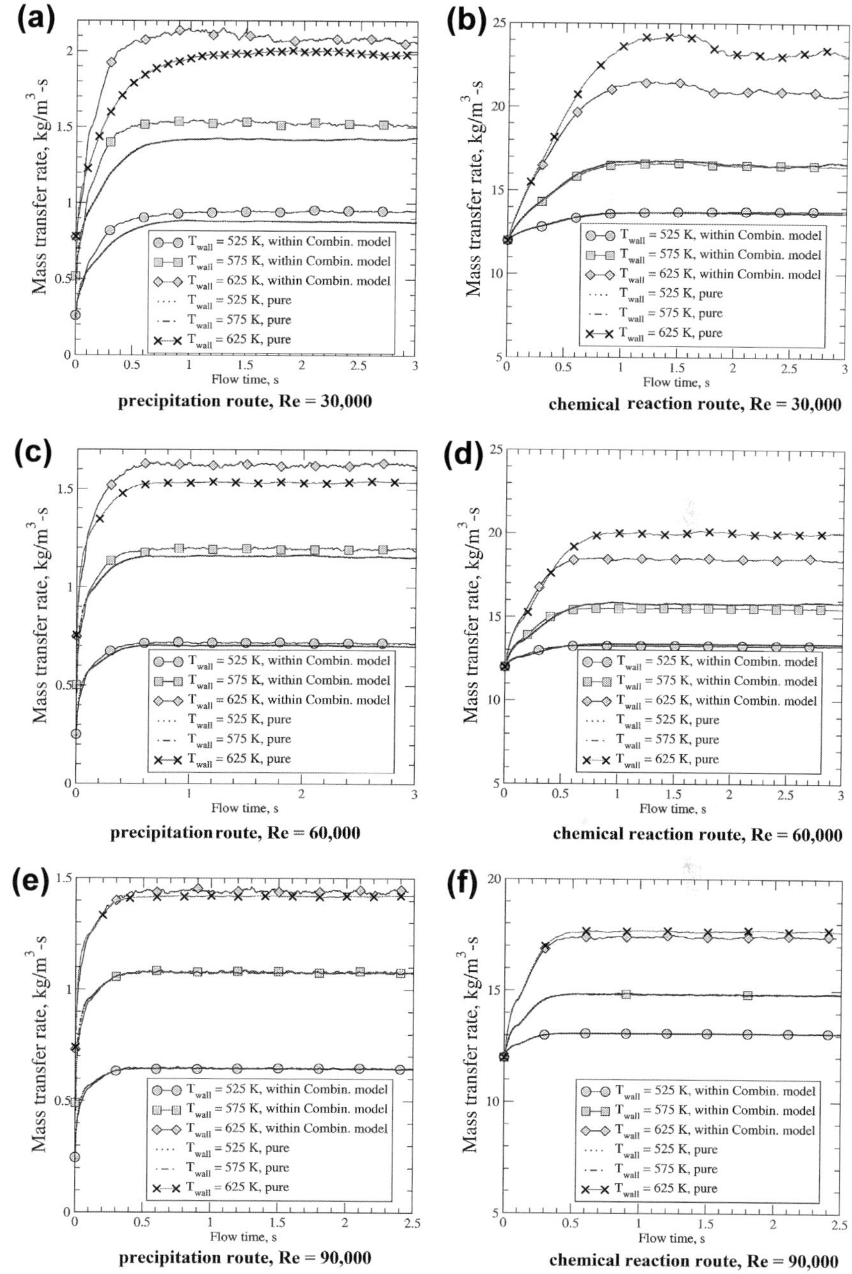

Figure 5.34 Comparison of predicted mass transfer rate contributed by asphaltene precipitation route and chemical reaction fouling route.

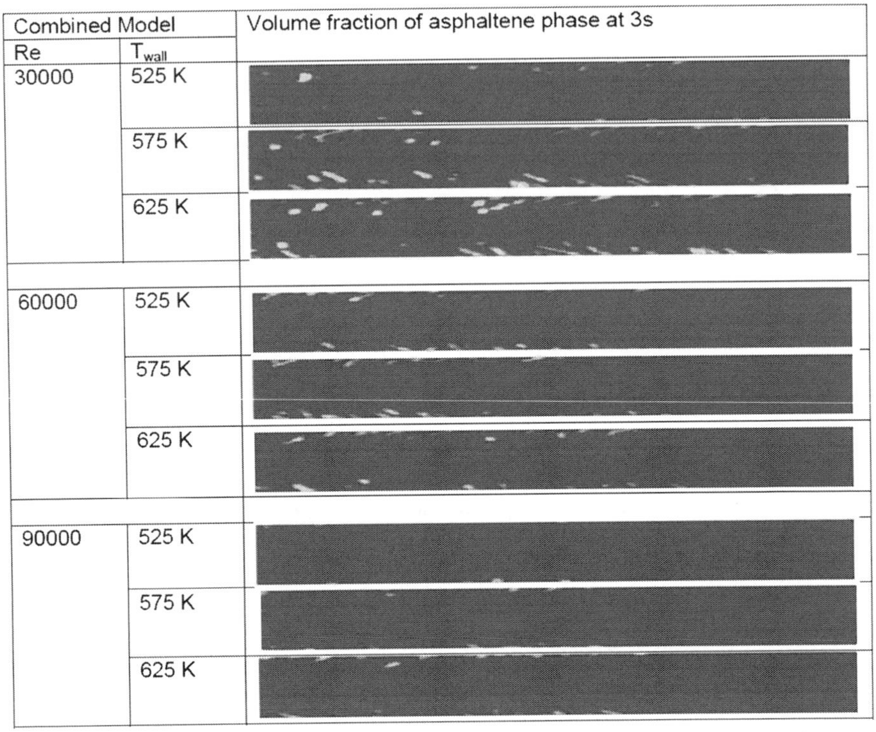

Figure 5.35 Comparison of VOF of asphaltene under different operating conditions.

5.3.4.4 *Aging Effect*

The aging effect on fouling formation was investigated within the framework of the combined model under the same conditions as those in the previous section. The predicted total asphaltene mass and mass transfer rate contributed by chemical reaction fouling are illustrated in Figure 5.36. As shown, the aging effect on the fouling formation is relatively minor. Under relatively low-temperature conditions, the mass plots with and without aging are nearly identical. At higher temperatures, aging leads to slightly lower foulant formation rates. This result is consistent with the DNS prediction for laminar flow (see Figure 5.15). Note that all the calculations are based on the clean-wall configuration.

The averaged viscosity and thermal conductivity of asphaltene were plotted versus the flow time, for the case with wall surface temperature 525 K and Re 15,000 (see Figure 5.37). As shown, the viscosity and thermal conductivity gradually increase with time. The thermal conductivity approached its maximum value $2\ \mathrm{W\ m^{-1}\ K^{-1}}$ after 3 s, however, the

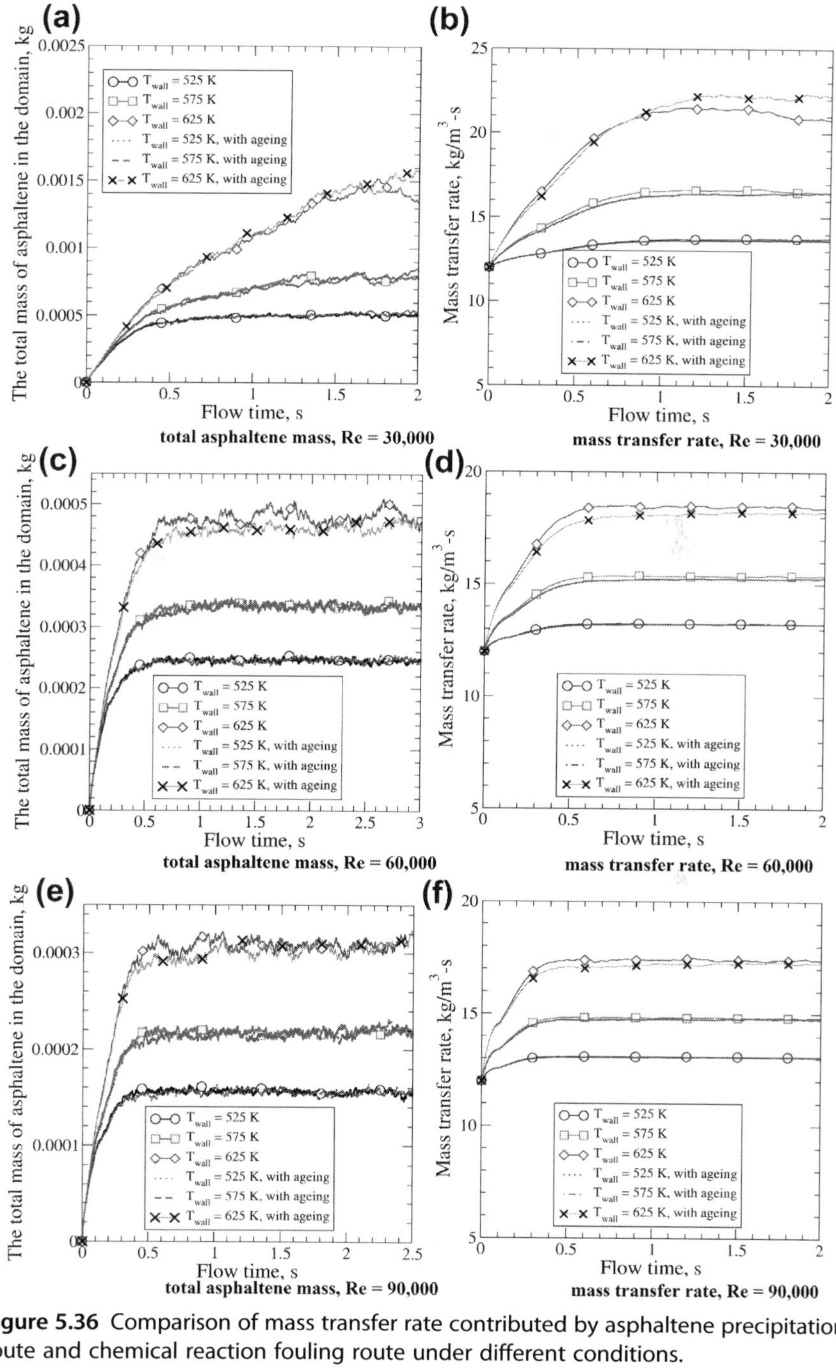

Figure 5.36 Comparison of mass transfer rate contributed by asphaltene precipitation route and chemical reaction fouling route under different conditions.

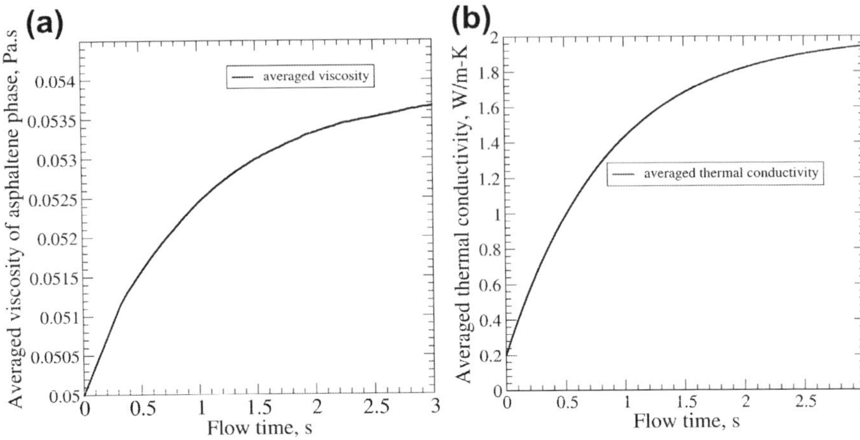

Figure 5.37 The time history of averaged asphaltene properties, viscosity (a), thermal conductivity (b) for the case of wall surface temperature 525 K and Re 15,000.

viscosity only increased by c. 7%. This may explain why aging was shown to play a minor role on fouling removal at the low-temperature and low Re conditions.

5.3.5 Conclusion

In the present work, the dynamics and evolution of crude oil fouling phenomena under laminar regime and turbulent regime have been studied numerically using 2D DNS and 3D LES techniques, respectively. Based on these simulation results, several conclusions can be drawn. For the laminar regime, the mathematical model is a combination of the nondimensionalized Navier–Stokes equations for conservation of mass and momentum for the flow dynamics plus a convective–diffusive Cahn–Hilliard equation to account for the concentration of the more viscous fluid. Also, an energy balance equation has been employed to account for heat transfer and convection and a chemical equilibrium model based on the Gibbs free energy is used to capture phase change phenomena. The deposited fluid has a time-dependent viscosity used to model aging effects and a structuration parameter *a*, Θ, to keep track of its change in time. These governing equations were solved via the diffuse interface method to capture the dynamics and evolution of fouling process in a 2D rectangular domain. By varying the diffusive rate of the component phase separating, D_s, which is inversely proportional to the deposition rate, it was inferred that for high values of D_s the buildup of deposit starts from the inlet and slowly moves downstream, contrary to the other cases with lower D_s values. This may be related to the presence of high recirculating diffusive areas downstream of the deposit. For intermediate values of D_s the presence of several droplets of oil trapped inside the deposit has been noted, due to the formation and coalescence of microligaments. The case with the lowest D_s is associated with a very fast and uniform buildup of deposit. Comparing mass plots in the presence and absence of aging it can be inferred that aging promotes the buildup of a thicker layer of deposit, inhibiting the formation of waves and the displacement of material downstream of the domain. An exception to this was found for the largest D_s values examined, for which the amount of deposit decreased with aging. This behavior may be linked to the interaction between the vortices ahead of the deposit and the fresh deposit near the inlet. A follow-up study is needed to explore whether this trend is inverted at longer time scales in the simulation. Also, the formation of ligaments and droplet pinch-off was not present in the case without aging due to the increase in the viscosity of the deposit layer which promotes the development of instabilities in the upper layer of deposit, leading to entrainment phenomena. By varying the Reynolds number (Re), it was found that an increase in Re resulted in an increase in interfacial waviness and a more effective displacement of the deposit along the domain. The interfacial instabilities observed at high Re are also responsible for the thinner highly viscous sublayer of deposit at the wall.

For the turbulent regime, a comprehensive model for crude oil fouling has been proposed that can be used to predict the evolution of the crude oil fouling process in typical industrial heat exchanger tubes. The overall model has several components: a chemical reaction fouling model in which sparingly soluble precursors present in the crude oil form an insoluble foulant via a heterogeneous reaction; an asphaltene precipitation model in which a chemical equilibrium process (asphaltene precipitation) based on the Gibbs free energy is used to account for the mass transfer from the oil phase to the foulant phase; a rheological model to describe the effects of aging on the fouling layer's viscosity; and a first-order kinetics model for describing changes in thermal conductivity. This model has been incorporated into an advanced CFD tool in a way that takes advantage of established species diffusion models as well as the LES and VOF methods.

A set of CFD simulations was then performed to assess the model's ability to predict foulant formation, removal, and aging. The results of these simulations indicated that chemical reaction fouling accounts for far more of the overall foulant formation than asphaltene precipitation and thus dominates the foulant mass generation process. In addition, the rate of the combined fouling route was increased by raising the wall surface temperature. The bulk flow velocity determines the degree of interfacial shear stress between the crude oil and the fouling layer. Higher bulk flow velocities tend to strengthen interface wave propagation and breaking, resulting in severe foulant droplet entrainment. The impact of aging on the removal process seems to be minimal for the set of simulations performed. This is probably due to competition between the structuration and the destructuration terms of the aging model. In addition, the impairment of chemical reaction fouling and enhancement of the asphaltene precipitation route were observed in the combined model. This could be relevant to the competition mechanism between of them.

The results presented herein demonstrate that the model by Yang et al. (2014) can be used to perform CFD simulations that should be useful in the design of improved strategies for mitigating the impact of crude oil fouling. Moreover, the CFD modeling provides a promising future to bridge the molecular-scale simulation (Section 5.2) and plant-scale simulation (Section 5.4) and is key in achieving the vision described in Section 5.1 of a multiscale framework for modeling of crude oil fouling.

5.4 INDUSTRIAL SCALE HIGH-FIDELITY MODELING

F. Coletti and S. Macchietto

In Section 5.1 the current heat exchanger design methodologies and their limitations have been reviewed. This has highlighted the importance of integrating fouling models with thermal ones to capture adequately the different phenomena involved in the deposition of unwanted material on thermal surfaces. As recognized by Bott (1995b) the purpose of any fouling model is

> *"to assist the designer or indeed the operator of heat exchangers, to make an assessment of the impact of fouling on heat exchanger performance given certain operating conditions."*

To accomplish this goal, Schreier and Fryer (1995) noted that fouling models should take into account:

- The rates of the processes that lead to deposition.
- The temperature distribution and deposit thickness profile.
- The effect of flow on deposition and, if relevant, reentrainment (removal).

While Section 5.2 and Section 5.3 were concerned with improving underlying fouling models at a thermodynamic and fluid-dynamic level, this section focuses on the development and use of an industrial-level model for shell-and-tube heat exchangers undergoing crude oil fouling by Coletti and Macchietto (2011). This model integrates all the three aspects above and captures the thermal and hydraulic effects of fouling within refinery heat exchangers and preheat trains and is now implemented in Hexxcell Studio™, commercially available via Hexxcell Ltd.

The main features of the model are summarized as follows:

- Distributed heat balances are written in cylindrical coordinates where appropriate (i.e., for tube wall and fouling layer domains), which makes it possible to overcome the thin slab approximation—often used in the past—accounting for curvature effects of the heat exchanger tubes on the heat flux.
- The heat exchanger configuration is accounted for (e.g., number of tube-side passes, tube diameter and length, baffle spacing, pitch arrangement, etc.). The heat transfer coefficients and pressure drops on the shell-side are calculated using the Bell–Delaware method (Taborek, 2008a,b) or the flow stream analysis (Hewitt, 2008).
- Physical properties for both shell-side and tube-side fluids are considered as a function of temperature. API-based relationships are used as default model but other thermophysical property models (including company proprietary ones) can also be used if available.
- The Ebert–Panchal model (Equation (5.30)) used in a distributed (rather than lumped) way allows calculating the local value of fouling resistance at each point along the exchanger length as opposed to an average value for the whole exchanger.
- The growth of the fouling layer over time at any given point across the tube length is captured using a moving boundary approach. The corresponding reduction in flow radius, R_{flow} (the tube radius available for flow in actual fouled conditions), and restriction in cross-sectional flow area are calculated as a function of the fouling thickness, δ. Tube-side velocity (thus shear stress) and heat transfer coefficient are calculated as a function of R_{flow}.
- The structural changes of the fouling deposit over time are captured with the aging model by Coletti et al. (2010). This describes the changes in the thermal conductivity of the fouling layer at any point in the axial and radial direction as a function of the specific temperature history. More details on this model have been given in Section 5.1.5.1.
- An advanced equation-oriented process simulator that features an easy-to-use flowsheeting environment (gPROMS by PSE Ltd.) is used to

solve the equations and estimate necessary model parameters. This framework has several advantages:

- It allows an easy setup of heat exchanger networks by interconnecting of individual heat exchanger units dragged-and-dropped in a flowsheet.
- The equation oriented architecture supports the solution of complex numerical problems and allows multiple types of calculations utilizing the same model.
- It is not a "black box." In contrast to other tolls implemented in procedural programing language (e.g., c++ or Matlab), the architecture chosen here allows the user to easily access and modify model equations (e.g., fouling model), thus giving full control of the model.
- It allows the easy integration with other process models (e.g., distillation columns, controllers).
- It provides a unified framework that helps preventing "silos" between company functions allowing easy sharing of assumptions, validations, and developments among R&D, engineering design, operations.

The following sections will provide a summary of the model, its software implementation, the results, and applications that derive from its use.

5.4.1 Governing Equations

The model equations are reported in the original publication (Coletti and Macchietto, 2011). Here are briefly summarized the main equations of the thermal, hydraulic, and fouling models and their integration.

5.4.1.1 Thermal Model

One key feature of the Hexxcell model is that the underlying governing equations do not rely on averaged models such as the logarithmic mean temperature or the ε-NTU methods (Section 5.1.1). In the model by Coletti and Macchietto (2011), distributed heat balances are solved along the axial direction in each pass, n, of the heat exchanger:

$$\frac{\partial}{\partial t}\left(\rho c_p T A_{flow}\right) = -dir_n \frac{\partial}{\partial z}\left(\rho c_p T u A_{flow}\right) + \frac{\partial}{\partial z}\left(\lambda A_{flow}\frac{\partial T}{\partial z}\right) + \alpha P\left(T_L\big|_{r=R_{flow}} - T\right) \tag{5.68}$$

where dir_n is a variable direction term, introduced to take into account the direction of the flow ($dir_n = 1$ in case of an odd pass, $dir_n = -1$ in case of an even pass). The fluid temperature, T, and its physical properties, ρ, c_p, λ, are a function of time, t, and spatial position, z, along the exchanger. The change in cross-sectional area of a tube caused by fouling is accounted for by considering the area of flow, A_{flow}, as a function of the fouling thickness, δ:

$$A_{flow} = \pi R_{flow}^2 = \pi(R_i - \delta)^2 \tag{5.69}$$

where R_{flow} denotes the flow radius and R_i the internal tube radius (Figure 5.38). The changes over time of the fouling layer thickness, δ, are calculated as a function of the local deposition rate, which, in turn, depends on the local value of the fouling resistance:

$$\frac{d\delta}{dt} = \lambda_L^0 \frac{dR_f}{dt} \tag{5.70}$$

In Equation (5.70), λ_L^0 denotes the value of the thermal conductivity of the fresh deposits accumulating on the thermal surface whereas R_f is calculated according to Equation (5.77) below.

The effects of the restrictions caused by fouling buildup on the convective heat transfer coefficient, α, are also accounted for in Equation (5.68) by writing α as function of R_{flow}:

$$\alpha = \frac{\lambda Nu}{2R_{flow}} \tag{5.71}$$

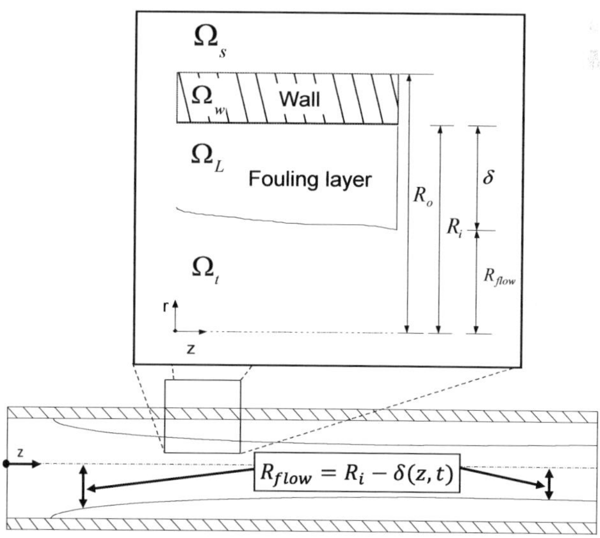

Figure 5.38 Main model domains and key geometrical quantities.

where λ denotes the fluid thermal conductivity and *Nu* is the Nusselt number, calculated through the Dittus–Boelter relationship (Hewitt et al., 1994). It can be noted that in Equation (5.71) the heat transfer coefficient is enhanced by the progressive reduction in cross-sectional area.

The fouling layer domain is represented as a two-dimensional conductive solid with a moving boundary defined a function of the thickness of the deposit growing in the heat exchanger tubes. Assuming negligible temperature gradients in the axial and angular coordinates and negligible heat of reaction from the aging process, the heat balance on differential segment of the fouling layer is

$$\delta^2 \rho_L c_{p,L} \frac{\partial T_L}{\partial t} = \frac{\partial \lambda_L}{\partial \tilde{r}} \frac{\partial T_L}{\partial \tilde{r}} - \frac{\lambda_L \delta}{R_i - \tilde{r}\delta} \frac{\partial T_L}{\partial \tilde{r}} + \lambda_L \frac{\partial^2 T_L}{\partial \tilde{r}^2} \tag{5.72}$$

Equation (5.72) is written with respect to a dimensionless radial coordinate, $\tilde{r}$:

$$\tilde{r} = \frac{R_i - r}{\delta} \tag{5.73}$$

The above formulation in cylindrical coordinates written with respect to the dimensionless variable makes it possible to overcome the thin slab approximation—often used in the past—by accounting for curvature effects in the heat flux:

$$q''_L = \frac{\lambda_L}{\delta} \frac{\partial T_L}{\partial \tilde{r}} \tag{5.74}$$

In both Equations (5.72) and (5.73), the thermal conductivity, λ_L, is a function of time, space, and temperature and varies according to the aging model by Coletti et al. (2010) described in Section 5.1.5.1. This way, each point in the deposit has a distinct value of thermal conductivity that reflects the specific temperature history experienced by that point.

The shell-side model is also treated as a distributed domain:

$$\frac{\partial}{\partial t}\left(\rho_S c_{p,S} T_S\right) = -dir_S \frac{\partial}{\partial z}\left(\rho c_{p,S} T_S u_S\right) + \frac{\partial}{\partial z}\left(\lambda_S \frac{\partial T}{\partial z}\right) + \frac{1}{A_S} \sum_{n=1}^{N_p} \alpha_S P_S \left(T_s - T_w\big|_{r=R_o}\right) \tag{5.75}$$

where dir_S is a variable direction term, introduced to take into account the internal arrangement, with respect to the direction of the flow in the first tube pass. The shell-side heat transfer coefficient, α_S (and pressure drop), can be calculated either with the Bell–Delaware method (Taborek, 2008a) or with the flow stream analysis (Hewitt, 2008). In the results presented in later sections, the former method is used, where the convective heat transfer coefficient, α_S, is calculated as

$$\alpha_S = \alpha_{id} J_c J_l J_b J_s J_r \tag{5.76}$$

In Equation (5.76), α_{id} denotes the heat transfer coefficient for ideal cross-flow whereas J_c, J_l, J_b, J_s, and J_r are correction factors for, respectively, the segmental baffle window, the baffle leakage, the bypass tube bundle to shell, the laminar heat transfer, and the nonequal inlet/outlet baffle spacing. The only difference from the standard calculation is that the ideal heat transfer coefficient is explicitly considered as a function of the spatial position (i.e., $\alpha_{id} = \alpha_{id}(z)$).

5.4.1.2 Fouling Model

The thermal model described above is coupled with the fouling rate model proposed by Ebert–Panchal:

$$\frac{dR_f(z,t)}{dt} = A\,\mathrm{Re}(z,t)^{-0.66}\mathrm{Pr}(z,t)^{-0.33}\exp\left(\frac{-E_f}{R_g T_f(z,t)}\right) - \gamma\tau(z,t) \tag{5.77}$$

While the original model was developed to fit lumped fouling resistance calculations, here it is used in a distributed way to describe the net deposition rate at each axial point. The dependency of all variables on space and time is written explicitly in Equation (5.77) to highlight that the local fouling rate is calculated for each pass at each axial coordinate, z. It should be noted however, that the calculations of the fouling resistance via Equation (5.77) are only used to determine the thickness of the fouling layer at any point in the heat exchanger as a function of process conditions via Equation (5.70) which in turn determines the performance of the heat exchanger. Moreover, any fitting of key parameters such as the activation energy, E_f, the preexponential factor, A, and the suppression parameters, γ, is only performed on primary measurements (e.g., temperature) and not on derived ones (e.g., fouling resistance calculations from experimental or plant data). Details of the parameter estimation procedure are reported in Section 5.4.2.

Since a common measure of fouling in industry is the average fouling resistance, typically calculated with one of the methods described in

Section 5.1.3, a similar, average quantity, $\overline{R_f}$, can be calculated with the model from the distributed quantities:

$$\overline{R_f} = \frac{1}{N_S} \sum_{i=1}^{N_p} \frac{\sum_{n=1}^{N_p} \frac{1}{L} \int_0^L \left(\int_0^1 \frac{\delta}{\lambda_L} d\tilde{r} \right) dz}{N_p L} \tag{5.78}$$

where N_S, is the shell number in the heat exchanger for a multishell unit.

5.4.1.3 Hydraulics

The velocity inside the tubes, u, on which Re and, ultimately, the heat transfer coefficient and pressure drops depend, is written as a function of the flow area, A_{flow} (which is a function of the fouling thickness via Equation (5.69)):

$$u = \frac{\dot{M}}{\rho A_{flow} N_T / N_p} \tag{5.79}$$

where N_p denotes the number of tube-side passes and N_T the number of tubes in the heat exchanger. The definition of velocity as a function of the flow area allows to capture the effects of fouling growth on the hydraulic behavior of the exchanger. Quantities related to the velocity, such as the wall shear stress, τ, which governs the suppression term in Equation (5.77), are therefore also affected by fouling:

$$\tau = C_f \left(\frac{\rho u^2}{2} \right) \tag{5.80}$$

where C_f is the Fanning friction factor for rough tubes (Yeap et al., 2004):

$$C_f = 0.0035 + 0.264 \mathrm{Re}^{-0.42} \tag{5.81}$$

The physical properties appearing in the equations above are calculated using API relationships (Riazi, 2005) as a function of temperature and space. This allows to capture the effects of the variation in heat transfer coefficient along the length of the heat exchanger. Other thermophysical property models can be used if available.

5.4.2 Parameter Estimation Procedure

The model equations summarized above comprise a number of parameters. These can be divided in a set of two parameters:

- Set A, which includes the physical properties (i.e., °API, MeABP, and $\nu_{100°F}$) of the fluids on both tube-side and shell-side. The values for these can, sometimes, be obtained from the refineries and may therefore be used as an input to the model rather than estimated.

- Set B, which includes the activation energy, E_f; preexponential deposition constant, A, and suppression constant, γ, in the fouling model, Equation (5.77). These quantities are not readily available and are difficult to measure experimentally.

The procedure developed by Coletti and Macchietto (2011) outlined below has been devised to use primary and readily available plant measurements (temperatures and flow rates) to estimate the necessary model parameters when these are not known:

1. Plant data are filtered according to a standardized procedure detailed in Coletti and Macchietto (2011) and summarized in Section 5.4.3. This procedures allow discarding from the estimation data set the data points that do not satisfy a minimum quality requirement.
2. The model is fitted by adjusting the values of the parameters to one set of plant data. A check is performed to assess whether it is able to reproduce those data.
3. The parameters obtained in step 2 are used to predict the outlet temperatures from the heat exchanger against a further set of plant data, different from the one used for the estimations. This step allows testing of the predictive capabilities of the model with the estimated value of the parameters.

Since parameters in Set A do not depend on fouling, they are estimated over the induction period when fouling has not yet started and the exchanger can be considered clean. This is defined as Period Ia or the "Clean period." On the contrary, parameter set B should be estimated over a longer period where fouling is indeed affecting the performance of the unit. The latter is defined as Period I or the "Estimation period" and includes Period Ia. Mode-based parameter estimation techniques are used to estimate the values of the parameters that give the best fit to the retained plant data. The choice of Period Ia is important: it must start when the unit is fully cleaned and if too long, the assumption of no fouling being present on the tubes may not be valid. Conversely, if the period is too short, fluid properties may be fitted to a single oil slate, which may not be representative of the typical crudes processed by the refinery.

A second period immediately following Period I, Period II, defined as the "Prediction period" is used to test the model for its predictive capabilities. The responses simulated using the parameter estimated in Period I are compared with plant measurements (outlet tube- and shell-side temperatures).

5.4.3 Plant Data

Raw plant measurements (rather than fouling resistances, derived from measurements with the use of simplistic heat balances) are used with model-based parameter estimation techniques to estimate values for the model parameters. This avoids incorporating in the analysis the errors resulting from any simplifying assumptions in calculations, such as the use of overall heat balances. Plant data are often considered unreliable by refinery operators. They typically trust temperature measurements to be within a ± 5 °C error. However, for the parameter estimation procedure used here a variance of ± 1 °C has been considered for all measurements as it will be justified later.

The measured outlet temperatures on both shell-side and tube-side are used to fit the model parameters in Period I following the procedure outlined above. These data are typically recorded at least on a daily basis, sometimes more frequently. Daily averages were used here. However, as highlighted by several authors (Crittenden et al., 1992) it is notoriously difficult to obtain reliable plant data, thus some judgment is required when analyzing such records.

As shown in Section 5.1, heat duties on both sides of the exchanger are typically calculated in refinery practice using a simple lumped heat balance model

$$Q = \dot{M} c_p \Delta T \tag{5.82}$$

that assumes constant mass flow rates, fluid physical properties, and heat transfer coefficients. Accuracy of the measurements is typically checked by calculating the percentage difference of the heat duty, Q, on the two sides of the heat exchanger. Here this difference is defined as the error φ:

$$\varphi = \frac{Q_S - Q_T}{Q_S} \% \tag{5.83}$$

where Q_S is the heat duty on calculated on the shell-side and Q_T is the heat duty calculated on the tube-side. Given the many assumptions described above, values of the error $\varphi = 10$–30% are often found when using raw plant measurements. Moreover, such errors typically have a significant nonzero average offset and wide variability (Crittenden et al., 1992).

Rigorous statistical methodologies existing in the literature could be used to detect instrument bias prior to the use of the following procedure. However, it is assumed here that the measurements do not suffer from systematic bias for the time periods considered and thus a simple procedure is sufficient to screen out grossly inaccurate measurements that would otherwise cause numerical problems in the estimation of model parameters and result in wrong parameter values. The procedure proposed by Coletti and Macchietto (2011) calculates φ from Equation (5.83) using measured plant temperature and flow rate data, and estimates its average over time $\overline{\varphi}$ and standard deviation σ. Daily data for which φ deviates more than a certain value from the mean are deemed to be grossly erroneous or outliers and are discarded from consideration for parameter fitting purposes. The remaining data are defined as *filtered* and are used in a proper statistically sound model-based parameter estimation, in conjunction with the full detailed dynamic model, using the procedure described in Section 5.4.2.

5.4.4 Implementation

One key characteristic of a good tool is its user friendliness. As shown in Figure 5.39 the setup of a single heat exchanger model is performed via a tailored interface through specially designed interactive forms and

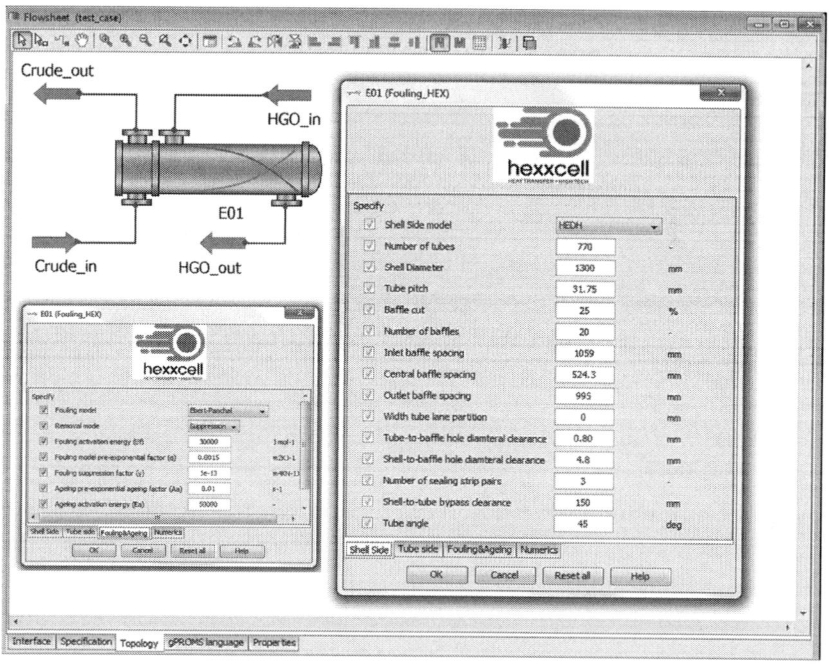

Figure 5.39 Specification of a shell-and-tube heat exchanger and fouling model parameters. *Copyright (2014) Hexxcell Ltd.*

drag-and-drop items. Key attributes to be specified include oil characteristics, heat exchanger characteristics, and parameters for fouling and aging. A standardized interface with Excel is also available for importing exchanger geometry data directly from refinery proprietary data sheets. Once specified, each particular exchanger may be added to the general Hexxcell Studio™ library. Individual library components may be reused alone or assembled in combination with other models to provide higher level models. The framework then allows a rapid replication of similar units and interconnecting them together, via simple drag-and-drop, to construct a heat exchanger network model (Figure 5.40). This allows to quickly match a current preheat train configuration, or alternative ones being assessed. The composite network model (e.g., an entire preheat train or a section of it) may itself be saved in the Hexxcell Studio™ library and reused, alone or with other models. This way, it is very easy to generate alternative configurations at various levels of aggregation, test and evaluate modifications under exactly the same conditions, etc. The cost parameters of an economic model (e.g., pumping costs, furnace fuel costs, production loss) may also be defined, to help assess performance in economic rather than purely technical terms.

Although the number of equations in the model is fairly large when an entire preheat network is simulated, it is possible to obtain a solution in a few minutes of computational time. It typically takes less than ca. 300 s to an Intel i7 processor at 3.4 GHz with 8 GB RAM to solve a 10 heat exchanger network. Moreover, the robustness of the model is ensured by the use of Hexxcell's proprietary initialization procedures which allow the solver to determine the initial point for the calculations transparently to the user. As a result, the model can be easily initialized and used by thermal and process engineers at all levels without the support of a specialist in numerical techniques.

5.4.5 Model Validation and Results for Individual Heat Exchangers

The details of the model validation are reported in Coletti and Macchietto (2010b, 2011). To show the model validation over a wide range of operating conditions a short summary is given here based on the results reported by Coletti and Macchietto (2010b) for four industrial units in two refineries operated by major oil companies:

- RA-E01, RA-E02. Two double shell units in Refinery A.
- RB-E01, RB-E02. Two single shell units in Refinery B.

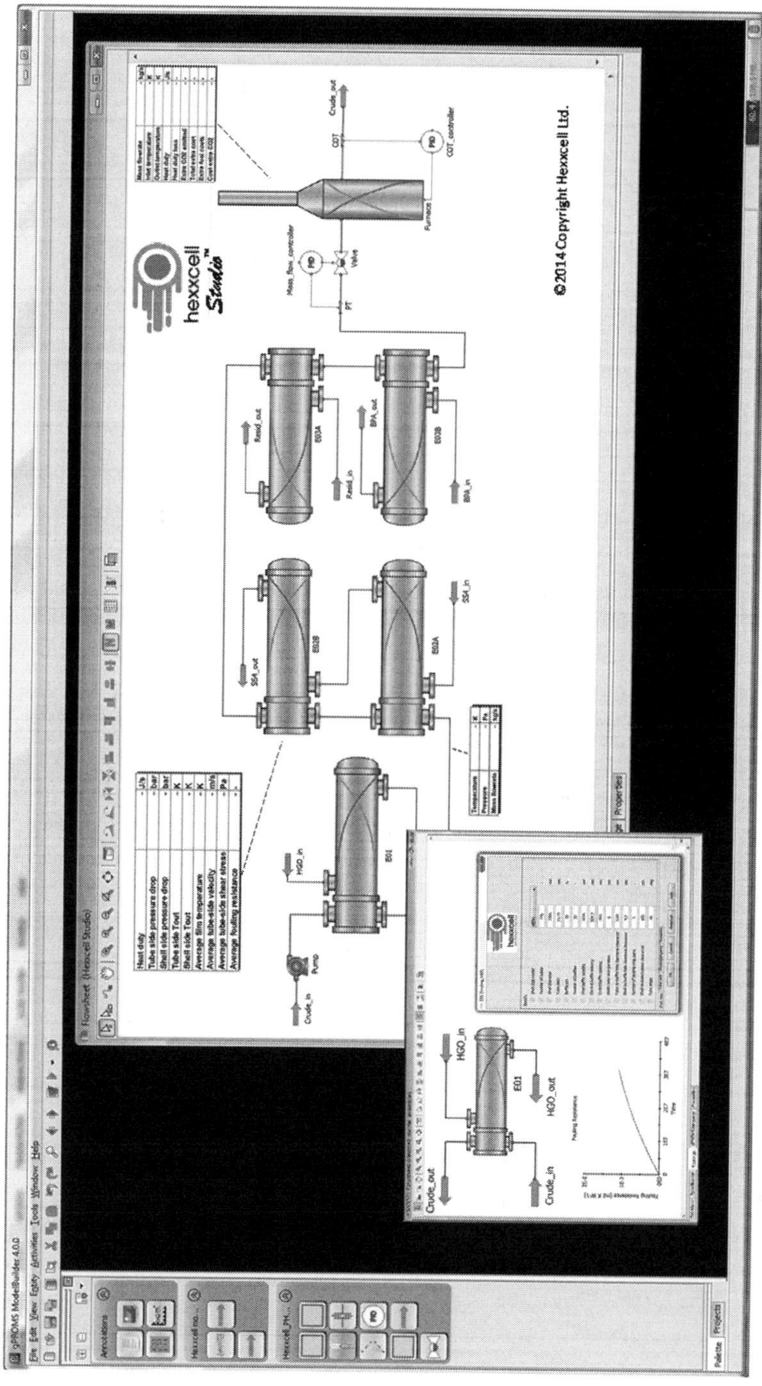

Figure 5.40 Screenshot of the graphical user interface of a preheat train model using the Hexxcell Studio™ library. *Copyright (2014) Hexxcell Ltd.*

Table 5.12 Summary of data used. Reproduced with permission from Coletti and Macchietto (2010b).

HEX name	N_s	N_p	T_s [°C]	T_t [°C]	Design u_t [m s^{-1}]	Run length [d]
RA-E01	2	4	220–350	190–250	2.5	349
RA-E02	2	4	220–320	170–250	1	162
RB-E01	1	4	240–300	180–210	1.7	550
RB-E02	1	2	155–270	130–160	2.7	310

Temperature range refers to the minimum coldest and the maximum hottest temperature on record for a given side of the unit. Run length is the period of time between two mechanical cleanings considered for the analysis.

The use of data from different refineries is particularly important to ensure that the model capabilities in predicting fouling behavior are not refinery specific or oil specific. While refineries typically change crude blends every 2–3 days, the crude types processed by one refinery are usually from a limited number of geographic origins. As a result, testing the model on data from different refineries has the significance of generalizing its validity to a large number of crude blends. Moreover, considering units from different refineries also means dealing with different design and operating philosophies. The geometries considered vary with respect to shell size, number of tube passes, tube diameter, pitch angle, baffle number, spacing, etc. The range of operating temperatures considered is 150–350 °C on the shell-side and 130–245 °C on the tube-side (Table 5.12).

In particular, one of the units, RB-E02, operated at a relatively low temperature (crude is between c. 130 and 150 °C) just below the range in which chemical reaction fouling is believed to be the dominant fouling mechanism. The hydraulic conditions on the tube-side considered also vary significantly. The design velocity of the heat exchanger considered ranges between 1 and 2.5 m s^{-1}.

For all the units considered, Period Ia was set to include the first 15 accurate measurements. Depending on the quality of the data, φ, it ends at different dates for each unit. Period I was 60 days for all units. The length of the prediction period, Period II, is related to the length of the run specific for each exchanger, which is between 162 and 550 days (last column of Table 5.12). The prediction periods considered here therefore range from 5.5 months for unit RA-E2 in this study up to 16 months for unit RB-E01.

The procedure used for parameter estimation and results of the model predictions are illustrated in the following section in detail for exchanger RA-E01; the results for other cases are reported in Section 3.4.5.2. Section 5.4.6.2 shows how the model can be used to propose effective retrofits that minimize fouling and maximize energy efficiency.

5.4.5.1 Model Results and Predictions for RA-E01

The heat exchanger considered comprised two identical shells with four tube passes per shell. The crude was allocated on the tube-side, whereas residuum from the crude distillation column was on the shell-side. Refinery observation when dismantling the unit for cleaning indicated that, although some shell-side fouling was detected, the dominant fouling resistance was on the tube-side.

Parameters in Set A and Set B were estimated over Period I using plant data (inlet temperatures and volumetric flow rates) starting after a mechanical cleaning. Over this period, 30% of the data points were discarded as deemed inaccurate by the procedure described above. The estimated parameters are reported in Table 5.13.

To test whether the model can be used in a predictive mode (i.e., to extrapolate its results beyond the estimation period, without reestimating its parameters), the model was then used to simulate the exchanger for the rest of the year. Measured inlet temperatures and flow rates were input to the simulation, but all adjustable parameters were kept fixed. The simulated performance, in terms of output temperatures over the whole time horizon considered (347 days) is shown by the continuous lines in Figures 5.41 and 5.42, together with plant measurements. The exit temperatures on both sides were predicted well over the entire period, even in the presence of rather major excursions, including the period between 60–100 days.

The contribution of fouling and aging to the overall behavior of the unit is illustrated by the two extra lines plotted in the same figures:

1. The dashed line was obtained using the same parameter Set A as before but setting the fouling rate to 0 (i.e., $dR_f/dt = 0$), corresponding to a case in which fouling had not occurred and the heat exchanger operated in clean conditions.

Table 5.13 Estimates for parameter Set A and B. Parameter Set B shows estimates of the fouling model in each shell comprising the heat exchanger. Reproduced with permission from Coletti and Macchietto (2011).

Parameter Set A			Parameter Set B		
Parameter	**Optimal estimate**	**Units**	**Parameter**	**Optimal estimate**	**Units**
Resid. °API	17.4	°API	E01A α	0.00165	$K\,m^2\,W^{-1}\,s^{-1}$
Resid. $\nu_{100°F}$	28.8	Cs	E01A E	28,491	$J\,mol\,s^{-1}$
Resid. MeABP	736.7	°C	E01A γ	9.28×10^{-13}	$N\,m^4\,K\,W^{-1}\,s^{-1}$
Crude °API	37.5	°API	E01B α	0.00165	$K\,m^2\,W^{-1}\,s^{-1}$
Crude $\nu_{100°F}$	13.9	Cs	E01B E	28,523	$J\,mol\,s^{-1}$
Crude MeABP	350.2	°C	E01B γ	9.35×10^{-13}	$N\,m^4\,K\,W^{-1}\,s^{-1}$

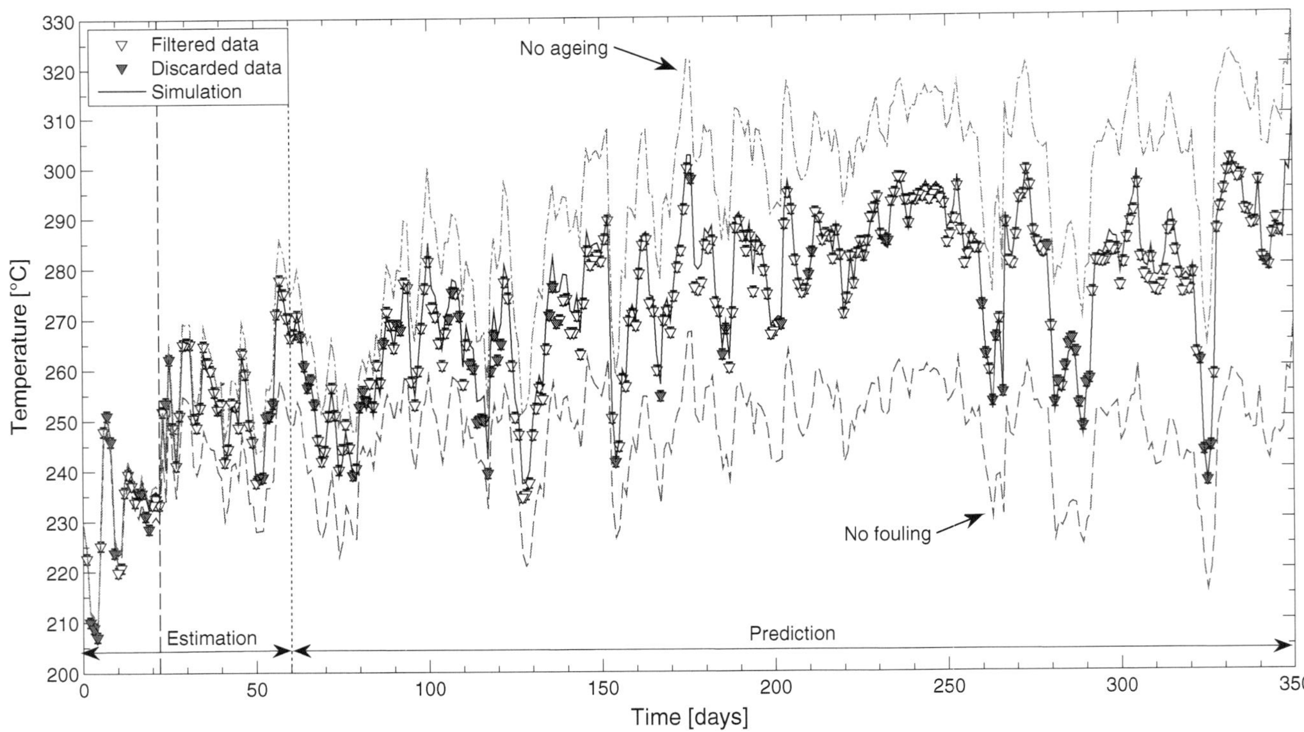

Figure 5.41 Overlay plot over the 350 days after cleaning of model simulations with tube-side outlet temperature. Dashed line shows simulation run setting $R_f = 0$ whereas the dashed–dotted line shows simulation results with no aging. White markers show data filtered with the procedure developed whereas colored markers show data deemed inaccurate. The error bars indicate ± 1 °C. The vertical dashed line shows the end of Period Ia whereas the dotted line indicates the end of Period I and the beginning of Period II. *Reproduced with permission from Coletti and Macchietto, A dynamic, distributed model of shell-and-tube heat exchangers undergoing crude oil fouling, 2011. Ind. Eng. Chem. Res. 50(8): 4515–4533. Copyright (2011) American Chemical Society.*

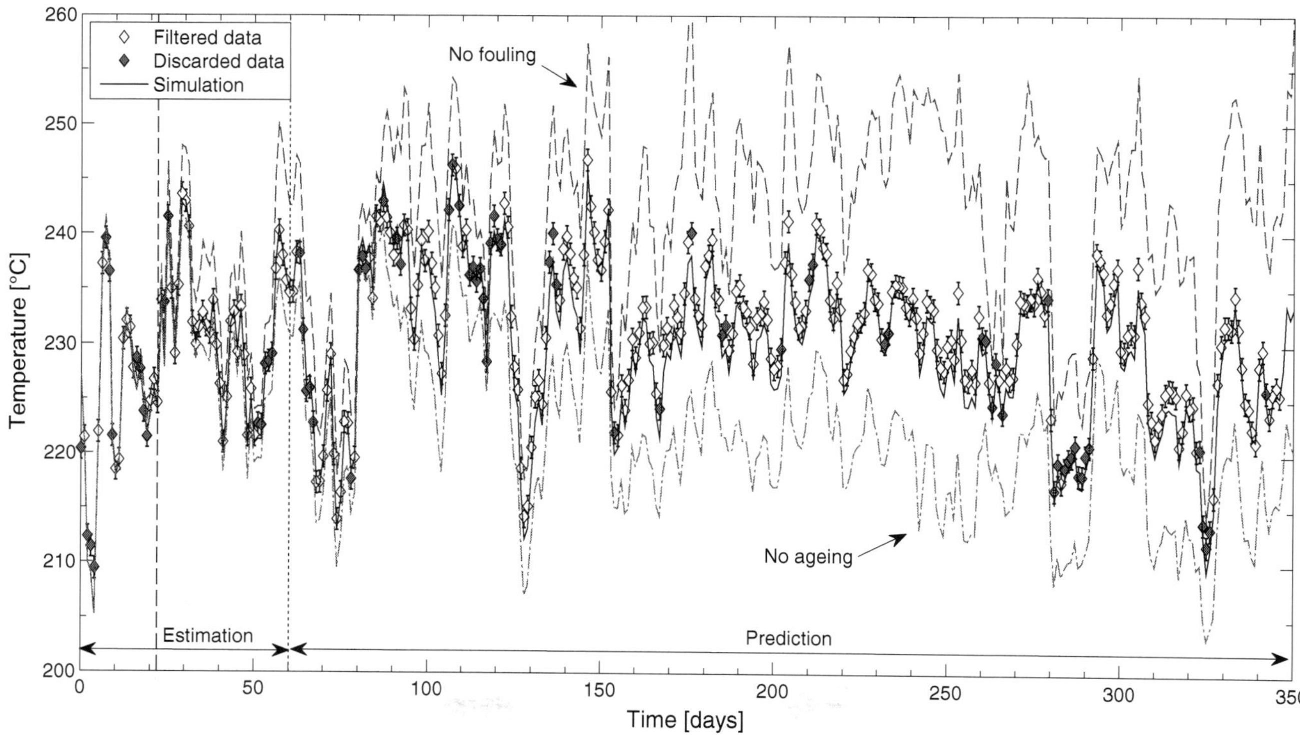

Figure 5.42 Overlay plot over the 350 days after cleaning of model simulations with shell-side outlet temperature. Dashed line shows simulation run setting $R_f = 0$ whereas the dashed-dotted line shows simulation results with no aging. White markers show data filtered with the procedure developed whereas colored markers show data deemed inaccurate. The error bars indicate ± 1 °C. The vertical dashed line shows the end of Period Ia whereas the dotted line indicates the end of Period I and the beginning of Period II. *Reproduced with permission from Coletti and Macchietto, A dynamic, distributed model of shell-and-tube heat exchangers undergoing crude oil fouling, 2011. Ind. Eng. Chem. Res. 50(8): 4515–4533. Copyright (2011) American Chemical Society.*

2. The dashed–dotted line was obtained using same parameter Sets A and B but setting the preexponential term in the aging model to 0, corresponding to a case in which fouling occurred but the deposits did not age.

Figures 5.41 and 5.42 provide an insight on the opposite effects of fouling and aging on the crude outlet temperature. Whilst fouling reduced the heat transfer efficiency thus decreasing the crude outlet temperature of the heat exchanger, aging acted in the opposite way, by increasing the thermal conductivity of the deposits, thus enhancing the overall heat transfer. Temperature estimated without fouling or aging in the model did not match the plant data, but did so when both phenomena were included.

The contribution of each of the two shells, E01A and E01B to the overall average fouling resistance $\overline{R_f}$, calculated from Equation (5.78), is shown in Figure 5.43. As expected for a clean exchanger, the value of the fouling resistance increased from an initial zero value. The rate of increase was high initially but reduced after about 150 days to an approximately constant value. Fouling in shell E01A was slightly higher than in shell E01B but

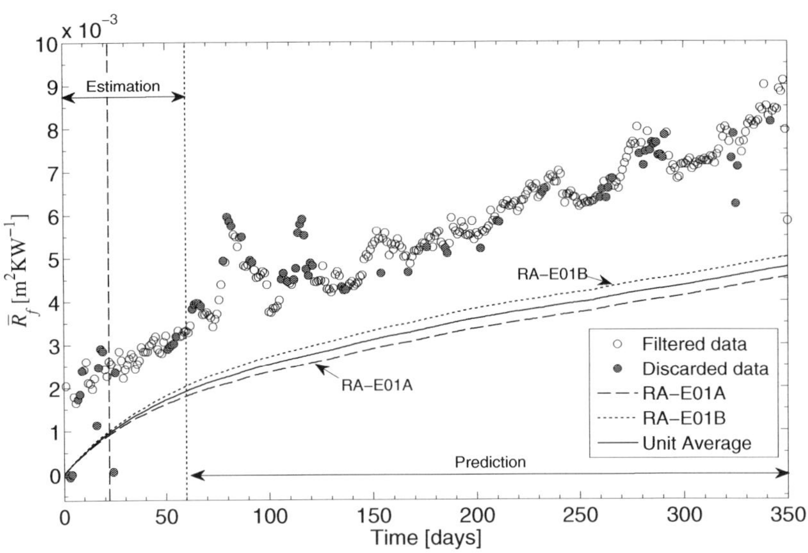

Figure 5.43 Refinery-calculated fouling resistance and model-calculated overall average R_f. The dotted and dashed lines indicate average R_f calculated for RA-E01A and RA-E01B respectively. *Reproduced with permission from Coletti and Macchietto, A dynamic, distributed model of shell-and-tube heat exchangers undergoing crude oil fouling, 2011. Ind. Eng. Chem. Res. 50(8): 4515–4533. Copyright (2011) American Chemical Society.*

the difference (in this case) was not large. Model outputs for $\overline{R_f}$ were also compared with refinery calculations. The latter showd a zero fouling resistance for the first few days, after which a sharp jump to an offset value of ca. 2×10^{-3} m^2KW^{-1} occurd.

Given the high degree of accuracy of the exit temperatures predicted by the model, it is likely that the overall fouling resistance calculated by the refinery merely reflects the gross approximations (e.g., in physical properties) used in its calculation. The overall trends shown substantially agree, and when the offset is subtracted, numerical values are also close. In particular, Figure 5.43 shows that the (perceived) sudden increase in fouling resistance around 80–100 days can be fully explained by the measured changes in inlet flow rates, temperatures, and complex interactions arising.

One of the main benefits of the model is that it enables the identification of the effects of fouling on the performance of the heat exchanger. Figure 5.44(a) reports the heat duty from model simulations in the case of fouling and no fouling over time for the heat exchanger RA-E01. It can be noted that the two curves diverge significantly over time. The difference between the two heat duties shown in Figure 5.44(b) represents the energy lost instantly due to fouling. Figure 5.44(b) also shows the integral of the heat duty loss over time, which represents the cumulative energy loss, calculated as 22.4 GWh after 347 days of operations. If this loss were to be compensated entirely at the furnace (i.e., no dumping effects given by the interaction with other exchangers in the network), the cost in extra fuel burnt would be in excess of US$670,000 (based on fuel a cost of US$27 MWh^{-1} and 90% efficiency of the furnace).

5.4.5.2 Model Results and Predictions for Other Heat Exchangers

To test the range of applicability of the model in a variety of cases, the same procedure described for RA-E01 was applied to the other three heat exchangers in Table 5.12.

The residuals (Figure 5.45) for each individual heat exchanger show some weak systematic trend, as opposed to purely random scatter, indicating there is possibly some underlying model mismatch (not surprising, for example, in view of the many oil slate changes over the long periods of time considered). Even so, 93% of all data are within ±2% error on the shell-side (ca. ±4 °C) and ±1% on the tube-side (ca. ±2 °C). The major deviation is given by RB-E02, which is expected as this is the lowest temperature unit and small absolute errors translate into large percentage errors. It should be noted that the values of the residuals calculated also depend on the temperature scale used.

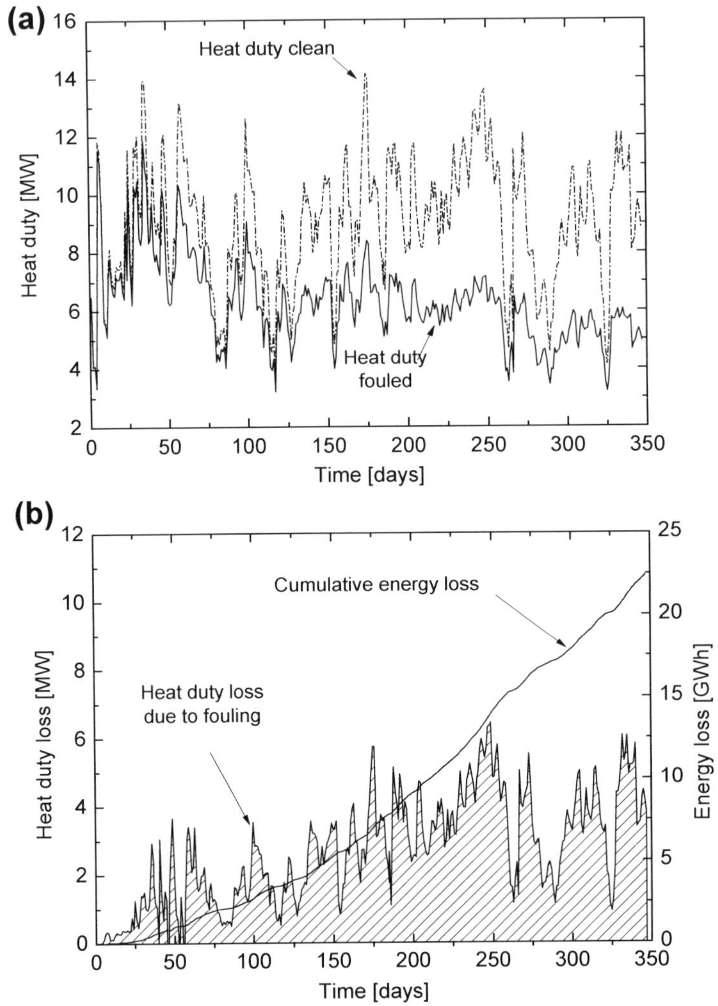

Figure 5.44 Heat duty from model simulations in case of fouling and no fouling over time in unit RA-E01 (a). The energy lost due to fouling, represented by the shaded area, and the cumulative energy loss over time are given in (b).

It is interesting to note that the quality of model predictions remained good (within a narrow error band) over extended periods of time (up to 16 months). These are industrially relevant time horizons as required for planning of cleanings to optimize a refinery economic performance.

Figure 5.46 reports parity plots of the predicted versus measured temperatures for the shell (a) and tube (b) sides, for all four units, thus covering a wide temperature range. The dashed lines show a deviation of ±1% from

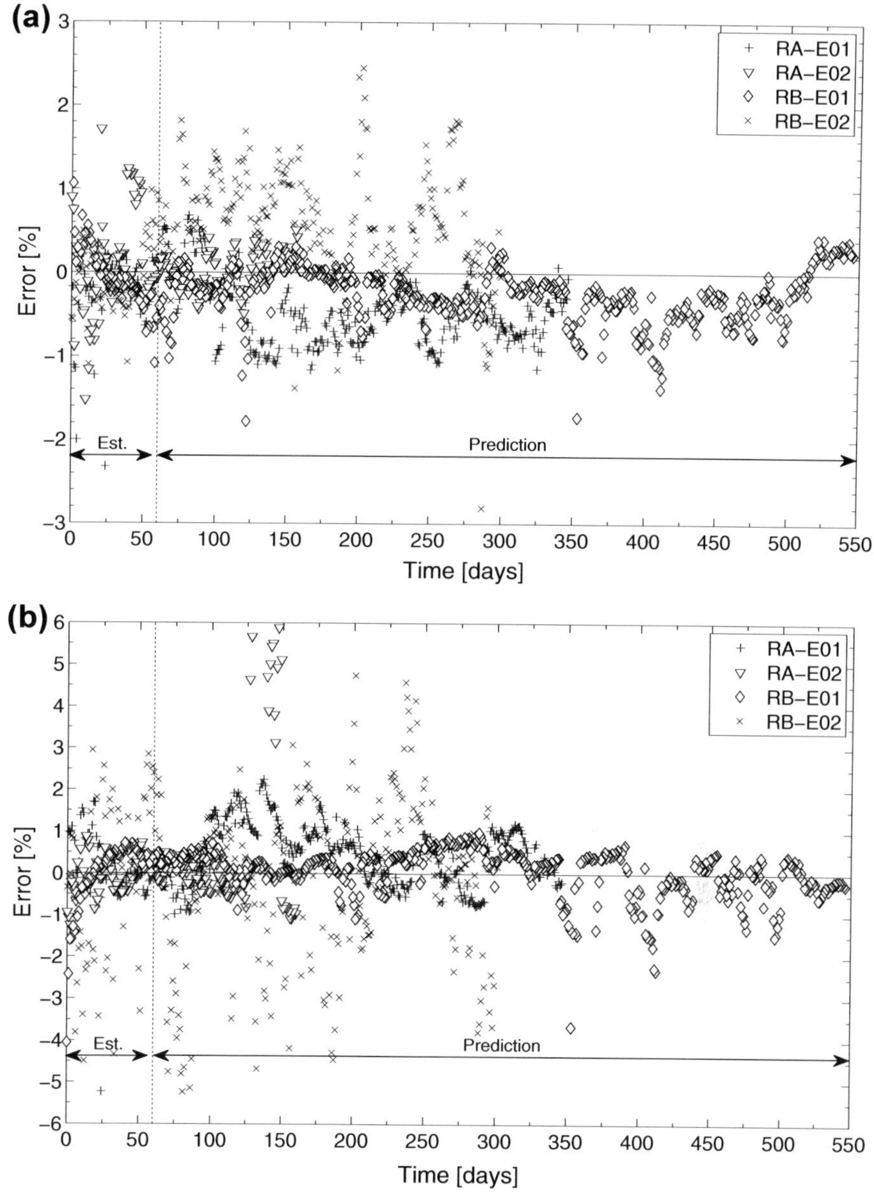

Figure 5.45 Percentage residuals between plant measurements of fluid outlet temperatures (in °C) and model simulations on the shell (a) and tube (b) side for all units examined. The vertical, dashed line indicates the end of the estimation period (Period I) and the horizontal dashed line indicates zero error. *Reproduced with permission from Coletti and Macchietto, Validation of a novel model for Shell-and-Tube heat exchangers undergoing crude oil fouling, 2010. In: Proc. 14th Int. Heat Transfer Conf. (IHTC14); Vol. 4, Washington, DC, USA, August 8–13. ASME, pp. 741–750. Copyright (2010) ASME.*

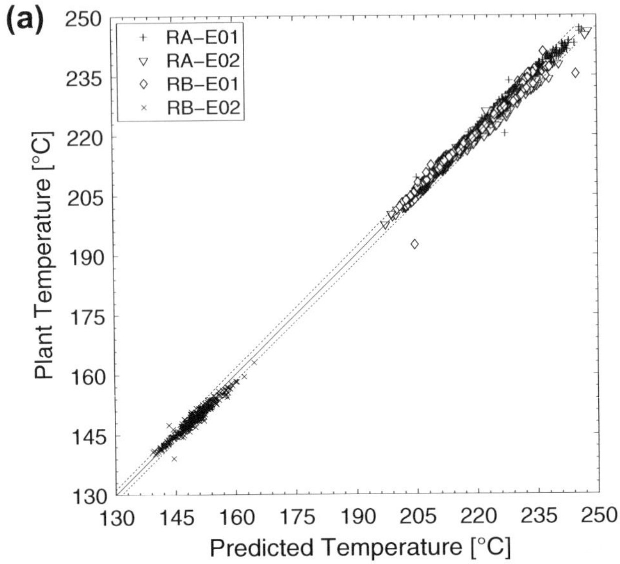

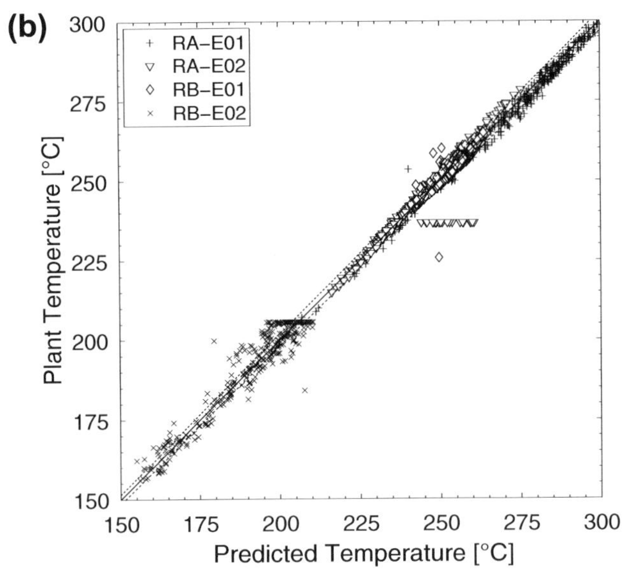

Figure 5.46 Parity plot of fluid outlet temperature on the shell (a) and tube (b) side for all units examined. The dashed lines indicate ±1% deviation from the parity line on the diagonal. *Reproduced with permission from Coletti and Macchietto, Validation of a novel model for Shell-and-Tube heat exchangers undergoing crude oil fouling, 2010. In: Proc. 14th Int. Heat Transfer Conf. (IHTC14); Vol. 4, Washington, DC, USA, August 8–13. ASME, pp. 741–750. Copyright (2010) ASME.*

the parity line. Analysis of these plots reveals that over 80% of the points are within a ±1% error for both sides of the unit. It should be noted that this 80% includes points which are given by clearly erroneous measurements (e.g., the previously noted flat plant temperatures at ca. 235 °C, evident in Figure 5.46(a)). This analysis provides an a posteriori justification of the use of ±1 °C variance in the measurements, which was used in the parameter estimation step.

The estimated fouling activation energy, E_f, shows a remarkable consistency, varying between 28.5 and 32.1 kJ mol^{-1} and compares well with typical values reported in the literature for crude oil reaction fouling.

5.4.5.3 Retrofit of Unit RA-E01

The model validation with plant data performed in the previous section gives the necessary confidence to exploit its capabilities to assess alternative design that might mitigate fouling and reduce energy losses. The detailed analysis carried out in the previous section unveiled that energy losses produced by fouling in unit RA-E01 added up to 22.4 GWh after 347 days of operation. In this section, a study of a simple retrofit option which aimed at reducing these losses is reported.

It has already been noted in previous Chapters of this book that increasing velocity—thus wall shear stress—has a beneficial effect on fouling and it is the process variable on which the designer has most control. It seems reasonable therefore to choose this variable first to improve the unit's performance. For a given flow rate, a way of achieving higher velocities within the tubes is to increase the number of tube-side passes, N_p. This is a relatively inexpensive option to be implemented in existing units as it requires only replacement of the heat exchanger headers, leaving in place existing tube bundle and shell.

In the case considered here, the retrofit of unit RA-E01 was performed by increasing the number of tube-side passes of the existing configuration from 4 to 6. All other geometric parameters (e.g., flow arrangement, number of tubes) were kept at the original values. Values of physical properties and the fouling model parameters were fixed at the values estimated with the procedure described in Section 5.4. Simulations were then performed under the same plant conditions, over 347 days after a mechanical cleaning, used for the 4 pass unit. This allowed for testing how the retrofitted unit would have performed in the refinery if the proposed design were actually used under the process conditions experienced in the plant. Comparison with simulation results reported in the previous section for the existing exchanger allows assessment of the potential saving achievable with the new configuration.

Figure 5.47(a) shows the reduction in fouling resistance produced by the higher velocities within the 6 pass exchanger. The value of the overall fouling resistance was reduced by ca. 25% after a year of operations, at the expenses of an increased pumping power required to counter the larger pressure drop (Figure 5.47(b)). While the extra pressure drop generated by the new configuration, ca. 0.5 bar, seemed acceptable under clean conditions, at the end of the operating period, it reached ca. 3 bar. This may be accepted, depending on the hydraulic flexibility of the refinery (considerations on mechanical design are excluded in this preliminary analysis).

Figure 5.48(a) shows that the difference in outlet temperatures of the tube-side fluid (crude) between operations with the existing unit and

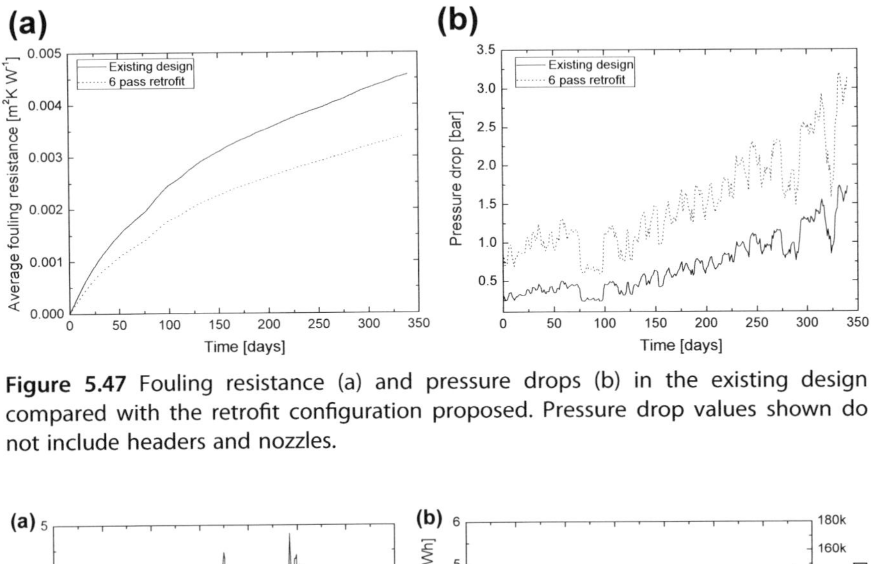

Figure 5.47 Fouling resistance (a) and pressure drops (b) in the existing design compared with the retrofit configuration proposed. Pressure drop values shown do not include headers and nozzles.

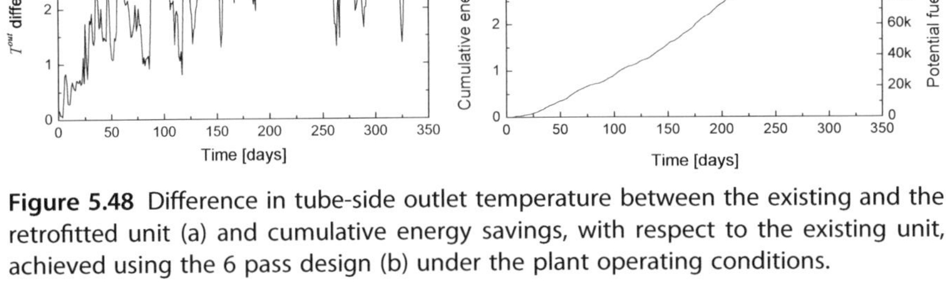

Figure 5.48 Difference in tube-side outlet temperature between the existing and the retrofitted unit (a) and cumulative energy savings, with respect to the existing unit, achieved using the 6 pass design (b) under the plant operating conditions.

the retrofitted one was as large as 4 °C. Cumulative energy savings were estimated to add up to almost 5 GWh (Figure 5.48(b)). This is equivalent to a 22% reduction in the heat duty losse generated by fouling in the first 347 days of operation and to ca. US$150,000 savings in furnace fuel alone.

5.4.6 Whole Preheat Train Case Study

The detailed dynamic model described above is for a single shell-and-tube heat exchanger. However, in a PHT, no heat exchanger exists on its own and the distinct fouling behavior of each heat exchanger affects the overall performance of the network. Complex interactions among the heat exchangers can only be unveiled by a simulation of the entire network. The implementation of the model equations as detailed in Section 5.4.4 greatly facilitates the setup of a network simulation. All the necessary equations that govern the single heat exchanger model are wrapped in an object that can be easily instantiated through a graphic user interface and readily replicated in a flowsheet environment to generate the topology of a specific network.

The operation of the network is simulated via the simultaneous solution of equations for all exchangers. Each exchanger in the network is represented using the detailed dynamic model with suitably instantiated parameters for geometry, fouling, aging, etc. Simulating each unit at this level of detail allows gaining insights on how each heat exchanger individually contributes to the overall network performance.

5.4.6.1 Assessment of Network Performance: Key Performance Indicators

Variable flow conditions, temperatures, and crude slates as well as geometry of single HEXs and structure of the network affect different ways of the fouling behavior of a PHT, thus the energy recovered and, ultimately, the refinery bottom line. However, given the large number of variables involved, it may be difficult to isolate causes of inefficiencies and identify different variables contributing to the overall performance. The task of identifying dynamic interactions among units is particularly challenging when considering different structures for network retrofit. To assess in a consistent way the performance of a PHT network undergoing fouling and to be able to evaluate potential benefits in restructuring the network, the set of key performance indicators proposed by Coletti et al. (2011) is summarized in the following section with a modification introduced to also take into account possible hydraulic limitations of the heat exchanger network. In what follows, each network structure is denoted by Cj where j is the number of the configuration considered.

Given that the phenomena involved are intrinsically dynamic, some of the KPIs presented in the following sections (and summarized in Table 5.14) are referred to a reference time, t_{KPI}, and used to provide a snapshot of network conditions and history up to that point in time. Some of these indicators refer to clean conditions which can be easily calculated at time 0 if all the heat exchangers in the network can be assumed clean at that time and the inputs (flow rate, temperature, and pressure) are kept constant. However, if real plant data are to be used as input, it is not easy to determine a reference clean conditions performance. In this case model simulations can be exploited to provide reference clean conditions by setting $R_f = 0$ in each heat exchanger.

5.4.6.1.1 Fuel Costs

The decline in CIT with respect to clean conditions over time for any network j can be used as a measure of the overall thermal performance of the network:

$$\Delta CIT = CIT_{clean} - CIT \tag{5.84}$$

Equation (5.84) evaluated at $t = t_{KPI}$ provides KPI-01, the first key performance indicator considered here. The energy loss associated with the extra fuel burnt at the furnace to compensate the decrease in CIT, E_{Qloss}^{Cj}, is calculated as the integral over time of the difference between the total actual heat supplied by the furnace to the crude, $Q_{furnace}^{Cj}$, and the total heat duty under clean conditions, $Q_{furnace}^{Cj,clean}$:

$$E_{Qloss}^{Cj} = \int_0^t \left(Q_{furnace}^{Cj} - Q_{furnace}^{Cj,clean} \right) dt \tag{5.85}$$

Taking into account overall efficiency of the furnace, $\eta_{furnace}$, the cost of this extra energy requirement is a function of the cost of the fuel being burnt, P_{fuel}:

$$C_{FF}^{Cj} = \frac{E_{Qloss}^{Cj}}{\eta_{furnace}} \times P_{fuel} \tag{5.86}$$

C_{FF}^{Cj} evaluated at time $t = t_{KPI}$ provides KPI-02. To evaluate different retrofit options, it is useful to compare this cost for given network structure, configuration Cj, with that in a reference configuration, C1. To do this, first

Table 5.14 Summary of key performance indicators (KPI) for assessment of network performance

	KPI	Symbol	Description	Units	Equation
Thermal performance	01	ΔCIT	Decline in CIT with respect to clean conditions at time t_{KPI}	°C	(5.84)
	02	C_{FF}	Cumulative cost at t_{KPI} associated with extra fuel burnt in furnace	US$	(5.86)
	02r	Sv_{FF}	Cumulative cost at t_{KPI} associated with different energy recovery at the furnace vs. reference configuration	US$	(5.89)
Environmental impact	03	C_{CO_2}	Cumulative cost at t_{KPI} of extra CO_2 released because of fouling	US$	(5.91)
	03r	Sv_{CO_2}	Cumulative cost difference at t_{KPI} of extra CO_2 released because of fouling vs. reference configuration	US$	(5.92)
Production loss	04	t_{PL}	Time at which production loss starts	Days	(5.95)
	04a	t_{FFL}	Time at which the furnace firing limit is reached	Days	(5.93)
	04b	t_{NHL}	Time at which the hydraulic limit is reached	Days	(5.94)
	05	C_{PL}	Cumulative cost of production loss at t_{KPI}	US$	(5.97)
	05r	Sv_{PL}	Cumulative difference of cost of production loss vs. reference configuration at t_{KPI}	US$	(5.98)
Hydraulic performance	06	ΔCIP	Increase in CIP at t_{KPI}	Bar	(5.99)
	07	C_{PP}	Cumulative cost of energy losses at the pump due to fouling	US$	(5.101)
	07r	Sv_{PP}	Cumulative difference of energy losses at the pump due to fouling vs. reference configuration at t_{KPI}	US$	(5.102)
Overall economics	08	C_{tot}	Cumulative total costs due to fouling at t_{KPI}	US$	(5.103)
	08r	Sv_{tot}	Cumulative difference of total costs due to fouling at t_{KPI}	US$	(5.104)

it is necessary to calculate the total (cumulative) energy required for each configuration considered:

$$E_{FF}^{Cj} = \int_0^t \dot{M}^{Cj} c_p \left(COT - CIT_{Cj} \right) dt; \quad for\, j = 1, \ldots, N_c \tag{5.87}$$

where N_c denotes the number of retrofit configurations explored and CIT_j is the coil inlet temperature in each configuration j, and COT the fixed coil outlet temperature, required for distillation. The difference in performance (extra energy recovered at the furnace) of each retrofit provides the comparison between the different configurations considered:

$$\Delta E_{FF}^{C1-Cj} = E_{FF}^{C1} - E_{FF}^{Cj}; \quad for\, j = 1, \ldots, N_c \tag{5.88}$$

In economic terms, the furnace fuel savings of a particular configuration Cj with respect to $C1$, Sv_{FF}^{C1-Cj}, associated with ΔE_{FF}^{C1-Cj} are

$$Sv_{FF}^{C1-Cj} = \frac{\Delta E_{FF}^{C1-Cj}}{\eta_{furnace}} P_{fuel}; \quad for\, j = 1, \ldots, N_c \tag{5.89}$$

The cumulative fuel savings up to time $t = t_{KPI}$ provide the KPI-02r indicator for network retrofit.

5.4.6.1.2 Emission Costs

The combustion of extra fuel produces the release of greenhouse gases to the environment, which, under environmental laws (e.g., the Emissions Trading Scheme in Europe), adds economic penalties to the operations. In this study it is assumed that under clean conditions the refinery is just within its allocated allowance and that any extra ton of carbon dioxide caused by fouling, $M_{CO_2}^{Cj}$, must be paid for:

$$M_{CO_2}^{Cj} = \frac{E_{Qloss}^{Cj}}{\eta_{furnace}} \times m_{CO_2} \tag{5.90}$$

In Equation (5.90), m_{CO_2} denotes the carbon emission per Joule of energy produced in the combustion of a given fuel.

$$C_{CO_2}^{Cj} = M_{CO_2}^{Cj} \times P_{CO_2} \tag{5.91}$$

where P_{CO_2} is the price per ton of CO_2. $C_{CO_2}^{Cj}$ at $t = t_{KPI}$ provides KPI-03 for configuration Cj. If network retrofit is considered, the savings in CO_2

emissions for each configuration Cj with respect to the reference configuration C1 are calculated from the fuel energy savings in Equation (5.92):

$$Sv_{CO_2}^{C1-Cj} = \Delta E_{FF}^{C1-Cj} \times \frac{m_{CO_2}}{\eta_{furnace}} \times P_{CO2};\ for\, j = 2, \ldots, N_c \tag{5.92}$$

The cumulative CO_2 savings up to time $t = t_{KPI}$ provide the KPI-03r for network retrofit.

5.4.6.1.3 Production Loss

As noted in Chapter 1, the reduction in thermal and hydraulic efficiency caused by fouling is paid not only at the furnace or the pump as extra energy but also as loss of production. When the furnace hits its firing limit, the throughput must be reduced causing a cost in loss of production. The time when this happens, t_{FFL}, provides a key performance indicator KPI-04a:

$$t_{FFL} = t|_{Q=Q^{max}} \tag{5.93}$$

If the pressure drop across the preheat train reaches a maximum value beyond which any increase cannot be compensated for by the hydraulic systems, the hydraulic limit of the network is reached. The time at which this happens is KPI-04b

$$t_{NHL} = t|_{\Delta P=\Delta P^{max}} \tag{5.94}$$

Whether the thermal or the hydraulic limit is reached, the result is that production needs to be reduced:

$$t_{PL} = min(t_{FFL}, t_{NHL}) \tag{5.95}$$

The production loss, M_{PL}, associated with the reduction in throughput due to either the thermal or the hydraulic limit is

$$M_{PL}^{Cj} = \int_0^t \left(\dot{M}^{Cj,clean} - \dot{M}^{Cj}\right) dt \tag{5.96}$$

where $\dot{M}_{clean}$ is the mass flow rate under clean conditions and $\dot{M}$ is the actual throughput. The cost associated with the production loss, C_{PL}^{Cj}, is then calculated as

$$C_{PL}^{Cj} = \frac{M_{PL}^{Cj}}{0.1569\rho} \times P_{bbl} \tag{5.97}$$

where P_{bbl} is the operating margin per barrel of crude, ρ the crude density, and 0.1569 is a conversion factor. C_{PL}^{Cj} as calculated in Equation. (5.97) at time $t = t_{KPI}$ provides KPI-05.

To assess network retrofit performance with respect to production loss, the difference (excess) in production achieved by each configuration Cj with respect to the reference configuration C1 is calculated from the difference in mass flow rate over time. The associated production loss savings are

$$Sv_{PL}^{C1-Cj} = C_{PL}^{C1} - C_{PL}^{Cj}; \ for\, j = 2, \ldots, N_c \tag{5.98}$$

Sv_{PL}^{C1-Cj} evaluated at $t = t_{KPI}$ denotes KPI-05r.

5.4.6.1.4 Assessment of Hydraulic Efficiency

The reduction in cross-sectional area inside the heat exchanger tubes produces an increase in pressure drops that must be countered by increasing the energy supplied to the pump to maintain the largest throughput achievable within the furnace firing limit constraint. Pressure drops, shear stress, and velocity within individual heat exchangers can be used to evaluate the hydraulic impact of fouling. However, the hydraulic performance of the whole network can be assessed via the increase in coil inlet pressure (CIP) over time with respect to clean conditions:

$$\Delta CIP = CIP_{clean} - CIP \tag{5.99}$$

The value of ΔCIP at $t = t_{KPI}$ provides KPI-06. The cost of energy loss associated with the increase in pumping power required to counter the hydraulic effects of fouling is another useful indicator of the network performance. The integral over time of the difference between pumping power under clean conditions, W_{clean}, and the actual pumping power (i.e., under fouled conditions), W, gives the energy losses at the pump due to fouling, E_{PP}^{Cj}:

$$E_{PP}^{Cj} = \int_0^t \left(W^{Cj} - W^{Cj,clean} \right) dt \tag{5.100}$$

This translates in extra electricity costs due to fouling-related extra pumping power requirements as

$$C_{PP}^{Cj} = \frac{E_{PP}^{Cj}}{\eta_{pump}} \times P_{elec} \tag{5.101}$$

where P_{elec} denotes the price of electricity, η_{pump} denotes the pump efficiency. C_{PP}^{Cj} evaluated at $t = t_{KPI}$ provides KPI-07. In the case of a retrofit, the CIP in a given configuration C1 is compared to that in a reference configuration C1. The savings in pumping power can be estimated from the total (cumulative) pumping power required for each configuration Cj, and the price of electricity:

$$Sv_{PP}^{C1-Cj} = \left(E_{PP}^{C1} - E_{PP}^{Cj}\right) \times \frac{P_{elec}}{\eta_{pump}}; \ for\ j = 2, \ldots, N_c \tag{5.102}$$

which, calculated at $t = t_{KPI}$, indicates KPI-07r for network retrofit.

5.4.6.1.5 Overall Fouling Costs

The cumulative cost of fouling, C_{tot}, evaluated at $t = t_{KPI}$ denotes KPI-08 which comprehensively summarizes the economic impact of fouling in the CDU.

Starting at time 0 from clean conditions for the PHT, the extra costs due to fouling for a given network Cj are evaluated as:

$$C_{tot} = C_{FF} + C_{CO_2} + C_{PL} + C_{PP} \tag{5.103}$$

where C_{FF} is the cost of the additional fuel that must be burnt in the furnace to counter the decline over time in the oil temperature at its inlet, the coil inlet temperature (CIT), C_{PL} is the cost associated with the reduction in throughput, C_{CO_2} is the costs associated with the extra emission of CO_2 due to fouling, and C_{PP} is the electricity cost due to increase in pumping power required to maintain a constant throughput.

For network retrofit, KPI-08r is the sum of KPI-02r, KPI-03r, KPI-05r, and KPI-07r:

$$Sv_{tot}^{C1-Cj} = C_{tot}^{C1} - C_{tot}^{Cj} \tag{5.104}$$

Cleaning and shutdown costs are not considered here. The assessment of the costs is therefore based on no action having been taken to clean any unit. The following sections detail the calculations for the each of the terms in Equation (5.103).

5.4.6.2 *Case Study on Entire Preheat Train*

Coletti et al. (2011) showed that using high-fidelity thermo-hydraulic simulations to capture complex dynamic interactions in the network provides a more useful way to analyze alternative PHT retrofit options than currently available. For this purpose, a network was modeled based on an industrial case study of a small refinery (ca. 20,000 bbl day^{-1}). In order to protect proprietary information, the heat and mass balance for the existing network comprising seven exchangers were adjusted and some features of the plant were changed without affecting the validity of the conclusions. This network structure represented the base case for the study and was referred to as configuration C1 (Figure 5.49). The performance of the preheat train was monitored at startup (after cleaning) and after 8000 h of operation. While units E-01, E-02, and E-04 did not change significantly their performance over the operating period, the other four units in the network exhibited severe fouling. The observations of the refinery operators were that fouling in units E-05 and E-06 occurred mainly on the tube-side (were crude oil was allocated) and this was the only side of these units which was Periodically cleaned. Unit E-03 fouled on the shell-side (which handled column residues) but not on the tube-side. Unit E-07 fouled heavily on both sides.

For the analysis it was assumed that:

1. Crude was on the tube-side and hot fluids flow on the shell-side in all units.
2. The performance of the desalter did not affect the fouling behavior of the network and its temperature is optimally controlled (i.e., the temperature at the outlet of the desalter was constant).
3. The furnace firing limit was never reached (i.e., $Q_{furnace} \ll Q_{furnace}^{max}$).
4. The hydraulic limit of the network was never reached.

Parameters that characterize the fouling behavior in each unit were estimated to fit the performance of the base-case configuration C1. Once the fouling behavior was captured, the same values were used to assess alternative network retrofit configurations, with the goal of increasing overall energy recovery. By analyzing the stream temperatures reported in Coletti et al. (2011), it was decided to study the effect of adding an extra heat exchange area to the network to recover more energy from the residue stream (highlighted in red in Figure 5.49). This can be done, for example, by adding an extra unit E-07x, identical to E-07 in geometry. The question then becomes where the extra area should be added to maximize energy recovery while taking into account fouling dynamics. Three alternative network structures were considered by Coletti et al. (2011). The first broadly followed pinch rules to maximize energy recovery based on clean exchanger performance and the remaining two sought to improve the energy recovery

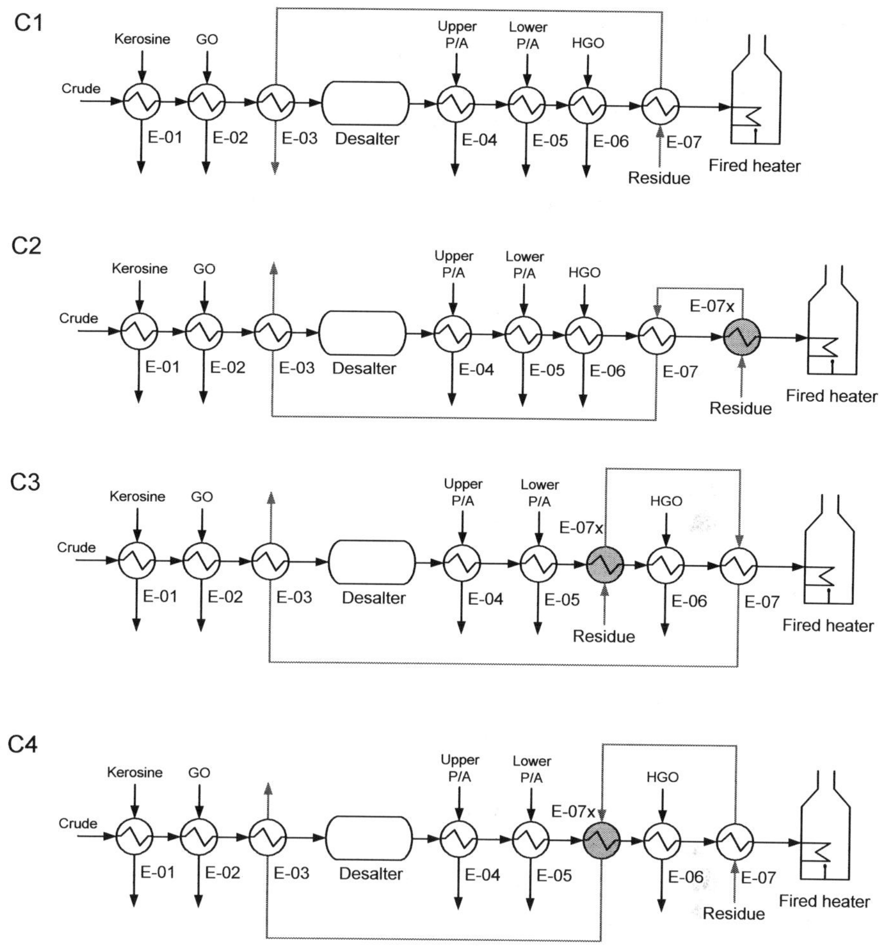

Figure 5.49 HEN2 network configurations considered. *Adapted from Coletti et al. (2011).*

while also taking into account the fouling behavior. In summary, the following alternative configurations were studied (Figure 5.49):

- Configuration C2, shown in. In this configuration, pinch rules for maximum energy recovery were applied. As a result, E-07x was added between the E-07 and the furnace, at the hottest position in the network. The hot residue stream is matched with the crude at its highest temperature. Although this ensures maximum heat recovery under clean conditions, over time fouling is expected to penalize the overall heat recovery of the network as the wall temperatures, on which fouling depends, are also maximized.

- Configuration C3 in. This configuration goes against traditional pinch rules by matching the hot residue stream with the crude at an intermediate temperature (at the exit of E-05). The residue stream enters the extra unit E-07x, placed between E-05 and E-06 and subsequently enters E-07 and E-03. As a result, heat recovery under clean conditions is expected to be less than that achievable with configuration C3; however, the crude at the highest temperature exchanges heat with a lower temperature residue stream.
- Configuration C4 in. This considers the residue entering first unit E-07 as in the base-case structure and then unit E-07x, which is placed between E-05 and E-06 as in structure C3.

Note that the design of the extra heat exchanger was not optimized for each configuration to provide a more direct comparison of the effect of adding the same area in different positions of the network. The performance given by the different designs is assessed below by the use of the KPI illustrated in the previous section.

By inspecting the values of the CIT for the base case and the three retrofit options (Figure 5.50) it can be seen that for configurations C2 and C4 the CIT under clean conditions (i.e., $t = 0$) is over 13 °C higher than that achieved in C1, indicating a good extra energy recovery. Conversely, the initial CIT in configuration C3 is 3.5 °C lower than the base-case C1, despite the extra heat transfer area available. This can be explained with the fall in performance of the existing units E6 and E7 caused by the reduced temperature driving force. Over time, however, things change significantly because of fouling. Fouling rates in E-07 are highest for configurations C3 and C4, and lowest for C1 and C2. Unit E-07x fouls more in C2 than in C3 and C4. As a result, after less than a month of operations, the CIT in C3 is maintained at a higher value compared to that of the base-case C1. After 150 days C3 starts recovering more energy than C2. After ca. 300 days, the CIT in the retrofit structure C4 also falls below that of C3. The structure generated according to pinch rules, C2, results in the worst performance over a long time.

Figure 5.51(a) shows that the drop in CIT over time from clean conditions is drastic for C1 and C2 whereas C4 performs better. Configuration C3 in particular is the one that suffers the least from the decrease in CIT ($\Delta\text{CIT}|_{\text{C1};t_{KPI=600\text{d}}} = -20.8°\text{C}$).

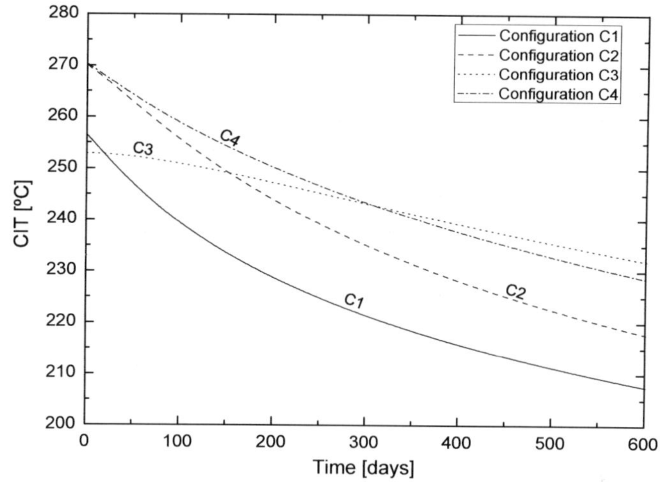

Figure 5.50 Coil inlet temperature (CIT) for the base-case and the three alternative network configurations. *Reproduced with permission from Coletti et al., Effects of fouling on performance of retrofitted heat exchanger networks; a thermo-hydraulic based analysis, 2011. Comput. Chem. Eng. 35 (5), 907–917.*

The networks perform very much differently also from a hydraulic point of view. Figure 5.51(b) shows the increase in pressure drops with respect to clean conditions for each configuration considered (ΔCIP, calculated with Equation (5.99)). Since the existing network C1 has one unit less (E-07x), it has a lower total pressure drop and a lower ΔCIP compared to the others ($\Delta\text{CIP}|_{\text{C1};t_{KPI=600\text{d}}} = 5.4$ bar). Among the three proposed retrofits, C4 performs the worst, with $\Delta\text{CIP}|_{\text{C4};t_{KPI=600\text{d}}} = 22.3$ bar.

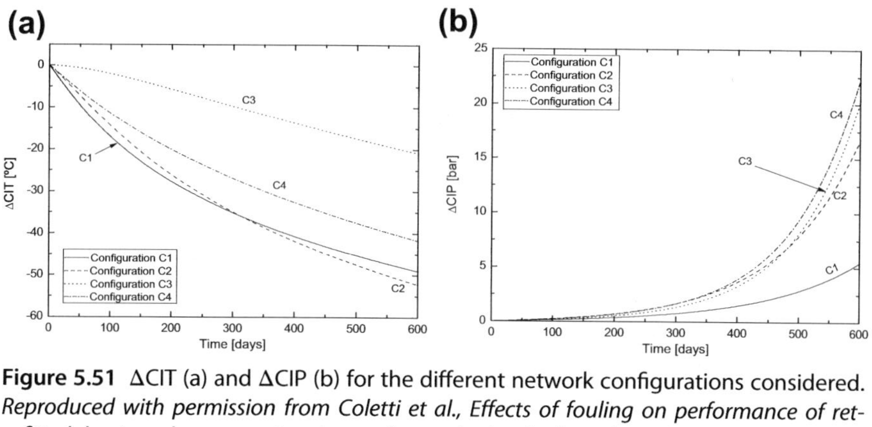

Figure 5.51 ΔCIT (a) and ΔCIP (b) for the different network configurations considered. *Reproduced with permission from Coletti et al., Effects of fouling on performance of retrofitted heat exchanger networks; a thermo-hydraulic based analysis, 2011. Comput. Chem. Eng. 35 (5), 907–917.*

Although the analysis of the CIT and CIP highlights the importance of considering fouling dynamics in the retrofit of PHT networks, it is not sufficient to assess which structure, among those proposed, provides the largest overall amount of energy recovered over time. Figure 5.52 shows the (cumulative) amount of extra energy recovered by the network through one of the three proposed retrofits, as variation in the energy required in the furnace, with respect to the base case. Configuration C3 has a negative impact on the overall energy recovered by the network under clean conditions as confirmed by the fact that ΔE_{FF}^{C1-C3} is negative for ca. the first 50 days of operations (Figure 5.52). However, after roughly a year of operations configuration C3 starts performing better than C2, the configuration proposed following pinch rules. By analyzing Figure 5.52 another important aspect can be unveiled. While the CIT in configuration C3 becomes larger than that in C4 after ca. 300 days, the cumulative extra energy recovered by the latter is constantly larger than that recovered by the former. One of the key observations enabled by the use of the model is that the position of the extra unit and its fouling behavior are paramount for the overall performance of the network. In configuration C4, the fouling resistance in E-07 increases with respect to the base case (the tube-side temperatures are higher) but it results in the lowest fouling resistance in E-07x. As a result,

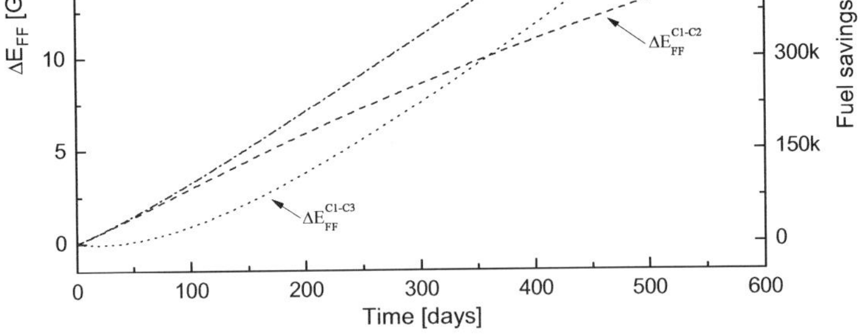

Figure 5.52 Extra energy recovered by each configuration proposed with respect to the base case. The right axis shows the equivalent savings in terms of fuel burnt in the furnace, Sv_{FF}^{C1-Cj}. *Reproduced with permission from Coletti et al., Effects of fouling on performance of retrofitted heat exchanger networks; a thermo-hydraulic based analysis, 2011. Comput. Chem. Eng. 35 (5), 907–917.*

Table 5.15 KPI summary at $t_{KPI} = 600$ days for the three proposed retrofits

#	KPI	Description	Units	C1	C2	C3	C4
01	ΔCIT	CIT drop	°C	−49.0	−52.2	−20.9	−41.7
02r	Sv_{FF}	Fuel savings	US$	–	451,000	616,000	961,000
03r	Sv_{CO_2}	CO_2 savings	US$	–	110,000	150,000	168,000
04	t_{PL}	Production loss start	Days	∞	∞	∞	∞
05	C_{PL}	Production loss	US$	0	0	0	0
05r	Sv_{PL}	Production loss savings	US$	0	0	0	0
06	ΔCIP	CIP increase	Bar	5.4	16.7	20.2	22.3
07r	Sv_{PP}	Pumping savings	US$	<1000	<1000	<1000	<1000
08r	Sv_{tot}	Total costs	US$	–	561,000	766,000	859,000

Adapted from Coletti et al. (2011).

configuration C4 is capable of saving over 22.5 GWh after 600 days in furnace fuel alone ($\Delta E_{FF}^{C1-C4}\big|_{t_{KPI=600d}} = 22.7$ GWh). In economic terms, savings in fuel costs alone are shown on the right axis of Figure 5.52.

Table 5.15 reports a summary of the KPIs used to comprehensively assess the fouling behavior of the different configurations at $t_{KPI=600d}$. With respect to CIT drop, network C2 is the worst performer ($\Delta CIT|_{C2;t_{KPI=600d}} = -52$ °C). KPI-02r, Sv_{FF}, estimated for the different configurations shows that of over US$450,000 in fuel savings can be achieved by using configuration C2 while c. US$600,000 using C3 and ca. US$680,000 using configuration C4. The extra economic benefit of C4 versus C2 is ca. 50%. While after 300 days of operation, the savings at the furnace ranks options are $Sv_{FF}^{C1-C4} > Sv_{FF}^{C1-C2} > Sv_{FF}^{C1-C3}$, after 600 days the ranking becomes $Sv_{FF}^{C1-C4} > Sv_{FF}^{C1-C3} > Sv_{FF}^{C1-C2}$. In fuel costs alone, these differences translate in ca. US$230,000 over 600 days between a network structures designed accordingly to pinch rules and one that takes into account fouling. This is considered large for the size of the refinery considered (20,000 bbl day^{-1}). It should be noted that in absolute energy and monetary terms, benefits are expected to be much higher for larger refineries. Moreover, other factors not included in this case study, such as reduction in throughput, could play a crucial role in the choice of the arrangement.

5.4.7 Concluding Remarks

At the beginning of Chapter 5, the vison for a multiscale model of refinery heat exchangers undergoing crude oil fouling was illustrated. This envisaged the integration of thermodynamic and molecular models with fluid-dynamic and industrial-scale models to enable accurate and detailed predictions of fouling deposition from crude oils. The following Section 5.2 and Section 5.3 discussed two of the building blocks of this vision (namely thermodynamic and fundamental transport modeling). Although significant progress has been made in these areas, more development is required to enable the overall vison to be achieved and to make use at industrial scale of the predictive features of such an approach.

This final section of the chapter presented a modeling approach which is in itself a multiscale one. Although it incorporates less descriptive submodels for thermodynamics and fluid dynamics, the model is able to capture some of the complex phenomena described in Chapter 2, such as deposit aging, to a level of detail and accuracy that enables immediate and practical use in industrial settings. The parameters for such submodels may be obtained directly from macro-level experiments (such as those described in Chapter 3) or refinery data.

The distinctive features of the model allow to calculate the thermal resistance and thickness of the fouling layer along the tube as well as its interactions with heat transfer and fluid flow. The distributed nature of the model allows to identify critical zones where deposition is particularly severe and proposing geometries that mitigate fouling. Such features together with the implementation in a commercial process simulator make the model, commercialized by Hexxcell Ltd., a user-friendly and powerful tool.

The excellent agreement (less than 2% error) between model predictions and primary plant measurements (i.e., temperatures) from different refineries over extended Periods of time (i.e., up to 16 months) gives the necessary confidence that the model is capable of representing the performance of individual heat exchangers and entire networks in a reliable way. As a result, the model can be used with confidence to precisely quantify energy losses caused by fouling and propose retrofit options to minimize them at both individual heat exchanger and network levels.

CHAPTER SIX

Concluding Remarks

F. Coletti[1], G.F. Hewitt[2]
[1]Chief Technology Officer, Hexxcell Ltd, UK
[2]Professor of Chemical Engineering, Imperial College London, UK

As seen from Chapters 1–5, crude oil fouling is a highly complex process, reflecting the complex nature of crude oil itself. Crude oil composition and behavior change from source to source, reflecting the wide variety of prehistoric events that led to the original formation of the crude oil. However, the supplier of products arising from crude oil wishes to make these products relatively independent of the source and the *crude oil distillation* process is an important tool in achieving product consistency. As demonstrated, fouling of heat exchangers in the preheat train leads to a significant increase in the cost of this distillation process.

There are many types of fouling (see Chapter 2), but it seems likely that the dominant mechanism in crude oil fouling is *chemical reaction* based. The deposits are subject to important *aging* processes that change the physical nature of the deposit with time and it is important to recognize these processes within any meaningful model of the fouling process.

The economic importance of crude oil fouling and its formidable complexity make it absolutely essential to continue research on this topic. Thus, in one sense, the work presented in this book is a kind of "progress report" on this research; it will be expected to continue in the future. However, what was new in the CROF project mentioned in Chapter 1, and the associated studies which led to most of the work reported here, was the *multidisciplinary* approach in which a wide variety of skills were brought together to attack the problem. The themes of this work can be briefly sated as follows:

- **Experiments on fouling.** Several new initiatives were taken in gathering experimental data on fouling. These ranged from the use of microbomb reactors (Chapter 3, Section 1), the development of a rotating cell device to study fouling under controlled conditions (Chapter 3, Section 2), the development of a novel method for fouling layer thickness measurement (Chapter 4, Section 4), and the construction of a major new facility for studying fouling under realistic conditions (Chapter 3, Section 3).

Crude Oil Fouling
ISBN: 9780128012567
http://dx.doi.org/10.1016/B978-0-12-801256-7.00006-3

- **Characterization of fouling deposits.** Here, methods were developed that allow determination of the chemical structure and molecular weight of deposits (Chapter 4, Section 1 and Section 2) and also allow the location of various species on the surface of a deposit (*chemical imaging*, Chapter 4, Section 3).
- **Modeling.** Here, the modeling efforts ranged from the *molecular scale* in which the determination of key parameters such as diffusion coefficients and fluid physical properties was facilitated and the development of new equations of state that accounted for the presence of large molecules such as asphaltenes (Chapter 5, Section 2), the modeling of the fundamental processes of foulant deposition and aging (Chapter 5, Section 3), and the modeling of and prediction of data from operating plants (Chapter 5, Section 4).

This multidisciplinary effort has already brought several benefits and provided useful insights in this complex field of study. However, to give heat transfer equipment engineers and refinery operators the possibility of optimizing their design and operations, it is important that the work continues. Key areas of focus for this ongoing work are as follows:

- The generation of data on fouling under conditions closely matching those encountered in industry in apparatus such as the HIPOR facility described in Chapter 3. Testing crude oils under controlled conditions allows the determination of the effects that individual variables have on the fouling behavior. The use of such equipment is essential not only for determining the fundamental parameters governing fouling but also for testing possible fouling mitigation strategies and technologies (e.g., coatings, chemicals, tube inserts).
- The further development of reliable analytical techniques described in Chapter 4 and protocols that allow the characterization of oils and samples of deposits from both experimental testing facilities and refinery heat exchangers. Understanding what the species deposited are, especially if this can be done with in situ heating. This will help to shed light on the underlying mechanisms dominating the fouling process.
- Achieving the vision of a truly predictive multiscale model for fouling in heat exchangers that does not rely on past data to simulate fouling behavior as a function of crude oil composition, process conditions, and equipment design. This means both sharpening the existing tools, presented in Chapter 5, that describe the different scales of investigation and integrating them in a coherent framework.

As a part of this multidisciplinary approach, the close collaboration between academia and industry is a pivotal activity. Industrial experience and practical needs should drive the overall direction of the fundamental research. Data and deposits from the plant should be carefully analyzed to provide actionable information for operating personnel and heat exchanger designers that enable them to manage, mitigate, and, ideally, eliminate fouling.

It is believed that the multidisciplinary approach described in this book represents a paradigm shift in the approach to crude oil fouling studies, and that it can (and should) be extended to other systems (e.g., pipelines and upstream equipment). The problem is an ongoing one and one may expect that research will continue!

APPENDIX 1

Table A.1 Summary of analytical techniques used in fouling studies

Short name	Name	Application	Information type	Sample state	Advantages	Disadvantages/ Limitations	Sample size	Resolution	Experiment duration
UV–F	Ultraviolet–fluorescence spectroscopy	Determine aromaticity of sample	Chemical structure	Solid	Fast and easy to use	Qualitative information	2 ml	1 nm	<1 min
FT–IR	Fourier transform–infrared spectroscopy	Chemical bond analysis, determining functional groups	Chemical structure	Liquid, solid	Fast and easy to use	Cannot analyse pure metals	<0.1 ml, <10 mg	$\pm 0.01\ cm^{-1}$	<1 min
NMR	Nuclear magnetic resonance spectroscopy	Determine aromaticity of sample, C/H ratio, functional groups, aromatic/aliphatic ratio	Chemical structure	Liquid, solid	High resolution	Requires high solubility in the solvent, longer time	0.7 ml	±0.01 ppm	5min–1 h
GC/MS	Gas chromatography/ Mass spectrometry	Determine molecular weight of crude oil volatile compounds with boiling point <500 °C	Molecular weight	Liquid	Molecular weight analysis of highly complex mixtures of volatile compounds	A considerable proportion of hydrocarbons cannot be observed (400 u for aromatics, 500–550 u for aliphatic)	<1 ml	GC: ±1 s MS: $\pm 0.01\ mz^{-1}$	1–2 h
SEC	Size exclusion chromatography	Molecular weight distribution Characterization of high-mass, polydisperse Hydrocarbon mixtures	Molecular weight	Liquid	Higher MM ceiling respect to the GC/MS	Can only detect just the soluble part of the sample	<1 ml	1 s	1–2 h
MALDI-MS	Matrix assisted laser desorption/ Ionization mass spectrometry	Determine molecular weight	Molecular weight	Solid	Soft mass analysis technique for large and sensitive materials. Higher mass ceiling.	Desorbed asphaltene degradation may still a problem with MALDI-MS		$\pm 0.01\ mz^{-1}$	30 min
TGA	Thermogravimetric analysis	Ash content, fixed carbon, % volatile lost	Reactivity	Liquid, solid	Analysis of thermophysical properties of samples and allows probing into thermodegradation reactions at high temperature range in excess of 1000 °C	Long analysis time ~2–5 h	<5 mg	Mass: ±0.0001 g Temp: ±0.01 °C	2–5 h

ST	Solubility test	Determine what solvant dissolves solid samples; oil incompatibility	Reactivity	Liquid, solid	Quick and straight forward	Subjective and imprecise; depends on physical conditions	>10 ml	–	~10 min
EA	Elemental analysis	Determine C, H, N, O, S content	Elemental composition	Liquid, solid	Well established; C, H, N, O difficult to determine by other means	Can be affected by other elements e.g., F	>100 mg	±0.01%	Samples sent for analysis
XRF	X-ray fluorescence	Determine Fe, S, Si, Na, Ca, Al, metals	Elemental composition	Liquid, solid	Fast and easy to use	Large sample size required	5–10 ml, >3 g	1 ppm	30 min
ICP	Inductively coupled plasma	Determine Fe, S, Si, Na, Ca, Al, metals	Elemental composition	Liquid, solid, gas	accurate for trace metal analysis	Long preparation required for samples in solution	~100 mg	1 ppm	1 h
OM	Optical microscopy	Morphology analysis	Morphology	Solid	Shows whether the foulant is isotropic or not, which suggests the main reason of coke formation	Long sample preparation	–	1 μm	N/A
SEM-EDX	Scanning electron microscopy	Topology/Morphology analysis	Morphology	Solid	Identification of elements and their association with each other	Long sample preparation	–	1 nm	N/A
XRD	X-ray diffraction	Phase identification, crystallinity	Morphology	Solid	Detection of metal abundance, dispersion, and oxidation state	No spatial information	~100 mg	N/A	~30 min
XPS	X-ray photoelectron spectroscopy	Composition analysis	Morphology	Solid, liquid	Elemental composition of the surface	–	~100 mg	N/A	N/A
–	Spectrometer in a transmission configuration	Possibility of taking chemical maps of materials and systems at the microscopic level	Chemical structure	Liquid and solid					
Macro-ATR	Spectrometer combined with macro ATR Accessories and focal plane array infrared detector	Large fields of view for chemical imaging and characterization; studies dynamic systems in situ, studies precipitation and deposition in situ, in situ experiments at high temperatures and pressures; depth profile measurements.	Chemical structure	Gases, liquids, and solids can be studied (including supercritical fluids)	Versatility and chemical specificity	–		10–15 μm	seconds

(Continued)

Table A.1 Summary of analytical techniques used in fouling studies—cont'd

Short name	Name	Application	Information type	Sample state	Advantages	Disadvantages/ Limitations	Sample size	Resolution	Experiment duration
ATR-FTIR	FTIR spectroscopic imaging system combined with microscope, for use in Micro-ATR imaging studies	Chemical imaging of dynamic processes	Chemical structure	Solid deposits, crude oil	Chemical specificity and high spatial resolution	Good contact with ATR crystal is required		1–4 μm	seconds to minutes
ATR-FTIR	FTIR spectrometer combined with an infared microscope and a macro chamber	Chemical imaging of dynamic processes	Chemical structure	Solid deposits, crude oil	Versatility and chemical specificity	Good contact with ATR crystal is required		1–4 μm or 200 nm	seconds to minutes
Raman	Raman spectrometers	Chemical analysis of samples, including depth profile of microscopic samples, high spatial resolution	Chemical structure	Solid and liquid	Noncontact	Thin layers studied			minutes
FT-Raman	FT-Raman spectroscopic system	Chemical characterization of challenging carbonaceous samples and deposits	Chemical structure	Solid	Noncontact	–			
TERS	Tip-enhanced Raman system	Nanoscale imaging of samples as it combines the high spatial resolution of atomic force microscopy with the chemical information of Raman spectroscopy	Chemical structure	Solid	Provides unique nanoscale insight a very high spatial resolution	A very challenging and expensive experiment		10 nm	minutes to hours

REFERENCES

Aguilera-Mercado, B., Herdes, C., Murgich, J., Müller, E.A., 2006. Mesoscopic simulation of aggregation of asphaltene and resin molecules in crude oils. Energy Fuels 20 (1), 327–338.

Ahmed, T., 1989. Hydrocarbon Phase Behaviour. Gulf Publishing Company, Houston.

Akınç, H.E., Becer, M., Bakır, M., Özçelik, Y., Balkan, F., 2013. Fouling monitoring in crude oil pre-heat trains. In: Malayeri, M.R., Müller-Steinhagen, H., Watkinson, A.P. (Eds.), Proc. Int. Conf. On Heat Exchanger Fouling and Cleaning X, June 9–14, 2013, Budapest, Hungary, pp. 18–21. Published online. www.heatexchanger-fouling.com.

Akrami, H.A., Yardim, M.F., Akar, A., Ekinci, E., 1997. FTIR characterization of pitches derived from Avgamasya asphaltite and Raman-Dincer heavy crude. Fuel 76 (14–15), 1389–1394.

Al-Atar, E., Watkinson, A.P., 2001. Effects of resins/asphaltenes ratio on heat exchanger fouling of mixtures containing heavy oil. In: Bott, T.R., Watkinson, A.P., Panchal, C.B. (Eds.), Proc. Int. Conf. Mitigation of Heat Exchanger Fouling and its Economic and Environmental Implications, July 18–23, 1999, Banff, Canada. Begell House, New York, pp. 359–366.

Al-Bagawi, J.J., Said, S., 2001. Mitigation of heat exchanger fouling in the oil industry. In: Bott, T.R., Watkinson, A.P., Panchal, C.B. (Eds.), Proc. Int. Conf. Mitigation of Heat Exchanger Fouling and its Economic and Environmental Implications, July 18–23, 1999, Banff, Canada. Begell House, New York, pp. 193–197.

Albert, F., Augustin, W., Scholl, S., 2009. Enhancement of heat transfer in crystallization fouling due to surface roughness. In: Müller-Steinhagen, H., Malayeri, M.R., Watkinson, A.P. (Eds.), Int. Conf. On Heat Exchanger Fouling and Cleaning VIII, June 14–19, 2009, Schladming, Austria, pp. 303–310. Published online. www.heatexchanger-fouling.com.

Alboudwarej, H., Akbarzadeh, K., Beck, J., Svrewk, W.Y., Yarranton, H.W., 2003. Regular solution model for asphaltene precipitation from bitumens and solvents. AIChE J. 49 (11), 2948–2956.

Ali, A., Chapman, G.J., Chew, Y.M.J., Gu, T., Paterson, W.R., Wilson, D.I., 2013. A fluid dynamic gauging device for measuring fouling deposit thickness in opaque liquids at higher temperature and pressure. Exp. Therm. Fluid Sci. 48, 19–28.

Altgelt, K.H., Boduszynski, M.M., 1994. Compositional Analysis of Heavy Petroleum Fractions. Marcel Dekker Inc., New York.

Álvarez, P., Granda, M., Sutil, J., Menendez, R., Fernández, J.J., Viña, J.A., Morgan, T.J., Millan, M., Herod, A.A., Kandiyoti, R., 2008. Characterization and pyrolysis behavior of novel anthracene oil derivatives. Energy Fuels 22 (6), 4077–4086.

Álvarez, P., Menendez, J.L., Berrueco, C., Rostani, K., Millan, M., 2012. Determination of crude oil incompatibility regions by ellipsometry. Fuel Process. Technol. 96, 16–21.

American Petroleum Institute, 1992. Technical Data Book-Petroleum Refining, Eleventh ed. API, New York.

Ancheyta, J., Trejo, F., Rana, M.S., 2010. Asphaltenes: Chemical Transformation during Hydroprocessing of Heavy Oils. CRC Press.

Andersson, E., Quah, J., Polley, G.T., 2009. Experience in application of compabloc heat exchangers in refinery pre–heat trains. In: Müller-Steinhagen, H., Malayeri, M.R., Watkinson, A.P. (Eds.), Int. Conf. On Heat Exchanger Fouling and Cleaning VIII, June 14–19, 2009, Schladming, Austria, pp. 39–43. Published online. www.heatexchanger-fouling.com.

ANSYS® Academic Research, 2013. Release 14.5, Help System, ANSYS FLUENT Manual. ANSYS, Inc.

Aquino-Olivos, M.A., Andersen, S.I., Lira-Galeana, C., 2003. Comparisons between asphaltenes from the dead and live-oil samples of the same crude oils. Petroleum Sci. Technol. 21 (5–6), 1017–1041.

Artola, P.-A., Pereira, F.E., Adjiman, C.S., Galindo, A., Müller, E.A., Jackson, G., Haslam, A.J., 2011. Understanding the fluid phase behaviour of crude oil: asphaltene precipitation. Fluid Phase Equilibr. 306 (1), 129–136.

Asomaning, S., Watkinson, A.P., 1992. Heat exchanger fouling by Olefin-Kerosene mixtures. Can. J. Chem. Eng. 70 (3), 444–451.

Asomaning, S., Watkinson, A.P., 1997. Deposit formation by asphaltene–rich heavy oil mixtures on heat transfer surfaces. In: Bott, T.R., Melo, L.F., Panchal, C.B., Somerscales, E.F.C. (Eds.), Proc. Int. Conf. On Understanding Heat Exchanger Fouling and its Mitigation, May 1997, Castelvecchio Pascoli, Italy. Begell House, New York, pp. 283–290.

Asomaning, S., Watkinson, A.P., 2000. Petroleum stability and heteroatom species effects in fouling of heat exchangers by asphaltenes. Heat. Transf. Eng. 21 (3), 10–16.

Asomaning, S., Panchal, C.B., Liao, C.F., 2000. Correlating field and laboratory data for crude oil fouling. Heat Transf. Eng. 21 (3), 17–23.

Atkins, G.T., 1962. What to do about high coking rates. Petro/Chem. Eng. 34 (4), 20–25.

ASTM, 2007. Standard Test Method for N-heptane Insolubles. Standard D3279. ASTM International, West Conshohocken, PA.

ASTM, 2009. Standard Test Method for Determination of Total, Combustible and Carbonate Carbon in Solid Residues from Coal and Coke. Standard D6316. ASTM International, West Conshohocken, PA.

ASTM, 2012. Standard Test Method for Forms of Sulfur in Coal. Standard D2492. ASTM International, West Conshohocken, PA.

ASTM, 2014a. Standard Test Methods for Determination of Carbon, Hydrogen and Nitrogen in Analysis Samples of Coal and Carbon in Analysis Samples of Coal and Coke. Standard D5373. ASTM International, West Conshohocken, PA.

ASTM, 2014b. Standard Test Method for Sulfur in the Analysis Sample of Coal and Coke Using High-temperature Tube Furnace Combustion. Standard D4239. ASTM International, West Conshohocken, PA.

Avaulle, L., Neau, E., Jaubert, J.-N., 1997. Thermodynamic modeling for petroleum fluids II. Prediction of PVT properties of oils and gases by fitting one or two parameters to the saturation pressures of reservoir fluids. Fluid Phase Equilibr. 139 (1–2), 171–203.

Avendaño, C., Lafitte, T., Galindo, A., Adjiman, C.S., Jackson, G., Müller, E.A., 2011. SAFT-γ force field for the simulation of molecular fluids: 1. A single-site coarse grained model of carbon dioxide. J. Phys. Chem. B 115 (38), 11154–11169.

Avendaño, C., Lafitte, T., Adjiman, C.S., Galindo, A., Müller, E.A., Jackson, G., 2013. SAFT-γ force field for the simulation of molecular fluids: 2. coarse-grained models of greenhouse gases, refrigerants and long alkanes. J. Phys. Chem. B 117 (9), 2717–2733.

Bagajewicz, M., Soto, J., 2000. Rigorous procedure for the design of conventional atmospheric crude fractionation units. Part II: heat exchanger network. Ind. Eng. Chem. Res. 40 (2), 627–634.

Bagajewicz, M., 2000. Process Plant Instrumentation: Design and Upgrade. CRC Press.

Bai, Z.S., Wang, H.L., 2007. Crude oil desalting using hydrocyclones. Chem. Eng. Res. Des. 85 (A12), 1586–1590.

Bansal, B., Chen, X.D., 2006. A critical review of milk fouling in heat exchangers. Compre. Rev. Food Sci. F 5 (2), 27–33.

Barbaro, A., Bagajewicz, M., 2005. New rigorous one-step MILP formulation for heat exchanger network synthesis. Comput. Chem. Eng. 29 (9), 1945–1976.

Bard, 1974. Nonlinear Parameter Estimation. Academic Press, London.

Basak, K., Abhilash, K.S., Ganguly, S., Saraf, D.N., 2002. On-line optimization of a crude distillation unit with constraints on product properties. Ind. Eng. Chem. Res. 41 (6), 1557–1568.

Baudelet, C.A., Krueger, A.W., 1998. A practical approach to fouling mitigation in refinery units: the SPIRELF® system. In: AIChE Spring National Meeting, New Orleans, LA, USA.

Bayat, M., Aminian, J., Bazmi, M., Shahhosseini, S., Sharifi, K., 2012. CFD modeling of fouling in crude oil pre-heaters. Energy Convers. Manage 64, 344–350.

Béhar, E., Mougin, P., Pina, A., 2003. Integration of asphaltenes flocculation modeling into Athos reservoir simulator. Oil Gas Sci. Technol. 58 (2), 625–634.

Bell, K.J., 2008. Introduction to heat exchanger design – fundamental concepts. In: Hewitt, G.F. (Ed.), Heat Exchanger Design Handbook. Begell House, New York.

Bennett, C.A., 2012. A Theory Describing Asphaltene Adhesion Fouling Inside Heat Exchanger Tubes. Heat Transfer Eng. 33 (15), 1246–1250.

Bennett, C.A., Appleyard, S., Gough, M., Hohmann, R.P., Joshi, H.M., King, D.C., Lam, T.Y., Rudy, T.M., Stomierowski, S.E., 2006. Industry–recommended procedures for experimental crude oil preheat fouling research. Heat Transf. Eng. 27 (9), 28–35.

Bennett, C.A., Kistler, R.S., Lestina, T.G., King, D.C., 2007. Improving heat exchanger designs. Chem. Eng. Prog. 103 (4), 40–45.

Bennett, C.A., Kistler, R.S., Nangia, K., Al-Ghawas, W., Al-Hajji, N., Al-Jemaz, A., 2009. Observation of an isokinetic temperature and compensation effect for high-temperature crude oil fouling. Heat Transf. Eng. 30 (10), 794–804.

Berrueco, C., Venditti, S., Morgan, T.J., Alvarez, P., Millan, M., Herod, A.A., Kandiyoti, R., 2008. Calibration of size-exclusion chromatography columns with 1-methyl-2-pyrrolidinone (NMP)/chloroform mixtures as eluent: applications to petroleum-derived samples. Energy Fuels 22 (5), 3265–3274.

Beuf, M., Rizzo, G., Leuliet, J.C., Müller-Steinhagen, H., Yiantsios, S., Karabelas, A., Benezech, T., 2003. Fouling and cleaning of modified stainless steel plate heat exchangers processing milk products. In: Watkinson, A.P., Müller-Steinhagen, H., Malayeri, M.R. (Eds.), ECI Conf. on Heat Exchanger Fouling and Cleaning: Fundamentals and Applications, Volume RP1. Berkeley Electronic Press, Santa Fe, NM, USA, pp. 99–106.

Bjork, K.M., Nordman, R., 2005. Solving large–scale retrofit heat exchanger network synthesis problems with mathematical optimization methods. Chem. Eng. Process. 44 (8), 869–876.

Blas, F.J., Vega, L.F., 1998. Prediction of binary and ternary diagrams using the statistical associating fluid theory (SAFT) equation of state. Ind. Eng. Chem. Res. 37 (2), 660–674.

Bohnet, M., 1987. Fouling of heat transfer surfaces. Chem. Eng. Technol. 10 (1), 113–125.

Bohnet, M., Augustin, W., Hirsh, H., 1997. Influence of fouling layer shear strength on removal behaviour. In: Bott, T.R., Melo, L.F., Panchal, C.B., Somerscales, E.F.C. (Eds.), Proc. Int. Conf. On Understanding Heat Exchanger Fouling and its Mitigation, May 1997, Castelvecchio Pascoli, Italy. Begell House, New York, pp. 201–208.

Bott, T.R., 1988a. Crystallisation fouling basic science and models. In: Melo, L.F., Bott, T.R., Bernardo, C. (Eds.), Fouling Science and Technology, NATO Science Series: E, May 18–30, 1987, Alvor, Portugal. Kluwer Academic, Dordrecht, pp. 251–260.

Bott, T.R., 1988b. General fouling problems. In: Melo, L.F., Bott, T.R., Bernardo, C. (Eds.), Fouling Science and Technology, NATO Science Series: E, May 18–30, 1987, Alvor, Portugal. Kluwer Academic, Dordrecht, pp. 3–14.

Bott, T.R., 1990. Fouling Notebook. IChemE, Rugby (UK).

Bott, T.R., 1995a. Design and operation of heat exchange equipment for fouling conditions. In: Panchal, C.B. (Ed.), Fouling Mitigation of Industrial Heat–exchange

Equipment, June 18–23, 1995, San Luis Obispo, CA, USA. Begell House, New York, pp. 3–11.

Bott, T.R., 1995b. Fouling of Heat Exchangers. Elsevier Science, Amsterdam.

Bott, T.R., 2001. To foul or not to foul. That is the question. Chem. Eng. Prog. 97 (11), 30–37.

Boukir, A., Guiliano, M., Doumenq, P., El Hallaoui, A., Mille, G., 1998. Structural characterization of crude oil asphaltenes by infrared spectroscopy (FTIR). Application to photo-oxidation. Comptes Rendus De L Acad. Des. Sci. Ser. Ii Fasc. C-Chimie 1 (10), 597–602.

Breure, B., Subramanian, D., Leys, J., Peters, C.J., Anisimov, M.A., 2013. Modeling asphaltene aggregation with a single compound. Energy Fuels 27 (1), 172–176.

Broeck, H.T., 1944. Economic selection of exchanger sizes. Ind. Eng. Chem. 36 (1), 64–67.

Brodowicz, K., Markowski, M., 2003. Calculation of heat exchanger networks for limiting fouling effects in the petrochemical industry. Appl. Therm. Eng. 23 (17), 2241–2253.

Buenrostro-González, E., Groenzin, H., Lira-Galeana, C., Mullins, O.C., 2001. The overriding chemical principles that define asphaltenes. Energy Fuel 9 (2), 225–230.

Buenrostro-Gonzalez, E., Lira-Galeana, C., Gil-Villegas, A., Wu, J., 2004. Asphaltene precipitation in crude oils: theory and experiments. AIChE J. 50 (10), 2552–2570.

Brackbill, J.U., Kothe, D.B., Zemach, C., 1992. A continuum method for modeling surface tension. J. Comput. Phys. 100 (2), 335–354.

Brons, G., Brown, L.D., Joshi, H., Kennedy, R.J., Bruno, T. and Rudy, T.M., 2010. Method for Refinery Foulant Deposit Characterization, United States Patent, Patent No.: 7,708,864 B2, May 4.

Brugma, A.J., 1941. The brugma process. Ref. Nat. Gasol. Manuf. 20, 86–88.

Bulasara, V.K., Uppaluri, R., Ghoshal, A.K., 2009. Revamp study of crude distillation unit heat exchanger network: Energy integration potential of delayed coking unit free hot streams. Appl. Therm. Eng. 29 (11–12), 2271–2279.

Burke, N.E., Hobbs, R.E., Kashow, S.F., 1990. Measurement and modeling of asphaltene precipitation. J. Pet. Tech. 42 (11), 1440–1446.

Burton, H., 1968. Section G. Deposits from whole milk in heat treatment plant - a review and discussion. J. Dairy Res. 35 (2), 317–330.

Butterworth, D., 1978a. The development of a model for three–dimensional flow in tube bundles. Int. J. Heat Mass Transfer 21 (2), 253–256.

Butterworth, D., 1978b. A model for heat transfer during three–dimensional flow in tube bundles. In: 6th Int. Heat Transfer Conf., Toronto, on, Canada. Hemisphere, Washington, pp. 219–224.

Butterworth, D., 1996. Visualize your design of shell–and–tube heat exchangers. Chem. Tech. Eur. 3 (4), 2024.

Butterworth, D., 2000. Designing shell–and–tube heat exchangers with velocity–dependant fouling. In: 34th US National Heat Transfer Conf., Pittsburgh, PA, USA. August 20–22. American Society of Mechanical Engineers, pp. 433–436.

Butterworth, D., 2002. Design of shell–and–tube heat exchangers when the fouling depends on local temperature and velocity. Appl. Therm. Eng. 22 (7), 789–801.

Butterworth, D., 2004. Process heat transfer 2010. Appl. Therm. Eng. 24 (89), 1395–1407.

Calemma, V., Iwanski, P., Nali, M., Scotti, R., Montanari, L., 1995. Structural characterization of asphaltenes of different origins. Energy Fuels 9 (2), 225–230.

Canapary, R.C., 1961. How to control refinery fouling. Oil Gas. J. 59 (41), 114–118.

Caputo, A.C., Pelagagge, P.M., Salini, P., 2011. Joint economic optimization of heat exchanger design and maintenance policy. Appl. Therm. Eng. 31, 1381–1392.

Carbognani, L., Espidel, J., 2003. Preparative subfractionation of petroleum resins and asphaltenes. II. Characterization of size exclusion chromatography isolated fractions. Pet. Sci. Technol. 21 (11–12), 1705–1720.

Castellanos Díaz, O., Modaresghazani, J., Satyro, M.A., Yarranton, H.W., 2011. Modeling the phase behaviour of heavy oil and solvent mixtures. Fluid Phase Equilibr. 304 (12), 74–85.

Cao, Q., Sarkar, K., Prasad, A.K., 2004. Direct numerical simulations of two-layer viscositystratified flow. Int. J. Multiphase Flow 30 (12), 1485–1508.

Cerda, J., Westerberg, A.W., Mason, D., Linnhoff, B., 1983. Minimum utility usage in heat exchanger network synthesis a transportation problem. Chem. Eng. Sci. 38 (3), 373–387.

Ceylan, K., Kelbaliyev, G., 2003. The roughness effects on friction and heat transfer in the fully developed turbulent flow in pipes. Appl. Therm. Eng. 23 (5), 557–570.

Chan, K.L.A., Hammond, S.V., Kazarian, S.G., 2003. Applications of attenuated total reflection infrared spectroscopic imaging to pharmaceutical formulations. Anal. Chem. 75 (9), 2140–2146.

Chan, K.L.A., Kazarian, S.G., 2003. New opportunities in micro- and macro-attenuated total reflection infrared spectroscopic imaging: spatial resolution and sampling versatility. Appl. Spectrosc. 57 (4), 381–389.

Chan, K.L.A., Kazarian, S.G., Mavraki, A., Williams, D.R., 2005. Fourier transform infrared imaging of human hair with a high spatial resolution without the use of a synchrotron. Appl. Spectrosc. 59 (2), 149–155.

Chan, K.L.A., Kazarian, S.G., 2006. Detection of trace materials with fourier transform infrared spectroscopy using a multi-channel detector. Analyst 131 (1), 126–131.

Chan, K.L.A., Kazarian, S.G., 2007. Attenuated total reflection fourier transform infrared imaging with variable angles of incidence: a three-dimensional profiling of heterogeneous materials. Appl. Spectrosc. 61 (1), 48–54.

Chan, K.L.A., Tay, F.H., Poulter, G., Kazarian, S.G., 2008. Chemical imaging with variable angles of incidence using a diamond attenuated total reflection accessory. Appl. Spectrosc. 62 (10), 1102–1107.

Chapman, W.G., Jackson, G., Gubbins, K.E., 1988. Phase equilibria of associating fluids: chain molecules with multiple bonding sites. Mol. Phys. 65 (5), 1057–1079.

Chapman, W.G., Gubbins, K.E., Jackson, G., Radosz, M., 1989. SAFT equation of state solution model for associating fluids. Fluid Phase Equilibr. 52, 31–38.

Chapman, W.G., Gubbins, K.E., Jackson, G., Radosz, M., 1990. New reference equation of state for associating liquids. Ind. Eng. Chem. Res. 29 (8), 1709–1721.

Chaudhuri, P.D., Diwekar, U.M., Logsdon, J.S., 1997. An automated approach for the optimal design of heat exchangers. Ind. Eng. Chem. Res. 36 (9), 3685–3693.

Chenoweth, J.M., 1988a. General design of heat exchangers for fouling conditions. In: Melo, L.F., Bott, T.R., Bernardo, C. (Eds.), Fouling Science and Technology, NATO Science Series: E, May 18–30, 1987, Alvor, Portugal. Kluwer Academic, Dordrecht, pp. 477–494.

Chenoweth, J.M., 1988b. Liquid fouling monitoring equipment. In: Melo, L.F., Bott, T.R., Bernardo, C. (Eds.), Fouling Science and Technology, NATO Science Series: E, May 18–30, 1987, Alvor, Portugal. Kluwer Academic, Dordrecht, pp. 49–65.

Chenoweth, J.M., 1990. Final report of the HTRI/TEMA joint committee to review the fouling section of the TEMA standards. Heat Transfer Eng. 11 (1), 73–107.

Chenoweth, J.M., 1995. The TEMA standards fouling section: another look. In: Panchal, C.B. (Ed.), Fouling Mitigation of Industrial Heat-exchange Equipment, June 18–23, 1995, San Luis Obispo, CA, USA. Begell House, New York, pp. 61–71.

Chew, J.Y.M., Paterson, W.R., Wilson, D.I., 2004a. Fluid dynamic gauging for measuring the strength of soft deposits. J. Food Eng. 65 (2), 175–187.

Chew, J.Y.M., Cardoso, S.S.S., Paterson, W.R., Wilson, D.I., 2004b. CFD studies of dynamic gauging. Chem. Eng. Sci. 59 (16), 3381–3398.

Chew, Y.M.J., Höfling, V., Augustin, W., Paterson, W.R., Wilson, D.I., 2005. A method for measuring the strength of scale deposits on heat transfer surfaces. Dev. Chem. Mineral Process. 13 (1/2), 21–30.

Chew, J.Y.M., Tonneijk, S.J., Paterson, W.R., Wilson, D.I., 2006. Solvent-based cleaning of emulsion polymerization reactors. Chem. Eng. J. 117 (1), 64–69.

Choi, D.W., 2005. How to buy and operate desalters. Hydrocarb. Process. 84 (3), 75–79.

Chouparova, E., Lanzirotti, A., Feng, H., Jones, K.W., Marinkovic, N., Whitson, C., Philp, P., 2004. Characterization of petroleum deposits formed in a producing well by synchrotron radiation-based microanalyses. Energy Fuels 18 (4), 1199–1212.

Christy, A.A., Dahl, B., Kvalheim, O.M., 1989. Structural features of resins, asphaltenes and kerogen studied by diffuse reflectance infrared-spectroscopy. Fuel 68 (4), 430–435.

Ciric, A.R., Floudas, C.A., 1991. Heat exchanger network synthesis without decomposition. Comput. Chem. Eng. 15 (6), 385–396.

Coelho, R.R., Hovell, I., Monte, M.B.D., Middea, A., de Souza, A.L., 2006. Characterisation of aliphatic chains in vacuum residues (VRs) of asphaltenes and resins using molecular modelling and FTIR techniques. Fuel Process. Technol. 87 (4), 325–333.

Coelho, R.R., Hovell, I., 2007. Characterization of functional groups of asphaltenes in vacuum residues using molecular modelling and FTIR techniques. Pet. Sci. Technol. 25 (1–2), 41–54.

Coelho, R.R., Hovell, I., Rajagopal, K., 2012. Elucidation of the functional sulphur chemical structure in asphaltenes using first principles and deconvolution of mid-infrared vibrational spectra. Fuel Process. Technol. 97, 85–92.

Colbert, R.W., 1982. Industrial heat exchange networks. Chem. Eng. Prog. 78 (7), 47–54.

Coletti, F., 2010. Multi-scale modelling of refinery pre-heat trains undergoing fouling for improved energy efficiency (Ph.D. Thesis). Department of Chemical Engineering and Chemical Technology, Imperial College London, UK.

Coletti, F., Ishiyama, E.M., Paterson, W.R., Wilson, D.I., Macchietto, S., 2010. Impact of deposit ageing and surface roughness on thermal fouling: distributed model. AIChE J. 56 (12), 3257–3273.

Coletti, F., Macchietto, S., 2010a. Refinery pre–heat train network simulation: assessment of energy efficiency and carbon emissions. Heat. Transf. Eng. 32 (3–4), 228–236.

Coletti, F., Macchietto, S., 2010b. Validation of a novel model for Shell–and–Tube heat exchangers undergoing crude oil fouling. In: Proc. 14th Int. Heat Transfer Conf. (IHTC14); Vol. 4, Washington, DC, USA, August 8–13. ASME, pp. 741–750.

Coletti, F., Macchietto, S., 2011. A dynamic, distributed model of shell–and–tube heat exchangers undergoing crude oil fouling. Ind. Eng. Chem. Res. 50 (8), 4515–4533.

Coletti, F., Macchietto, S., Polley, G.T., 2011. Effects of fouling on performance of retrofitted heat exchanger networks; a thermo–hydraulic based analysis. Comput. Chem. Eng. 35 (5), 907–917.

Coletti, F., 2014. Heat Exchanger Design Handbook. Begell House, New York, ISBN 978-1-56700-423-6.

Coletti, F., Diaz-Bejarano, E., Macchietto, S., Kulikov, A., Porsin, A., 2014. Heating Device for High Temperature Fouling Rig. Patent application: PCT/RU2014/000380 of 26.05.2014.

Corrieu, G., 1981. State-of-the-art of cleaning surfaces. In: Hallström, B., Lund, D.B., Trägårdh, A.C., Reprocentralen (Eds.), Proc. Fundamentals and Applications of Surface Phenomena Associated with Fouling and Cleaning in Food Processing, Lund, Sweden.

Costa, A.L.H., Tavares, V.B.G., Borges, J.L., Queiroz, E.M., Pessoa, F.L.P., Liporace, F.D.S., de Oliveira, S.G., 2013. Parameter estimation of fouling models in crude preheat trains. Heat. Transf. Eng. 34 (8–9), 683–691.

Cotterman, R.L., Prausnitz, J.M., 1985. Flash calculations for continuous or semicontinuous mixtures by use of an equation of state. Ind. Eng. Chem. Proc. Des. Dev. 24 (2), 434–443.

Cotterman, R.L., Bender, R., Prausnitz, J.M., 1985a. Phase equilibria for mixtures containing very many components. Development and application of continuous thermodynamics for chemical process design. Ind. Eng. Chem. Proc. Des. Dev. 24 (1), 194–203.

Cotterman, R.L., Dimitrelis, D., Prausnitz, J.M., 1985b. Design of supercritical fluid extraction processes using continuous Thermodynamics. In: AIChE Annual Meeting, San Francisco, CA.

Cousineau, P., Fulford, G., Simmons, P., 1988. Monitoring fouling using a novel technique. In: Melo, L.F., Bott, T.R., Bernardo, C. (Eds.), Fouling Science and Technology, NATO Science Series: E, May 18–30, 1987, Alvor, Portugal. Kluwer Academic, Dordrecht, pp. 77–84.

Coussot, P., 2007. Rheophysics of pastes: a review of microscopic modelling approaches. Soft Matter 3 (5), 528–540.

Crattelet, J., Ghnimi, S., Debreyne, P., Zaid, I., Boukabache, A., Esteve, D., Auret, L., Fillaudeau, L., 2013. On-line local thermal pulse analysis sensor to monitor fouling and cleaning: application to dairy product pasteurisation with an ohmic cell jet heater. J. Food Eng. 119 (1), 72–83.

Crittenden, B.D., 1988a. Basic science and models of reaction fouling. In: Melo, L.F., Bott, T.R., Bernardo, C. (Eds.), Fouling Science and Technology, NATO Science Series: E, May 18–30, 1987, Alvor, Portugal. Kluwer Academic, Dordrecht, pp. 293–313.

Crittenden, B.D., 1988b. Chemical reaction fouling of heat exchangers. In: Melo, L.F., Bott, T.R., Bernardo, C. (Eds.), Fouling Science and Technology, NATO Science Series: E, May 18–30, 1987, Alvor, Portugal. Kluwer Academic, Dordrecht, pp. 315–332.

Crittenden, B.D., Kolaczkowski, S.T., 1979a. Mass transfer and chemical kinetics in hydrocarbon fouling. In: Proc. Conf. Fouling - Science or Art? University of Surrey, Guildford. Institution of Corrosion Science and Technology and Institution of Chemical Engineers, London.

Crittenden, B.D., Kolaczkowski, S.T., 1979b. Energy savings through the accurate prediction of heat transfer fouling resistances. In: O'Callaghan, P.W. (Ed.), Energy for Industry. Pergamon, Oxford, pp. 257–266.

Crittenden, B.D., Hout, S.A., Alderman, N.J., 1987a. Model experiments of chemical reaction fouling. Chem. Eng. Res. Des. 65 (2), 165–170.

Crittenden, B.D., Kolaczkowski, S.T., Hout, S.A., 1987b. Modelling hydrocarbon fouling. Chem. Eng. Res. Des. 65 (2), 171–179.

Crittenden, B.D., Alderman, N.J., 1988. Negative fouling resistances: the effect of surface roughness. Chem. Eng. Sci. 43 (4), 829–838.

Crittenden, B.D., Kolaczkowski, S.T., Downey, I.L., 1992. Fouling of crude oil preheat exchangers. Chem. Eng. Res. Des. 70 (A6), 547–557.

Crittenden, B.D., Kolaczkowski, S.T., Takemoto, T., 1993. Use of in-tube inserts to reduce fouling from crude oils. AIChE Symp. Ser. 89 (295), 300–307.

Crittenden, B.D., Kolackzkowski, S.T., Phillips, D.Z., 1997. Chemical reaction fouling. In: Bott, T.R., Melo, L.F., Panchal, C.B., Somerscales, E.F.C. (Eds.), Proc. Int. Conf. On Understanding Heat Exchanger Fouling and its Mitigation, May 1997, Castelvecchio Pascoli, Italy. Begell House, New York, pp. 91–103.

Crittenden, B.D., Kolaczkowski, S.T., Takemoto, T., Phillips, D.Z., 2009. Crude oil fouling in a pilot-scale parallel tube apparatus. Heat. Transf. Eng. 30 (10–11), 777–785.

Crittenden, B.D., Yang, M.Y., Dong, L., Hanson, R., Jones, J., Kundu, K., Harris, J., Klochok, O., Arsenyeva, O., Kapustenko, P., 2013. Crystallization fouling with enhanced heat transfer surfaces. In: Malayeri, M.R., Müller-Steinhagen, H., Watkinson, A.P. (Eds.), Proc. Int. Conf. On Heat Exchanger Fouling and Cleaning X, June 9–14, 2013, Budapest. Hungary, pp. 379–385. Published online. www.heatexchanger-fouling.com.

Dabai, F., Paterson, N., Millan, M., Fennell, P., Kandiyoti, R., 2010. Tar formation and destruction in a fixed bed reactor simulating downdraft gasification: equipment development and characterization of tar cracking products. Energy Fuels 24 (10), 4560–4570.

Danesh, A., Xu, D., Todd, A.C., 1994. Comparative study of cubic equations of state for predicting phase behaviour and volumetric properties of injection gas-reservoir oil systems. Fluid Phase Equilib. 63 (3), 259–278.

Davies, T.J., Henstridge, S.C., Gillham, C.R., Wilson, D.I., 1997. Investigation of whey protein deposit properties using heat flux sensors. Food Bioprod. Process. 75 (C2), 106–110.

de Boer, R.B., Leerlooyer, K., Eigner, M.R.P., van Bergen, A.R.D., 1995. Screening of crude oils for asphaltene precipitation: theory, practice and the selection of inhibitors. SPE Prod. Facil. 10 (1), 55–61.

Deng, Y., Dixon, J.B., White, G.N., 2003. Molecular configurations and orientations of hydrazine between structural layers of kaolinite. J. Colloid Interf. Sci. 257 (2), 208–227.

Deshannavar, U.B., Rafeen, M.S., Ramasamy, M., Subbarao, D., 2010. Crude oil fouling: a review. J. Appl. Sci. 10 (24), 3167–3174.

Dickakian, G., Seay, S., 1988. Asphaltene precipitation: primary crude exchanger fouling mechanism. Oil Gas J. 86 (10), 147–150.

Dickakian, G., 1989. Crude oil fouling control by a fouling analyzer. In: Shah, R. (Ed.), National Heat Transfer Conf. On Heat Transfer Equipment Fundamentals, Design, Applications and Operating Problems, HTD-vol. 108, Philadelphia, PA, USA. American Society of Mechanical Engineers, Michigan, pp. 331–336.

Ding, L., Calhoun, R.J., Street, R.L., 2003. Numerical Simulation of Strongly Stratified Flow Over a Three-Dimensional Hill. Boundary-Layer Meteology 107 (1), 81–114.

Ding, H., Spelt, P.D.M., Shu, C., 2007. Diffuse interface model for incompressible two-phase flows with large density ratios. J. Comput. Phys. 226 (2), 2078–2095.

DOE, 2002. Martinez refinery Completes Plant–wide Energy Assessment. Technical report, Office of Industrial Technologies. U.S. Department of Energy.

DOE, 2006. Energy bandwidth for Petroleum Refining Processes. Technical report, Office of Energy Efficiency and Renewable Energy Industrial Technologies Program. U.S. Department of Energy.

Douda, J., Llanos, M.E., Alvarez, R., Bolanos, J.N., 2004. Structure of maya asphaltene-resin complexes through the analysis of soxhlet extracted fractions. Energy Fuels 18 (3), 736–742.

Dumas, P., Jamin, N., Teillaud, J.L., Miller, L.M., Beccard, B., 2004. Imaging capabilities of synchrotron infrared microspectroscopy. Faraday Discuss 126, 289–302.

Durand, E., Clemancey, M., Lancelin, J.M., Verstraete, J., Espinat, D., Quoineaud, A.A., 2010. Effect of chemical composition on asphaltenes aggregation. Energy Fuels 24 (2), 1051–1062.

Eaton, P., 1983. Fouling Test Apparatus, US patent 4383438.

Eaton, P., Lux, R., 1984. Laboratory fouling test apparatus for hydrocarbon feed-stocks. In: Suitor, J.W., Pritchard, A.M. (Eds.), ASME HTD, 35, pp. 33–42.

Ebert, W.A., Panchal, C.B., 1995. Analysis of exxon crude–oil–slip stream coking data. In: Panchal, C.B. (Ed.), Fouling Mitigation of Industrial Heat–Exchange Equipment, June 18–23, 1995, San Luis Obispo, CA. USA. Begell House, New York, pp. 451–460.

Ebrahim, M., Kawari, A., 2000. Pinch technology: an efficient tool for chemical-plant energy and capital-cost saving. Appl. Energy 65 (1–4), 45–49.

Economou, I.G., 2002. Statistical associating fluid theory: a successful model for the calculation of thermodynamic and phase equilibrium properties of complex fluid mixtures. Ind. Eng. Chem. Res. 41 (5), 953–962.

EIA, 2007. Annual Energy Outlook (Technical report, U.S. Energy Information Administration).

Ekgasit, S., Ishida, H., 1997. New optical depth-profiling technique by use of the multiple-frequency approach with single ATR FT-IR spectrum: theoretical development. Appl. Spectrosc. 51 (10), 1488–1495.

Epstein, N., 1981. Fouling in heat exchangers. In: Somerscales, E.F.C., Knudsen, J.G. (Eds.), Fouling of Heat Transfer Equipment. Hemisphere, Washington, pp. 701–734.

Epstein, N., 1983. Thinking about heat transfer fouling: a 5×5 matrix. Heat. Transf. Eng. 4 (1), 43–56.

Epstein, N., 1988. Particulate fouling of heat transfer surfaces: mechanism and models. In: Melo, L.F., Bott, T.R., Bernardo, C. (Eds.), Fouling Science and Technology, NATO Science Series: May 18–30, 1987, Alvor, Portugal. Kluwer Academic, Dordrecht, pp. 143–164.

Epstein, N., 1994. A model of the initial chemical reaction fouling rate for flow within a heated tube and its verification. In: Proc 10th International Heat Transfer Conference, Vol. 4. Institution of Chemical Engineers, Rugby, pp. 225–229.

Errico, M., Tola, G., Mascia, M., 2009. Energy saving in a crude distillation unit by a preflash implementation. Appl. Therm. Eng. 29 (8–9), 1642–1647.

Eryener, D., 2006. Thermoeconomic optimization of baffle spacing for shell and tube heat exchangers. Energy Convers. Manage 47 (11–12), 1478–1489.

ESDU, 1989. Fouling of Heat Exchanger Surfaces General Principles. data item 86038. Technical report. ESDU Int. plc.

ESDU, 1994. Selection and Costing of Heat Exchangers Shell–and–tube Type. data item 94042. Technical report. ESDU Int. plc.

ESDU, 2000. Heat Exchanger Fouling in the Pre–heat Train of a Crude Oil Distillation Unit. data item 00016. Technical report. ESDU Int. plc.

Everall, N.J., Bibby, A., 1997. Improvements in the use of attenuated total reflection fourier transform infrared dichroism for measuring surface orientation in polymers. Appl. Spectrosc. 51 (8), 1083–1091.

Fahim, M.A., Al-Sahhaf, T.A., Elkilani, A.S., 2001. Prediction of asphaltene precipitation for Kuwaiti crude using thermodynamic micellization model. Ind. Eng. Chem. Res. 40 (12), 2748–2756.

Faller, R., 2009. Coarse grained modeling of soft condensed matter (Themed Issued). Phys. Chem. Chem. Phys 11, 1867–1868.

Fan, Z., Watkinson, A.P., 2006. Aging of carbonaceous deposits from heavy hydrocarbon vapors. Ind. Eng. Chem. Res. 45 (18), 6104–6110.

Fan, Z., Watkinson, A.P., 2011. Kinetics and structural evolution during aging of coker vapor deposits. Heat. Transf. Eng. 32 (3–4), 237–247.

Feintuch, H.M., Peer, V., Bucukoglu, M.Z., 1985. A preflash drum can conserve energy in a crude preheat train. Energ. Prog. 5 (3), 165–172.

Fesanghary, M., Damangir, E., Soleimani, I., 2009. Design optimization of shell and tube heat exchangers using global sensitivity analysis and harmony search algorithm. Appl. Therm. Eng. 29 (5–6), 1026–1031.

Firoozabadi, A., 1988. Reservoir-fluid phase behavior and volumetric prediction with equations of state. J. Pet. Technol. 40 (4), 397–406.

Fitzer, E., Mueller, K., Schäffer, W., 1971. The chemistry of the pyrolytic conversion of organic compounds to carbon. In: Walker Jr., P.L. (Ed.), Chemistry and Physics of Carbon, Vol. 7. Marcel Dekker, New York.

Flory, P.J., 1941. Thermodynamics of high polymer solutions. J. Chem. Phys. 9 (8), 660–661.

Floudas, C.A., Ciric, A.R., Grossmann, I.E., 1986. Automatic synthesis of optimum heat exchanger network configurations. AIChE J. 32 (2), 276–290.

Floudas, C.A., Grossmann, I.E., 1987. Synthesis of flexible heat–exchanger networks with uncertain flowrates and temperatures. Comput. Chem. Eng. 11 (4), 319–336.

Fontein, H.J., Wassink, J.G., 1978. The economically optimal design of heat exchangers. Eng. Process. Econ. 3 (2), 141–149.

Förster, M., Bohnet, M., 2001. Modification of the characteristics of heat transfer surfaces as a fouling mitigation strategy. In: Bott, T.R., Watkinson, A.P., Panchal, C.B. (Eds.),

Proc. Int. Conf. Mitigation of Heat Exchanger Fouling and its Economic and Environmental Implications, July 18–23, 1999, Banff, Canada. Begell House, New York, pp. 229–238.

Forte, E., and Taylor, S. E., 2014. Thermodynamic Modelling of Asphaltene Precipitation and Related Phenomena. Submitted for publication.

Frausto-Hernndez, S., Rico-Ramrez, V., Jimnez-Gutirrez, A., Hernndez-Castro, S., 2003. MINLP synthesis of heat exchanger networks considering pressure drop effects. Comput. Chem. Eng. 27 (8–9), 1143–1152.

Frenkel, D., Louis, A.A., 1992. Phase separation in binary hard-core mixtures: an exact result. Phys. Rev. Lett. 68, 3363–3365.

Fernandez-Baujin, J.M., Solomon, S.M., 1976. An industrial application of pyrolysis technology: lummus SRT III Module. In: Albright, L.F., Crynes, B.L. (Eds.), Industrial and Laboratory Pyrolyses. American Chemical Society, Washington, D. C, pp. 345–372.

Fryer, P.J., Slater, N.K.H., 1985. A direct simulation procedure for chemical reaction fouling in heat exchangers. Chem. Eng. J. 31 (2), 97–107.

Fryer, P.J., Slater, N.K.H., 1986. The simulation of heat exchanger control with tube–side chemical reaction fouling. Chem. Eng. Sci. 41 (9), 2363–2372.

Fryer, P., 1988a. Basic concepts in heat exchanger network modelling. In: Melo, L.F., Bott, T.R., Bernardo, C. (Eds.), Fouling Science and Technology, NATO Science Series: E, May 18–30, 1987, Alvor, Portugal. Kluwer Academic, Dordrecht, pp. 495–511.

Fryer, P., 1988b. The fouling of heat exchanger networks. In: Melo, L.F., Bott, T.R., Bernardo, C. (Eds.), Fouling Science and Technology, NATO Science Series: E, May 18–30, 1987, Alvor, Portugal. Kluwer Academic, Dordrecht, pp. 511–527.

Fu, Y.-H., Sandler, S.I., 1995. A simplified SAFT equation of state for associating compounds and mixtures. Ind. Eng. Chem. Res. 34 (5), 1897–1909.

Furman, K.C., Sahinidis, N.V., 2002. A critical review and annotated bibliography for heat exchanger network synthesis in the 20th century. Ind. Eng. Chem. Res. 41 (10), 2335–2370.

Gabrienko, A.A., Lai, C.H., Kazarian, S.G., 2014. In situ chemical imaging of asphaltene precipitation from crude oil induced by n-heptane. Energy Fuels 28 (2), 964–971.

Gabrienko, A.A., Subramani, V., Martyanov, O.N., Kazarian, S.G., 2014. Correlation between asphaltene stability in n-heptane and crude oil composition revealed with in situ chemical imaging. Adsorpt. Sci. Technol. 32 (4), 243–255.

Gadalla, M., Jobson, M., Smith, R., 2003. Optimization of existing heat–integrated refinery distillation systems. Chem. Eng. Res. Des. 81 (1), 147–152.

Gadalla, M.A., Olujic, Z., Jansens, P.J., Jobson, M., Smith, R., 2005. Reducing CO_2 emissions and energy consumption of heat–integrated distillation systems. Environ. Sci. Technol. 39 (17), 6860–6870.

Gale, G.E., 1995. A thickness measuring device using pneumatic gauging to detect the sample. Meas. Sci. Technol. 6 (11), 447–451.

Gardner, K., Taborek, J., 1977. Mean temperature difference: a reappraisal. AIChE J. 23 (6), 777–786.

Georgiadis, M.C., 1998. Advanced Optimisation of Batch Plant Design and Operation (Ph.D. Thesis). Department of Chemical Engineering and Chemical Technology, Imperial College London, UK.

Georgiadis, M.C., Rotstein, G.E., Macchietto, S., 1998a. Modeling and simulation of shell and tube heat exchangers under milk fouling. AIChE J. 44 (4), 959–971.

Georgiadis, M.C., Rotstein, G.E., Macchietto, S., 1998b. Optimal design and operation of heat exchangers under milk fouling. AIChE J. 44 (9), 2099–2110.

Georgiadis, M.C., Papageorgiou, L., Macchietto, S., 2000. Optimal cleaning policies in heat exchanger networks under rapid fouling. Ind. Eng. Chem. Res. 39 (2), 441–454.

Gil-Villegas, A., Galindo, A., Whitehead, P.J., Mills, S.J., Jackson, G., Burgess, A.N., 1997. Statistical associating fluid theory for chain molecules with attractive potentials of variable range. J. Chem. Phys. 106 (10), 4168–4186.

Gilmour, C.H., 1965. No fooling – no fouling. Chem. Eng. Prog. 61 (7), 49–54.

Glen, N.F., Howarth, J.H., Jenkins, A.M., 1997. The estimation of errors and uncertainty in the determination of fouling resistances. In: Bott, T.R., Melo, L.F., Panchal, C.B., Somerscales, E.F.C. (Eds.), Proc. Int. Conf. On Understanding Heat Exchanger Fouling and its Mitigation, May 1997, Castelvecchio Pascoli, Italy. Begell House, New York, pp. 369–376.

Glen, N.F., Howarth, J.H., Jenkins, A.M., 2001. Fouling monitoring on process plant — field experience. In: Bott, T.R., Watkinson, A.P., Panchal, C.B. (Eds.), Proc. Int. Conf. Mitigation of Heat Exchanger Fouling and its Economic and Environmental Implications, July 18–23, 1999, Banff, Canada. Begell House, New York, pp. 70–76.

Gonçalves, C., de, O., Queiroz, E.M., Pessoa, F.L.P., Liporace, F.S., Oliveira, S.G., Costa, A.L.H., 2014. Heuristic optimization of the cleaning schedule of crude preheat trains. Appl. Therm. Eng. 73 (1), 1–12.

Gonzalez, D.L., Ting, P.D., Hirasaki, G.J., Chapman, W.G., 2005. Prediction of asphaltene instability under Gas injection with the PC-SAFT equation of state. Energy Fuels 19 (4), 1230–1234.

Gonzalez, D.L., Hirasaki, G.J., Creek, J., Chapman, W.G., 2007. Modeling of asphaltene precipitation due to changes in composition using the perturbed chain statistical associating fluid theory equation of state. Energy Fuels 21 (3), 1231–1242.

Gonzalez, D.L., Vargas, F.M., Hirasaki, G.J., Chapman, W.G., 2008. Modeling study of CO_2-induced asphaltene precipitation. Energy Fuels 22 (2), 757–762.

Gonzalez, M.F., Sosa-Stull, C., Lopez-Linares, F., Pereira-Almao, P., 2007. Comparing asphaltene adsorption with model heavy molecules over macroporous solid surfaces. Energy Fuels 21 (1), 234–241.

Gordon P.W., 2012. Development of a Scanning Fluid Dynamic Gauging for Cleaning Studies (Ph.D. Thesis). Department of Chemical Engineering and Biotechnology, Cambridge University, UK.

Gram, A.J., 1960. Mechanical design of heat exchangers. Ind. Eng. Chem. 52 (6), 468–473.

Gray, M.R., Tykwinski, R.R., Stryker, J.M., Tan, X., 2011. Supramolecular assembly model for aggregation of petroleum asphaltenes. Energy Fuels 25 (7), 3125–3134.

Greenfield, M.L., 2011. Molecular modelling and simulation of asphaltenes and bituminous materials. Int. J. Pavement Eng. 12 (4), 325–341.

Groenzin, H., Mullins, O.C., 1999. Asphaltene molecular size and structure. J. Phys. Chem. A 103 (50), 11237–11245.

Gross, J., Sadowski, G., 2001. Perturbed-chain SAFT: an equation of state based on a perturbation theory for chain molecules. Ind. Eng. Chem. Res. 40 (4), 1244–1260.

Gu, T., Chew, Y.M.J., Paterson, W.R., Wilson, D.I., 2009. Experimental and CFD studies of fluid dynamic gauging in annular flows. AIChE J. 55 (8), 1937–1947.

Gu, T., Albert, F., Augustin, W., Chew, Y.M.J., Mayer, M., Paterson, W.R., Scholl, S., Sheikh, I., Wang, K., Wilson, D.I., 2011. Application of fluid dynamic gauging to annular test apparatuses for studying fouling and cleaning. Exp. Therm. Fluid Sci. 35 (3), 509–520.

Gudmundsson, J.S., 1981. Particulate fouling. In: Somerscales, E.F.C., Knudsen, J.G. (Eds.), Fouling of Heat Transfer Equipment, August 1979, Troy, New York. Hemisphere, Washington, pp. 357–387.

Gudmundsson, O., Palsson, O.P., Palsson, H., Lalot, S., 2009. Fouling detection in a cross-flow heat exchanger based on physical modelling. In: Müller-Steinhagen, H., Malayeri, M.R., Watkinson, A.P. (Eds.), Int. Conf. On Heat Exchanger Fouling and

Cleaning VIII, June 14–19, 2009, Schladming, Austria, pp. 447–454. Published online. www.heatexchanger-fouling.com.

Gudmundsson, O., Lalot, S., Thorsen, J.E., 2013. Comparison of fouling detection methods using experimental data. In: Malayeri, M.R., Müller-Steinhagen, H., Watkinson, A.P. (Eds.), Proc. Int. Conf. On Heat Exchanger Fouling and Cleaning X, June 9–14, 2013, Budapest, Hungary, pp. 429–436. Published online. www.heatexchanger-fouling.com.

Gunness, R.C., Baker, J.G., 1938. Testing heat transfer equipment. Ind. Eng. Chem. 30 (4), 373–376.

Hardy, J., Fanni, J., Scher, J., 1985. Une méthode turbidimétrique de mesure de la coagulation du lait. Rev. Laitière Française 441, 13–18.

Haris, P.I., Chapman, D., 1995. The conformational-analysis of peptides using fourier-transform IR spectroscopy. Biopolymers 37 (4), 251–263.

Harrick, N.J., 1987. Internal Reflection Spectroscopy, third ed. Harrick Scientific Corporation, New York. p. 327.

Hasson, D., 1981. Precipitation fouling. In: Somerscales, E.F.C., Knudsen, J.G. (Eds.), Fouling of Heat Transfer Equipment, August 1979, Troy, New York. Hemisphere, Washington, pp. 527–568.

Haynes, H.W., Mathews, M.A., 1991. Continuous-mixture vapor-liquid equilibria computations based on true boiling point distillations. Ind. Eng. Chem. Res. 30 (8), 1911–1915.

Hays, G.F., 1989. Fouling allowances are self-fulfilling prophecies. In: Shah, R. (Ed.), National Heat Transfer Conf. On Heat Transfer Equipment Fundamentals, Design, Applications and Operating Problems. American Society of Mechanical Engineers, Philadelphia, PA, USA, pp. 325–330.

Heins, A., Veiga, R., Ruiz, C., Riera, A., 2008. Fouling monitoring and cleaning optimisation in a heat exchanger network of a crude distillation unit. In: Müller-Steinhagen, H., Malayeri, M.R., Watkinson, A.P. (Eds.), Int. Conf. On Heat Exchanger Fouling and Cleaning VII, June 14–19, 2007, Tomar, Portugal. Berkley Electronic Press, pp. 455–464.

Herod, A.A., Bartle, K.D., Morgan, T.J., Kandiyoti, R., 2012. Analytical methods for characterizing high-mass complex polydisperse hydrocarbon mixtures: an overview. Chem. Rev. 112 (7), 3892–3923.

Herod, A.A., Bartle, K.D. and Kandiyoti, R., 2008. Comment on a paper by Mullins, Martinez-Haya and Marshall "contrasting perspective on asphaltene molecular weight. This Comment versus Overview of A. A. Herod, K. D. Bartle R. Kandiyoti." Energy Fuels, 22(6), 4312–4317.

Hewitt, G.F., Shires, G.L., Bott, T.R., 1994. Process Heat Transfer. CRC Press, London.

Hewitt, G.F., 2007. Approximate design and costing methods for heat exchangers. Heat Transfer Eng. 28 (2), 76–86.

Hewitt, G.F., 2008. Heat Exchanger Design Handbook. Begell House, New York.

Hexxcell, 2014. Crude Oil Fouling Simulation Tool. Internal Report. Hexxcell Ltd, London.

Hildebrand, J.H., 1929. Solubility XII. Regular solutions. J. Am. Chem. Soc. 51 (1), 66–80.

Hildebrand, J.H., Prausnitz, J.M., Scott, J.M., 1970. Regular and Related Solutions; the Solubility of Gases, Liquids and Solids. Van Nostrand Reinhold Company, New York.

Hirschberg, A., deJong, L.N.J., Schipper, B.A., Meijer, J.G., 1984. Influence of temperature and pressure on asphaltene flocculation. Soc. Pet. Eng. J. 24 (3), 283–293.

Hirt, C.W., Nichols, B.D., 1981. Volume of fluid (VOF) method for the dynamics of free boundaries. J. Comput. Phys. 39 (1), 201–225.

Hohmann, E.C., 1971. Optimum Networks for Heat Exchange (Ph.D. Thesis). University of Southern California, CA, USA.

Hory, Y., 2000. Crude oil processing. In: Lucas, A.G. (Ed.), Modern Petroleum Technology, sixth ed., Volume 2 Downstream. Wiley & Sons, New York, pp. 24–38.

Howarth, J.H., Glen, N.F., Jekins, A.M., 1997. The use of fouling monitoring techniques. In: Bott, T.R., Melo, L.F., Panchal, C.B., Somerscales, E.F.C. (Eds.), Proc. Int. Conf. On Understanding Heat Exchanger Fouling and its Mitigation, May 1997, Castelvecchio Pascoli, Italy. Begell House, New York, pp. 309–314.

Huang, S.H., Radosz, M., 1990. Equation of state for small, large, polydisperse and associating molecules. Ind. Eng. Chem. Res. 29 (11), 2284–2294.

Huang, S.H., Radosz, M., 1991. Equation of state for small, large, polydisperse and associating molecules: extension to fluid mixtures. Ind. Eng. Chem. Res. 30 (8), 1994–2005.

Huang, J., 2006. Thermal degradation of asphaltene and infrared characterization of its degraded fractions. Pet. Sci. Technol. 24 (9), 1089–1095.

Huggins, M.L., 1941. Solutions of long chain compounds. J. Chem. Phys. 9 (5), 440.

Hwa, C.S., 1965. Mathematical Formulation and Optimization of Heat Exchanger Networks Using Separable Programming, Volume 4 of AIChE IChemE Symposium Series. AIChE, New York.

Ishiyama, E.M., Paterson, W.R., Wilson, D.I., 2008. Thermo-hydraulic channelling in parallel heat exchangers subject to fouling. Chem. Eng. Sci. 63 (13), 3400–3410.

Ishiyama, E.M., Paterson, W.R., Wilson, D.I., 2009a. The effect of fouling on heat transfer, pressure drop and throughput in refinery preheat trains: optimization of cleaning schedules. Heat Transfer Eng. 30 (10–11), 805–814.

Ishiyama, E.M., Paterson, W.R., Wilson, D.I., 2009b. Platform for techno-economic analysis of fouling mitigation options in refinery preheat trains. Energy Fuels 23 (3), 1323–1337.

Ishiyama, E.M., Coletti, F., Macchietto, S., Paterson, W.R., Wilson, D.I., 2010. Impact of deposit ageing on thermal fouling: lumped parameter model. AIChE J. 56 (2), 531–545.

Ishiyama, E.M., Paterson, W.R., Wilson, D.I., 2011. Optimum cleaning cycles for heat Transfer equipment undergoing fouling and ageing. Chem. Eng. Sci. 66 (4), 604–612.

Ishiyama, E.M., Pugh, S.J., Paterson, W.R., Polley, G.T., Kennedy, J., Wilson, D.I., 2013. Management of crude preheat trains subject to fouling. Heat Transfer Eng. 34 (8–9), 31–38.

Ishiyama, E.M., Paterson, W.R., Wilson, D.I., 2014. Aging is important: closing the fouling–cleaning loop. Heat Transfer Eng. 35 (3), 311–326.

Jamaluddin, A.K.M., Nighswander, J.N., Kohse, B.F., El Mahdi, A., 2000. Experimental and theoretical assessment of the asphaltene precipitation characteristics of the Sahil field under a proposed miscible gas injection scheme. In: Abu Dhabi International Petroleum Exhibition and Conference, October, 13-15, Abu Dhabi, UAE. Paper SPE 87292. Society of Petroleum Engineers.

Jamialahmadi, M., Soltani, B., Müller-Steinhagen, H., Rashtchian, D., 2009. Measurement and prediction of the rate of deposition of flocculated asphaltene particles from oil. Int. J. Heat Mass Transfer 52 (19–20), 4624–4634.

Jaubert, J.-N., Solimando, R., Paricaud, P., Barreau, A., 2000. Use of distribution functions: a useful tool to calculate the properties of condensate gases. Ind. Eng. Chem. Res. 39 (12), 5029–5036.

Jegla, Z., Kohoutek, J., Stehlík, P., 2011. Design and operating aspects influencing fouling inside the radiant coils of fired heaters operated in crude oil distillation plants. In: Müller-Steinhagen, H., Malayeri, M.R., Watkinson, A.P. (Eds.), Int. Conf. On Heat Exchanger Fouling and Cleaning VIII, June 14–19, 2009, Schladming, Austria, pp. 7–14. Published online. www.heatexchanger-fouling.com.

Jehlicka, J., Edwards, H.G.M., 2008. Raman spectroscopy as a tool for the non-destructive identification of organic minerals in the geological record. Org. Geochem. 39 (4), 371–386.

Ji, S., Bagajewicz, M., 2002a. Design of crude distillation plants with vacuum units. II. Heat exchanger network design. Ind. Eng. Chem. Res. 41 (24), 6100–6106.

Ji, S., Bagajewicz, M., 2002b. Design of crude fractionation units with preflashing or prefractionation: Energy targeting. Ind. Eng. Chem. Res. 41 (12), 3003–3011.

Jones, A.E., 2008. Fouling in spiral heat exchangers. In: Hewitt, G.F. (Ed.), Heat Exchanger Design Handbook. Begell House, New York.

Jones, D.S.J., 1995. Elements of Petroleum Processing. Wiley & Sons, New York.

Jones, G., Bott, T., 1999. Quantifying Petroleum Fouling of Refinery Heat Exchangers: A Literature Survey. Report F-9. Technical report.

Jones, G.M, Bott, T.R., 2001a. Designing petroleum processing heat transfer capital equipment subject to fouling. In: Müller-Steinhagen, H. (Ed.), Heat Exchanger Fouling - Fundamental Approaches and Technical Solutions. July 8–13, 2001, Davos, Switzerland. PP PUBLICO Publications, Germany, pp. 237–242.

Jones, G.M., Bott, T.R., 2001b. Monitors and models for the assessment of petroleum fouling of refinery heat exchangers. In: Bott, T.R., Watkinson, A.P., Panchal, C.B. (Eds.), Proc. Int. Conf. Mitigation of Heat Exchanger Fouling and its Economic and Environmental Implications, July 18–23, 1999, Banff, Canada. Begell House, New York, pp. 59–69.

Jones, S.A., Bird, M.R., Chew, Y.M.J., Wilson, D.I., 2012. Fluid dynamic gauging of microfiltration membranes fouled with sugar beet molasses. J. Food Eng. 108 (1), 22–29.

Joshi, H.M., 2001. The problem of heat exchanger fouling in petroleum refineries. In: Müller-Steinhagen, H. (Ed.), Heat Exchanger Fouling - Fundamental Approaches and Technical Solutions, July 8–13, 2001, Davos, Switzerland. PP PUBLICO Publications, Germany, pp. 213–217.

Joshi, H.M., Brons, G., 2003. Chemical cleaning of oil refinery heat exchangers - the need for a joint effort. In: Watkinson, A.P., Müller-Steinhagen, H., Malayeri, M.R. (Eds.), ECI Conf. On Heat Exchanger Fouling and Cleaning: Fundamentals and Applications, Volume RP1. USA. Berkeley Electronic Press, Santa Fe, New Mexico, pp. 219–220.

Joshi, H.M., Shilpi, N.B., Agarwal, A., 2009. Relate crude oil fouling research to field fouling observations. In: Müller-Steinhagen, H., Malayeri, M.R., Watkinson, A.P. (Eds.), Int. Conf. On Heat Exchanger Fouling and Cleaning VIII, June 14–19, 2009, Schladming, Austria, pp. 15–16. Published online. www.heatexchanger-fouling.com.

Joshi, H.M., 2013. Crude oil fouling data and a model for pilot-plant scale data. In: Malayeri, M.R., Müller-Steinhagen, H., Watkinson, A.P. (Eds.), Proc. Int. Conf. On Heat Exchanger Fouling and Cleaning X, June 9–14, 2013, Budapest, Hungary, pp. 22–26. Published online. www.heatexchanger-fouling.com.

Juyal, P., Merino-Garcia, D., Andersen, S.I., 2005. Effect on molecular interactions of chemical alteration of petroleum asphaltenes. I. Energy Fuels 19 (4), 1272–1281.

Karabelas, A.J., Yiantsios, S.G., Thonon, B., Grillot, J.M., 1997. Liquid-side fouling of heat exchangers. An integrated R&D approach for conventional and novel designs. Appl. Therm. Eng. 17 (810), 727–737.

Karaca, F., Islas, C.A., Millan, M., Behrouzi, M., Morgan, T.J., Herod, A.A., Kandiyoti, R., 2004. The calibration of size exclusion chromatography columns: molecular mass distributions of heavy hydrocarbon liquids. Energy Fuels 18 (3), 778–788.

Karimi, A., Qian, K., Olmstead, W.N., Freund, H., Yung, C., Gray, M.R., 2011. Quantitative evidence for bridged structures in asphaltenes by thin film pyrolysis. Energy Fuels 25 (8), 3581–3589.

Kakac, S., Agrawal, A., 1988. Basic relationships for heat exchangers and impact of fouling on design. In: Melo, L.F., Bott, T.R., Bernardo, C. (Eds.), Fouling Science and Technology, NATO Science Series: E, May 18–30, 1987, Alvor, Portugal. Kluwer Academic, Dordrecht, pp. 437–462.

Kawanaka, S., Park, S.J., Monsoori, G.A., 1991. Organic deposition from reservoir fluids: a thermodynamic predictive technique. SPE Res. Eng. 6 (2), 185–192.

Kays, W., London, A., 1964. Compact Heat Exchangers. McGraw–Hill, New York.

Kazarian, S.G., Chan, K.L.A., 2010. Micro- and macro-attenuated total reflection fourier transform infrared spectroscopic imaging. Appl. Spectrosc. 64 (7), 135–152.

Kazarian, S.G., Brantley, N.H., Eckert, C.A., 1999. Applications of vibrational spectroscopy to characterise poly(ethylene terephthalate) processed with supercritical CO_2. Vib. Spectrosc. 19 (2), 277–283.

Kazarian, S.G., Flichy, N.M.B., Coombs, D., Poulter, G., 2001. The potential of ATR-IR spectroscopy in applications to supercritical fluids and liquefied gases. Am. Lab 33 (16), 44–49.

Kazarian, S.G., Chan, K.L.A., Tay, F.H., 2009. ATR-FTIR imaging for pharmaceutical and polymeric materials: from micro to macro approaches. In: Salzer, R., Siesler, H.W. (Eds.), Infrared and Raman Spectroscopic Imaging. Wiley VCH, Weinheim, pp. 347–376.

Kazarian, S.G., Chan, K.L.A., 2013. ATR-FTIR spectroscopic imaging: recent advances and applications to biological systems. Analyst 138 (7), 1940–1951.

Kent, C.A., 1988. Biological fouling: basic science and models. In: Melo, L.F., Bott, T.R., Bernardo, C. (Eds.), Fouling Science and Technology, NATO Science Series: May 18–30, 1987, Alvor, Portugal. Kluwer Academic, Dordrecht, pp. 207–222.

Kern, D.Q., Seaton, R.E., 1959a. Surface fouling: how to calculate limits. Chem. Eng. Prog. 55 (6), 71–73.

Kern, D.Q., Seaton, R.E., 1959b. A theoretical analysis of thermal surface fouling. Brit. Chem. Eng. 4 (5), 258–262.

Kern, D.Q., 1966. Heat exchanger design for fouling service. Chem. Eng. Prog. 62 (7), 51–56.

Kern, D.Q., 1988. Process Heat Transfer. McGraw–Hill, New York.

Kho, T.T., Müller-Steinhagen, H., 1999. An experimental and numerical investigation of heat transfer fouling and fluid flow in flat plate heat exchangers. Chem. Eng. Res. Des. 77 (A2), 124–130.

Knudsen, J.G., Hays, G.F., 1995. Use of operating conditions to mitigate fouling of heat exchangers. In: Panchal, C.B. (Ed.), Fouling Mitigation of Industrial Heat–exchange Equipment, June 18–23, 1995, San Luis Obispo, CA. USA. Begell House, New York, pp. 41–52.

Knudsen, J.G., Lin, D.C., Ebert, W.A., 1997. The determination of the threshold fouling curve for a crude oil. In: Bott, T.R., Melo, L.F., Panchal, C.B., Somerscales, E.F.C. (Eds.), Proc. Int. Conf. On Understanding Heat Exchanger Fouling and its Mitigation, May 1997, Castelvecchio Pascoli, Italy. Begell House, New York, pp. 265–271.

Knudsen, J.G., 2002. Fouling in heat exchangers. In: Hewitt, G.F. (Ed.), Heat Exchanger Design Handbook. Begell House, New York.

Kohse, B.F., Nghiem, L.X., Maeda, H., Ohno, K., 2000. Modelling phase behaviour including the effect of pressure and temperature on asphaltene precipitation. In: SPE, Asia Pacific Oil and Gas Conference and Exhibition, October 16–18, Brisbane, Australia. Paper SPE-64465-MS. Society of Petroleum Engineers.

Kokal, S.L., Najman, J., Sayegh, S.G., George, A.E., 1992. Measurement and correlation of asphaltene precipitation from heavy oils by gas injection. J. Can. Pet. Tech. 31 (4), 24–30.

Kontogeorgis, G.M., Folas, G.K., 2010. Thermodynamic Models for Industrial Applications. From Classical and Advanced Mixing Rules to Association Theories. Wiley, United Kingdom.

Kontogeorgis, G.M., Voutsas, E.C., Yakoumis, I.V., Tassios, D.P., 1996. An equation of state for associating fluids. Ind. Eng. Chem. Res. 35 (11), 4310–4318.

Kontogeorgis, G.M., Yakoumis, I.V., Meijer, H., Hendriks, E.M., Moorwood, T., 1999. Multicomponent phase equilibrium calculations for water–methanol–alkane mixtures. Fluid Phase Equilibr. 158-160, 201–209.

Kraska, T., Gubbins, K.E., 1996a. Phase equilibria calculations with a modified SAFT equation of state. 1. pure alkanes, alkanols and water. Ind. Eng. Chem. Res. 35 (12), 4727–4737.

Kraska, T., Gubbins, K.E., 1996b. Phase equilibria calculations with a modified SAFT equation of state. 2. Binary mixtures of n-Alkanes, 1-Alkanols and water. Ind. Eng. Chem. Res. 35 (12), 4738–4746.

Kukulka, D.J., Czechowski, H., Kukulka, P.D., 2010. Factors associated with fouling in the process industry. Heat. Transf. Eng. 31 (9), 782–787.

Kuru, W.C., Panchal, C.B., Liao, C.F., Palen, J., Ebert, W.A., 1997. Field–fouling units for refinery experiments. In: AIChE/ASME National Heat Transfer Conference: Current Developments in Numerical Simulation of Heat and Mass Transfer, August 10–12, Baltimore, MD, USA, pp. 339–344.

Lafitte, T., Avendaño, C., Papaioannou, V., Galindo, A., Adjiman, C.S., Jackson, G., Müller, E.A., 2012. SAFT- γ force field for the simulation of molecular fluids: 3. Coarse-grained models of benzene and hetero-group models of n-decylbenzene. Mol. Phys. 110 (11–12), 1189–1203.

Lafitte, T., Apostolakou, A., Avendaño, C., Galindo, A., Adjiman, C.S., Müller, E.A., Jackson, G., 2013. Accurate statistical associating fluid theory for chain molecules formed from Mie segments. J. Chem. Phys. 139 (15), 154504.

Lambourn, G.A., Durrieu, M., 1986. Fouling in Crude Oil Preheat Trains. Hemisphere, Washington, DC, USA.

Lavaja, J.H., Bagajewicz, M.J., 2004. On a new MILP model for the planning of heat-exchanger network cleaning. Ind. Eng. Chem. Res. 43 (14), 3924–3938.

Lavaja, J.H., Bagajewicz, M.J., 2005a. On a new MILP model for the planning of heat exchanger network cleaning. Part II: throughput loss considerations. Ind. Eng. Chem. Res. 44 (21), 8046–8056.

Lavaja, J.H., Bagajewicz, M.J., 2005b. On a new MILP model for the planning of heat exchanger network cleaning. Part III: multiperiod cleaning under uncertainty with financial risk management. Ind. Eng. Chem. Res. 44 (21), 8136–8146.

Leekumjorn, S., Krejbjerg, K., 2013. Phase behavior of reservoir fluids: comparisons of PC-SAFT and cubic EOS simulations. Fluid Phase Equilibr. 359, 17–23.

Lewis, W.J.T., Chew, Y.M.J., Bird, M.R., 2012. The application of fluid dynamic gauging in characterising cake deposition during cross-flow microfiltration of a yeast suspension. J. Membr. Sci. 405-406, 113–122.

Li, Y., Motard, R.L., 1986. Optimal pinch approach temperature in heat-exchanger networks. Ind. Eng. Chem. Fund. 25 (4), 577–581.

Li, H., Kottke, V., 1998. Effect of baffle spacing on pressure drop and local heat transfer in shell–and–tube heat exchangers for staggered tube arrangement. Int. J. Heat Mass Transfer 41 (10), 1303–1311.

Li, W., 2005. Oscillatory asymptotic fouling model. J. Enhanc. Heat Transfer 12 (3), 239–248.

Li, Z., Firoozabadi, A., 2010a. Modeling asphaltene precipitation by n-alkanes from heavy oils and bitumens using cubic-plus-association equation of state. Energy Fuels 24 (2), 1106–1113.

Li, Z., Firoozabadi, A., 2010b. Cubic-plus association equation of state for asphaltene precipitation in live oils. Energy Fuels 24 (5), 2956–2963.

Liebmann, K., Dhole, V.R., 1995. Integrated crude distillation design. Comput. Chem. Eng. 19 (1), 119–124.

Liebmann, K., Dhole, V.R., Jobson, M., 1998. Integrated design of a conventional crude oil distillation tower using pinch analysis. Chem. Eng. Res. Des. 76 (3), 335–347.

Linnhoff, B., 1978. Thermodynamic Analysis in the Design of Process Networks (Ph.D. Thesis). Leeds University, UK.

Linnhoff, B., Flower, J.R., 1978a. Synthesis of heat exchanger networks: I. systematic generation of energy optimal networks. AIChE J. 24 (4), 633–642.

Linnhoff, B., 1993. Pinch analysis: a state–of–the–art overview: techno-economic analysis. Chem. Eng. Res. Des. 71 (5), 503–522.

Liporace, F.S., de Oliveira, S.G., 2007. Real time fouling diagnosis and heat exchanger performance. Heat Transfer Eng. 28 (3), 193–201.

Lister, D.H., 1981. Corrosion products in power generating systems. In: Somerscales, E.F.C., Knudsen, J.G. (Eds.), Fouling of Heat Transfer Equipment, August 1979, Troy, New York. Hemisphere, Washington, pp. 135–200.

Lin, R., Patrick Ritz, G., 1993. Studying individual macerals using I.R. microspectrometry and implications on oil versus gas/condensate proneness and "low-rank" generation. Org. Geochem. 20 (6), 695–706.

Liu, G., Xu, X., Gao, J., 2003. Study on the compatibility of high-paraffin crude oil with electric desalting demulsifiers. Energy Fuels 17 (3), 625–630.

Macchietto, S., Hewitt, G.F., Coletti, F., Crittenden, B.D., Dugwell, D.R., Galindo, A., Jackson, G., Kandiyoti, R., Kazarian, S.G., Luckham, P.F., Matar, O.K., Millan-Agorio, M., Muller, E.A., Paterson, W., Pugh, S.J., Richardson, S.M., Wilson, D.I., 2011. Fouling in crude oil preheat trains: a systematic solution to an old problem. Heat Transfer Eng. 32 (3–4), 197–215.

Mahdavi, S., Kharrat, R., 2011. Asphaltene precipitation prediction using a micellization model based on experimental data. Petrol. Sci. Technol. 29 (11), 1133–1146.

Mansoori, G.A., 2009. A unified perspective on the phase behaviour of asphaltene micelle in petroleum fluids. Int. J. Oil Gas Coal Technol. 2 (2), 141–167.

Markowski, M., 2000. Reconstruction of a heat exchanger network under industrial constraints – the case of a crude distillation unit. Appl. Therm. Eng. 20 (15–16), 1535–1544.

Markowski, M., Urbaniec, K., 2005. Optimal cleaning schedule for heat exchangers in a heat exchanger network. Appl. Therm. Eng. 25 (7), 1019–1032.

Mason, T.G., Lin, M.Y., 2003. Asphaltene nanoparticle aggregation in mixtures of incompatible crude oils. Phys. Rev. E 67, 050401.

Masso, A.H., Rudd, D.F., 1969. The synthesis of system designs II. Heuristic structuring. AIChE J. 15 (1), 10–17.

Master, B.I., Chunangad, K.S., 2008. Dynamic performance of heat exchangers. Heat Transfer Eng. 29 (3), 217–218.

McCabe, C., Gil-Villegas, A., Jackson, G., 1999. Gibbs ensemble computer simulation and SAFT-VR theory of non-conformal square-well monomer-dimer mixtures. Chem. Phys. Lett. 303 (1–2), 27–36.

McCabe, C., Galindo, A., 2010. SAFT associating fluids and fluid mixtures. In: Goodwin, A.R.H., Sengers, J.V., Peters, C.J. (Eds.), Applied Thermodynamics of Fluids. Royal Society of Chemistry, pp. 215–279.

Mejía, A., Herdes, C., Müller, E.A., 2014. Force Fields for coarse-grained molecular simulations from a corresponding states correlation. Ind. Eng. Chem. Res. 53 (10), 4131–4141.

Melo, L.F., 1997. An overview of bio-fouling form basic science to mitigation. In: Bott, T.R., Melo, L.F., Panchal, C.B., Somerscales, E.F.C. (Eds.), Proc. Int. Conf. On Understanding Heat Exchanger Fouling and its Mitigation, May 1997, Castelvecchio Pascoli, Italy. Begell House, New York, pp. 55–66.

Merdrignac, I., Espinat, D., 2007. Physicochemical characterization of petroleum fractions: the state of the art. Oil Gas Sci. Technol. – Rev. De. L'Institut Francais Du. Pet. 62 (1), 7–32.

Merry, H., Polley, G.T., 1981. Obtaining valid data on fouling resistance. In: Somerscales, E.F.C., Knudsen, J.G. (Eds.), Fouling of Heat Transfer Equipment, August 1979, Troy, New York. Hemisphere, Washington, pp. 83–94.

Meyers, G.E., Mitchell, J.W., Lindeman, J.C.F., 1970. The transient response of heat exchanger having an infinite capacitance rate fluid. J. Heat Transfer – T. ASME 92 (2), 269–275.

Middleman, S., 1998. An Introduction to Fluid Dynamics. Academic Press, New York.

Miquel, J., Hernandez, J., Castells, F., 1992. A new method for petroleum fractions and crude oil characterization. SPE Res. Eng. 7 (2), 265–270.

Mitrakirtley, S., Mullins, O.C., Vanelp, J., George, S.J., Chen, J., Cramer, S.P., 1993. Determination of the nitrogen chemical structures in petroleum asphaltenes using XANES spectroscopy. J. Am. Chem. Soc. 115 (1), 252–258.

Mizutani, F.T., Pessoa, F.L.P., Queiroz, E.M., Hauan, S., Grossmann, I.E., 2003. Mathematical programming model for heat-exchanger network synthesis including detailed heat-exchanger designs. 1. Shell-and-tube heat-exchanger design. Ind. Eng. Chem. Res. 42 (17), 4009–4018.

Morales-Fuentes, A., Picón-Núñez, M., Polley, G.T., Méndez-Díaz, S., 2014. Analysis of the influence of operating conditions on fouling rates in fired heaters. Appl. Therm. Eng. 62 (2), 777–784.

Morgan, T.J., Millan, M., Behrouzi, M., Herod, A.A., Kandiyoti, R., 2005. On the limitations of UV-fluorescence spectroscopy in the detection of high-mass hydrocarbon molecules. Energy Fuels 19 (1), 164–169.

Morton, D., 1960. Thermal design of heat exchangers. Ind. Eng. Chem. 52 (6), 474–478.

Müller, E.A., Gubbins, K.E., 2001. Molecular-based equations of state for associating fluids: a review of SAFT and related approaches. Ind. Eng. Chem. Res. 40, 2193–2211.

Müller, E.A., Jackson, G., 2014. Force field parameters from the SAFT-γ equation of state for use in coarse-grained molecular simulations. Ann. Rev. Chem. Biomol. Eng. 5, 405–427.

Müller-Steinhagen, H., 2000. Heat Exchanger Fouling: Mitigation and Cleaning Technologies. Institute of Chemical Engineers, Rugby, UK.

Müller-Steinhagen, H., Malayeri, M.R., Watkinson, A.P., 2009a. Heat exchanger fouling: environmental impacts. Heat. Transf. Eng. 30 (10), 773–776.

Müller-Steinhagen, H., Malayeri, M.R., Watkinson, A.P., 2009b. Preface to the proceedings of Int. Conf. On heat exchanger fouling and cleaning VIII. In: Müller-Steinhagen, H., Malayeri, M.R., Watkinson, A.P. (Eds.), Int. Conf. On Heat Exchanger Fouling and Cleaning VIII, June 14–19, 2009, Schladming, Austria. Published online. www.heatexchanger-fouling.com.

Mullins, O.C., Sheu, E.Y., Hammami, A., Marshall, A.G., 2007. Asphaltenes, Heavy Oils and Petroleomics. Springer, New York.

Mullins, O.C., 2008. Review of the molecular structure and aggregation of asphaltenes and petroleomics. SPE J. 13 (1), 48–57.

Mullins, O.C., Martinez-Haya, B., Marshall, A.G., 2008. Contrasting perspective on asphaltene molecular weight. This comment versus the overview of A. A. Herod, K. D. Bartle and R. Kandiyoti. Energy Fuels 22 (3), 1765–1773.

Mullins, O.C., 2010. The modified yen model. Energy Fuels 24 (4), 2179–2207.

Mullins, O.C., 2011. The asphaltenes. Annu. Rev. Anal. Chem. 4, 393–418.

Mullins, O.C., Sabbah, H., Eyssautier, J., Pomerantz, A.E., Barré, L., Andrews, A.B., Ruiz-Morales, Y., Mostowfi, F., McFarlane, R., Goual, L., Lepkowicz, R., Cooper, T., Orbulescu, J., Leblanc, R.M., Edwards, J., Zare, R.N., 2012. Advances in asphaltene science and the Yen–Mullins model. Energy Fuels 26 (7), 3986–4003.

Muralikrishna, K., Shenoy, U.V., 2000. Heat exchanger design targets for minimum area and cost. Chem. Eng. Res. Des. 78 (2), 161–167.

Murgich, J., Abanero, J.A., Strausz, O.P., 1999. Molecular recognition in aggregates formed by asphaltene and resin molecules from the Athabasca oil sand. Energy Fuels 13 (2), 278–286.

Murgich, J., 2002. Intermolecular forces in aggregates of asphaltenes and resins. Pet. Sci. Technol. 20 (9–10), 983–997.

Murphy, G., Campbell, J., 1992. Fouling in refinery heat exchangers: causes, effects, measurement and control. In: Bohnet, M. (Ed.), Fouling Mechanisms, GRETh Seminar, Grenoble, France. Editions Européennes Thermique et Industrie, Paris, pp. 249–261.

Nasr, M.R.J., Givi, M.M., 2006. Modeling of crude oil fouling in preheat exchangers of refinery distillation units. Appl. Therm. Eng. 26 (14), 1572–1577.

Nellensteyn, F.J., 1923. Bereiding en constitutie van asphalt (Manufacture and constitution of asphaltic bitumen). Dissertarie Technische Hoogescool, Delft.

Nelson, W.L., 1934a. Fouling of heat exchangers. Refin. Nat. Gasol. Manuf. 13 (7), 271–276.

Nelson, W.L., 1934b. Fouling of heat exchangers. Part II. Refin. Nat. Gasol. Manuf. 13 (8), 292–298.

Nesta, J., Bennett, C.A., 2004. Reduce fouling in shell-and-tube heat exchangers. Hydrocarb. Process. 83 (7), 77–82.

Nghiem, L.X., Hassam, M.S., Nutakki, R., George, A.E.D., 1993. Efficient modeling of asphaltene precipitation. In: SPE Annual Technical Conference and Exhibition. Society of Petroleum Engineers, Houston, TX, USA.

Nguyen, D.Q., Barbaro, A., Vipanurat, N., Bagajewicz, M., 2010. All-at-once and stepwise detailed retrofit of heat exchanger networks using an MILP model. Ind. Eng. Chem. Res. 49 (13), 6080–6103.

Orrego-Ruiz, J.A., Guzmań, A., Molina, D., Mejía-Ospino, E., 2011. Mid-infrared attenuated total reflectance (MIR-ATR) predictive models for asphaltene contents in vacuum residue: asphaltene structure—functionality correlations based on partial least-squares regression (PLS-R). Energy Fuels 25 (8), 3678–3686.

Ortega-Rodriguez, A., Duda, Y., Guevara-Rodriguez, F., Lira-Galeana, C., 2004. Stability and aggregation of asphaltenes in Asphaltene—Resin—Solvent mixtures. Energy Fuels 18 (3), 674–681.

Ostlund, J.A., Wattana, P., Nyden, M., Fogler, H.S., 2004. Characterization of fractionated asphaltenes by UV–vis and NMR self-diffusion spectroscopy. J. Colloid Interf. Sci. 271 (2), 372–380.

Pan, H., Firoozabadi, A., 1998. A thermodynamic micellization model for asphaltene precipitation: Part I: micellar size and growth. SPE Prod. Facil. 13 (2), 118–127.

Pan, M., Bulatov, I., Smith, R., Kim, J.K., 2013. Optimisation for the retrofit of large scale heat exchanger networks with different intensified heat transfer techniques. Appl. Therm. Eng. 53 (2), 373–386.

Panchal, C.B., 1988. Experimental investigation of marine bio-fouling and corrosion for tropical seawater. In: Melo, L.F., Bott, T.R., Bernardo, C. (Eds.), Fouling Science and Technology, NATO Science Series: E, May 18–30, 1987, Alvor, Portugal. Kluwer Academic, Dordrecht, pp. 241–247.

Panchal, C.B., Watkinson, A.P., 1993. Chemical Reaction Fouling Model for Single-phase Heat Transfer. Technical report. Argonne National Lab., IL (United States).

Panchal, C.B., Watkinson, A.P., 1994. Development of an Analytical Model for Organic Fluid Fouling. Technical report. Argonne National Laboratory, U.S. Department of Energy.

Panchal, C.B., Huangfu, E.P., 2000. Effects of mitigating fouling on the energy efficiency of crude–oil distillation. Heat Transf. Eng. 21 (3), 3–9.

Panchal, C.B., Kuru, W.C., Liao, C.F., Ebert, W.A., Palen, J.W., 1999. Threshold conditions for crude oil fouling. In: Bott, T.R., Melo, L.F., Panchal, C.B., Somerscales, E.F.C. (Eds.), Proc. Int. Conf. On Understanding Heat Exchanger Fouling and its Mitigation, May 1997, Castelvecchio Pascoli, Italy. Begell House, New York, pp. 273–282.

Panchal, C.B., 2001. Review of fouling mechanisms. In: Bott, T.R., Watkinson, A.P., Panchal, C.B. (Eds.), Proc. Int. Conf. Mitigation of Heat Exchanger Fouling and its Economic and Environmental Implications, July 18–23, 1999, Banff, Canada. Begell House, New York, pp. 8–15.

Panuganti, S.R., Tavakkoli, M., Vargas, F.M., Gonzalez, D.L., Chapman, W.G., 2013. SAFT model for upstream asphaltene applications. Fluid Phase Equilibr. 359, 2–16.

Papaioannou, V., Lafitte, T., Avendaño, C., Adjiman, C.S., Jackson, G., Müller, E.A., Galindo, A., 2014. Group contribution methodology based on the statistical associating fluid theory for heteronuclear molecules formed from Mie segments. J. Chem. Phys. 140, 054107.

Papalexandri, K., Pistikopoulos, E.N., 1993. An MINLP retrofit approach for improving the flexibility of heat exchanger networks. Ann. Oper. Res. 42 (1), 119–168.

Papalexandri, K.P., Pistikopoulos, E.N., 1994a. Synthesis and retrofit design of operable heat exchanger networks. 1. Flexibility and structural controllability aspects. Ind. Eng. Chem. Res. 33 (7), 1718–1737.

Papalexandri, K.P., Pistikopoulos, E.N., 1994b. Synthesis and retrofit design of operable heat exchanger networks. 2. Dynamics and control structure considerations. Ind. Eng. Chem. Res. 33 (7), 1738–1755.

Papoulias, S.A., Grossmann, I.E., 1983. A structural optimization approach in process synthesis–II: heat recovery networks. Comput. Chem. Eng. 7 (6), 707–721.

Paso, K.G., Fogler, H.S., 2003. Influence of n-paraffin composition on the ageing of wax-oil gel deposits. AIChE J. 49 (12), 3241–3252.

Patankar, S.V., 1980. Numerical Heat Transfer and Fluid Flow. CRC Press.

Patel, N.C., Teja, A.S., 1982. A new cubic equation of state for fluids and fluids mixtures. Chem. Eng. Sci. 77 (3), 463–473.

Paterson, W.R., Fryer, P.J., 1988. A reaction engineering approach to the analysis of fouling. Chem. Eng. Sci. 43 (7), 1714–1717.

Payne, F.A., Hicks, C.L., Pao-Sheng, S., 1993. Predicting optimal cutting time of coagulated milk using diffuse reflectance. J. Dairy Sci. 76 (1), 48–61.

Pedersen, K.S., Christensen, P.L., 2006. Phase Behaviour of Petroleum Reservoir Fluids. Taylor & Francis, Boca Raton.

Peneloux, A., Rauzy, E., Freze, R.A., 1982. Consistent correction for Redlich-Kwong-Soave volumes. Fluid Phase Equilibr. 8 (1), 7–23.

Peng, D.Y., Robinson, D.B., 1976. A new two-constant equation of state. Ind. Eng. Chem. Fundam. 15 (1), 59–64.

Peralta, J.M., Chew, Y.M.J., Wilson, D.I., 2011a. An analytical method for selecting the optimal nozzle external geometry for fluid dynamic gauging. Chem. Eng. Sci. 66 (16), 3579–3591.

Peralta, J.M., Chew, Y.M.J., Wilson, D.I., 2011b. Effect of nozzle external geometry on the pressure and shear stress exerted on the surface being gauged in fluid dynamic gauging. Chem. Eng. Res. Des. 88 (12), 2540–2551.

Pereira, A., Rosmaninho, R., Mendes, J., Melo, L.F., 2006. Monitoring deposit build-up using a novel mechatronic surface sensor (MSS). Food Bioprod. Process. 84 (4), 366–370.

Pereira, J.C., López, I., Salas, R., Silva, F., Fernández, C., Urbina, C., López, J.C., 2007. Resins: the molecules responsible for the stability/instability phenomena of asphaltenes. Energy Fuels 21 (3), 1317–1321.

Pereira, A., Mendes, J., Melo, L.F., 2009. Monitoring cleaning-in-place of shampoo films using nanovibration technology. Sens. Actuat. B-Chem. 136 (2), 376–382.

Pfeiffer, J.P.H., Saal, R.N.J., 1939. Asphaltic bitumen as colloid system. J. Phys. Chem. 44 (2), 139–149.

Phillips, D. Z., 1999. Mitigation of crude oil fouling by the use of hiTRAN inserts (Ph.D. Thesis). University of Bath, UK.

Pilavachi, P.A., Isdale, J.D., 1993. European community R&D strategy in the field of heat exchanger fouling: projects. Heat Recov Syst. CHP 13 (2), 133–138.

Pina, A., Mougin, P., Béhar, E., 2006. Characterisation of asphaltenes and modelling of flocculation – state of the art. Oil Gas Sci. Technol. 61 (3), 319–343.

Pinilla, J.L., Arcelus-Arrillaga, P., Puron, H., Millan, M., 2013. Selective catalytic steam cracking of anthracene using mesoporous Al_2O_3 supported Ni-based catalysts doped with Na, Ca or K. Appl. Catal. A-Gen. 459, 17–25.

Poddar, T.K., Polley, G.T., 1996. Heat exchanger design through parameter plotting. Chem. Eng. Res. Des. 74 (8), 849–852.

Pogiatzis, T., Ishiyama, E.M., Paterson, W.R., Vassiliadis, V.S., Wilson, D.I., 2012. Identifying optimal cleaning cycles for heat exchangers subject to fouling and ageing. Appl. Energy 81 (1), 60–66.

Polley, G.T., Wilson, D.I., Yeap, B.L., Pugh, S.J., 2002a. Evaluation of laboratory crude oil threshold fouling data for application to refinery pre–heat trains. Appl. Therm. Eng. 22 (7), 777–788.

Polley, G.T., Wilson, D.I., Yeap, B.L., Pugh, S.J., 2002b. Use of crude oil fouling threshold data in heat exchanger design. Appl. Therm. Eng. 22 (7), 763–776.

Polley, G.T., Wilson, D.I., Petitjean, E., Derouin, C., 2005. The fouling limit in crude oil preheat train retrofits. Hydrocarb. Process. 84 (7), 1–7.

Polley, G., Morales Fuentes, A., Pugh, S.J., 2009a. Design of shell–and–tube heat exchangers to achieve a specified operating period in refinery pre–heat trains. In: Müller-Steinhagen, H., Malayeri, M.R., Watkinson, A.P. (Eds.), Int. Conf. On Heat Exchanger Fouling and Cleaning VIII, June 14–19, 2009, Schladming, Austria, pp. 346–351. Published online. www.heatexchanger-fouling.com.

Polley, G.T., Morales-Fuentes, A., Wilson, D.I., 2009b. Simultaneous consideration of flow and thermal effects of fouling in crude oil preheat trains. Heat. Transf. Eng. 30 (10), 815–821.

Polley, G.T., 2010. Review of the development of models for prediction of fouling rates in exchangers heating crude oil. In: 11th Annual International Conference, Petroleum Phase Behaviour and Fouling, June 13–17, Jersey City, NJ.

Ponce-Ortega, J.M., Serna-González, M., Salcedo-Estrada, L.I., Jiménez-Gutiérrez, A., 2006. Minimum–investment design of multiple shell and tube heat exchangers using a MINLP formulation. Chem. Eng. Res. Des. 84 (10), 905–910.

Ponce-Ortega, J.M., Serna- González, M., Jiménez-Gutiérrez, A., 2007. Heat exchanger network synthesis including detailed heat exchanger design using genetic algorithms. Ind. Eng. Chem. Res. 46 (25), 8767–8780.

Pope, W.L., Pines, H.S., Fulton, R.L., Doyle, P.A., 1978a. Heat exchanger design: why guess a design fouling factor when it can be optimized. In: ASME Energy Technology Conference, Houston, TX, USA.

Pope, W.L., Pines, H.S., Silvester, L.F., Doyle, P.A., Fulton, R.L., Green, M.A., 1978b. Optimizations of geothermal cycle shell and tube exchangers of various configurations with variable fluid properties and site specific fouling. In: Geothermal Resources Council Transactions, Hilo, HI, USA. Geothermal Resources Council, Davis, California, USA, pp. 539–542.

Pritchard, A.M., 1979. Heat Exchanger Fouling in British Industry. Fouling prevention research digest, 1(1/4), IV-VI.

Pritchard, A.M., 1988a. Deposition of hardness salts. In: Melo, L.F., Bott, T.R., Bernardo, C. (Eds.), Fouling Science and Technology, NATO Science Series: May 18–30, 1987, Alvor, Portugal. Kluwer Academic, Dordrecht, pp. 261–273.

Pritchard, A.M., 1988b. The economics of fouling. In: Melo, L.F., Bott, T.R., Bernardo, C. (Eds.), Fouling Science and Technology, NATO Science Series: E, May 18–30, 1987, Alvor, Portugal. Kluwer Academic, Dordrecht, pp. 31–45.

Prakash, S., Datta, N., Deeth, H.C., 2005. Methods of detecting fouling caused by heating of milk. Food Rev. Int. 21 (3), 267–293.

PSE, 2009. Parameter estimation in gPROMS. In: gPROMS Advanced User Guide 3.2. Process Systems Enterprise, London.

Punnapala, S., Vargas, F.M., 2013. Revisiting the PC-SAFT characterization procedure for an improved asphaltene precipitation prediction. Fuel 108, 417–429.

Purón, H., Pinilla, J.L., Berrueco, C., Montoya de la Fuente, J.A., Millan, M., 2013. Hydrocracking of maya vacuum residue with NiMo catalysts supported on mesoporous alumina and silica alumina. Energy Fuels 27 (7), 3952–3960.

Rabas, T.J., Panchal, C.B., 2000. Fouling rates, not fouling resistances. Heat. Transf. Eng. 21 (2), 1–3.

Radhakrishnan, V.R., Ramasamy, M., Zabiri, H., Do Thanh, V., Tahir, N.M., Mukhtar, H., Hamdi, M.R., Ramli, N., 2007. Heat exchanger fouling model and preventive maintenance scheduling tool. Appl. Therm. Eng 27 (17-18), 2791–2802.

Rahmani, S., McCaffrey, W., Gray, M.R., 2002. Kinetics of solvent interactions with asphaltenes during coke formation. Energy Fuels 16 (1), 148–154.

Rao, K.R., Shrinivasa, U., Srinivasan, J., 1991. Synthesis of cost–optimal shell–and–tube heat exchangers. Heat Transfer Eng. 12 (3), 47–55.

Rao, K.R., Shrinivasa, U., Srinivasan, J., 1996. Constrained optimization of heat exchangers. Chem. Eng. 103 (6), 112–118.

Ratel, M., Kapoor, Y., Anxionnaz, Z., Seminel, I., Vinet, B., 2013. Investigation of fouling rates in a heat exchanger using an innovative fouling rig. In: Malayeri, M.R., Müller-Steinhagen, H., Watkinson, A.P. (Eds.), Proc. Int. Conf. On Heat Exchanger Fouling and Cleaning X, June 9–14, 2013, Budapest, Hungary, pp. 36–41. Published online. www.heatexchanger-fouling.com.

Ravagnani, M.A.S.S., Caballero, J.A., 2007a. A MINLP model for the rigorous design of shell and tube heat exchangers using the TEMA standards. Chem. Eng. Res. Des. 85 (10), 1423–1435.

Ravagnani, M.A.S.S., Caballero, J.A., 2007b. Optimal heat exchanger network synthesis with the detailed heat transfer equipment design. Comput. Chem. Eng. 31 (11), 1432–1448.

Ravagnani, M.A.S.S., Silva, A.P., Biscaia, E.C., Caballero, J.A., 2009. Optimal design of shell–and–tube heat exchangers using particle swarm optimization. Ind. Eng. Chem. Res. 48 (6), 2927–2935.

Redlich, O., Kwong, J.N.S., 1949. On the thermodynamics of solutions V. An equation of state. Fugacities of gaseous solutions. Chem. Rev. 44 (1), 233–244.

Reppich, M., Kohoutek, J., 1994. Optimal design of shell–and–tube heat exchangers. Comput. Chem. Eng. 18 (1), S295–S299.

Reppich, M., Zagermann, S., 1995. A new design method for segmental baffled heat–exchangers. Comput. Chem. Eng. 19 (1), 137–142.

Ritchie, J.M., Droegemueller, P., Simmons, M.J.H., 2009. hiTRAN wire matrix inserts in fouling applications. Heat Transfer Eng. 30 (10), 876–884.

Rodriguez, C., Smith, R., 2007. Optimization of operating conditions for mitigating fouling in heat exchanger networks. Chem. Eng. Res. Des. 85 (6), 839–851.

Rostani, K., 2014. Effect of crude oil Vanadium Porphyrin Content and blending on heat exchanger fouling (Ph.D. Thesis). Department of Chemical Engineering and Chemical Technology, Imperial College London, UK.

Russell, G.A., 1956. Oxidation of unsaturated compounds. III. Products of the reaction of indene and oxygen; stereochemistry of the addition of a peroxy radical and oxygen to a double bond. J. Am. Chem. Soc. 79 (5), 1035–1040.

Sabbagh, O., Akbarzadeh, K., Badamchi-Zadeh, A., Svrcek, W.Y., Yarranton, H.W., 2006. Applying the PR-EOS to asphaltene precipitation from n-alkane dilute heavy oils and bitumens. Energy Fuels 20 (2), 625–634.

Sabbah, H., Pomerantz, A.E., Wagner, M., Müllen, K., Zare, R.N., 2012. Laser desorption single-photon ionization of asphaltenes: mass range, compound sensitivity and matrix effects. Energy Fuels 26 (6), 3521–3526.

Saffaravval, M., Damangir, E., 1995. A general correlation for determining optimum bae spacing for all types of shell and tube exchangers. Int. J. Heat Mass Transfer 38 (13), 2501–2506.

Saikhwan, P., Chew, Y.M.J., Paterson, W.R., Wilson, D.I., 2007. Fluid dynamic gauging: a technique for studying the cleaning of food process surfaces. Food Manuf. Effic. 1 (2), 34–41.

Sanaye, S., Niroomand, B., 2007. Simulation of heat exchanger network (HEN) and planning the optimum cleaning schedule. Energy Convers. Manage 48 (5), 1450–1461.

Saleh, Z.S., Sheikholeslami, R., Watkinson, A.P., 2003. Fouling characteristics of a light Australian crude oil. In: Watkinson, A.P., Müller-Steinhagen, H., Malayeri, M.R. (Eds.), Proc. Heat Exchanger Fouling and Cleaning: Fundamentals and Applications, Engineering Conferences International, Santa Fe, New Mexico. Engineering Conferences International Digital Archives, pp. 226–233.

Saleh, Z.S., Sheikholesalami, R., Watkinson, A.P., 2005a. Blending effects on fouling of four crude oils. In: Müller-Steinhagen, H., Malayeri, M.R., Watkinson, A.P. (Eds.), 6th Int. Conf. On Heat Exchanger Fouling and Cleaning — Challenges and Opportunities, Volume RP2 of ECI Symposium Series, June 5–10, 20 Kloster Irsee, Germany. Engineering Conferences International Digital Archives, pp. 37–46.

Saleh, Z.S., Sheikholeslami, R., Watkinson, A.P., 2005b. Fouling characteristics of a light Australian crude oil. Heat Transfer Eng. 26 (1), 15–22.

Saghatoleslami, N., Salooki, M.K., Armin, M.A., 2010. Prediction of thickness and fouling rate in pulsating flow heat exchangers, using fluent simulator. Korean J. Chem. Eng. 27 (1), 96–103.

Santos, O., Nylander, T., Rosmaninho, R., Rizzo, G., Yiantsios, S., Andritsos, N., Karabelas, A., Müller-Steinhagen, H., Melo, L., BoulangPetermann, L., Gabet, C., Braem, A., Trgårdh, C., Paulsson, M., 2004. Modified stainless steel surfaces targeted to reduce fouling–surface characterization. J. Food Eng. 64 (1), 63–79.

Saxe, J.G., 1872. The blind men and the elephant, retrieved from en.wikisource.org/wiki/The_Blindmen_and_the_Elephant (14.05.2014).

Schenk, H.J., Witte, E.G., Littke, R., Schwochau, K., 1990. Structural modifications of vitrinite and alginite concentrates during pyrolitic maturation at different heating rates. A combined infrared, 13C NMR and microscopical study. Org. Geochem. 16 (4–6), 943–950.

Schöler, M., Fuchs, T., Helbig, M., Augustin, W., Scholl, S., Majschak, J.-P., 2009. Monitoring of the local cleaning efficiency of pulsed flow cleaning procedures. In: Müller-Steinhagen, H., Malayeri, M.R., Watkinson, A.P. (Eds.), Int. Conf. On Heat Exchanger Fouling and Cleaning VIII, June 14–19, 2009, Schladming, Austria, pp. 455–463. Published online. www.heatexchanger-fouling.com.

Schreier, P.J.R., Fryer, P.J., 1995. Heat exchanger fouling: a model study of the scaleup of laboratory data. Chem. Eng. Sci. 50 (8), 1311–1321.

Scott, R.L., 1945. The thermodynamics of high-polymer solutions: II. The solubility and fractionation of a polymer of heterogeneous distribution. J. Chem. Phys. 13 (5), 178–187.

Scott, R.L., Magat, M., 1945. The thermodynamics of high-polymer solutions: I. The free energy of mixing of solvents and polymers of heterogeneous distribution. J. Chem. Phys. 13 (5), 172–177.

Sedghi, M., Goual, L., 2014. PC-SAFT modeling of asphaltene phase behavior in the presence of nonionic dispersants. Fluid Phase Equilibr. 369, 86–94.

Selbaş, R., Kizilkan, O., Reppich, M., 2006. A new design approach for shell–and–tube heat exchangers using genetic algorithms from economic point of view. Chem. Eng. Process. 45 (4), 268–275.

Serna-González, M., Ponce-Ortega, J.M., Jiménez-Gutiérrez, A., 2004. Two–level optimization algorithm for heat exchanger networks including pressure drop considerations. Ind. Eng. Chem. Res. 43 (21), 6766–6773.

Serna-González, M., Ponce-Ortega, J.M., Castro-Montoya, A.J., Jiménez-Gutiérrez, A., 2006. Feasible design space for shell-and-tube heat exchangers using the Bell-Delaware method. Ind. Eng. Chem. Res. 46 (1), 143–155.

Shah, R.K., Sekulic, D.P., 2003. Fundamentals of Heat Exchanger Design, first ed. Wiley & Sons, New York.

Shanrang, Y., Shengxian, C., Yanling, C., Hua, Y., Kangle, Z., Xu, Z., 2009. An online real-time monitoring of fouling layer thickness based on AC impedance technology. In: Presented at: International Conference on Heat Exchanger Fouling and Cleaning, June 14–19, 2009, Schladming, Austria. Published online. www.heatexchanger-fouling.com.

Sheikh, A.K., Zubair, S.M., Haq, M.U., Budair, M.O., 1996. Reliability–based maintenance strategies for heat exchangers subject to fouling. J. Energ. Resour–ASME 118 (4), 306–312.

Sheikh, A.K., Zubair, S.M., Younas, M., Budair, M.O., 1997. Some remarks on the probabilistic basis for characterizing fouling data. In: Bott, T.R., Melo, L.F., Panchal, C.B., Somerscales, E.F.C. (Eds.), Proc. Int. Conf. On Understanding Heat Exchanger Fouling and its Mitigation, May 1997, Castelvecchio Pascoli, Italy. Begell House, New York, pp. 393–400.

Sheikh, A.K., Zubair, S.M., Younas, M., Budair, M.O., 2000. A risk based heat exchanger analysis subject to fouling. Part II: economics of heat exchangers cleaning. Energy 25 (5), 445–461.

Sheikh, A.K., Raza, M.K., Zubair, S.M., Budair, M.O., 2001. Predicting level of fouling using neural networks approach. In: Bott, T.R., Watkinson, A.P., Panchal, C.B. (Eds.), Proc. Int. Conf. Mitigation of Heat Exchanger Fouling and its Economic and Environmental Implications, July 18–23, 1999, Banff, Canada. Begell House, New York, pp. 27–41.

Sheremata, J.M., Gray, M.R., Dettman, H.D., McCaffrey, W.C., 2004. Quantitative molecular representation and sequential optimization of Athabasca asphaltenes. Energy Fuels 18 (5), 1377–1384.

Shick, R.A., Koenig, J.L., Ishida, H., 1996. Depth profiling of stratified layers using variable-angle ATR. Appl. Spectrosc. 50 (8), 1082–1088.

Shilling, R.L., Rudy, M.P., Rudy, T.M., 2009. Risk-based design margin selection for heat exchangers. In: Müller-Steinhagen, H., Malayeri, M.R., Watkinson, A.P. (Eds.), Int. Conf. On Heat Exchanger Fouling and Cleaning VIII, June 14–19, 2009, Schladming, Austria, pp. 340–345. Published online. www.heatexchanger-fouling.com.

Shirani, B., Nikazar, M., Mousavi-Dehghani, S.A., 2012. Prediction of asphaltene phase behaviour in live oil with CPA equation of state. Fuel 97, 89–96.

Siddiqui, M.N., 2003. Infrared study of hydrogen bond types in asphaltenes. Pet. Sci. Technol. 21 (9–10), 1601–1615.

Sikos, L., Klemes, J., 2010. Reliability, availability and maintenance optimisation of heat exchanger networks. Appl. Therm. Eng. 30 (1), 63–69.

Sileri, D., Sahu, K., Ding, H., Matar, O.K., 2009. Mathematical modelling of asphaltenes deposition and removal in crude distillation units. In: Müller-Steinhagen, H., Malayeri, M.R., Watkinson, A.P. (Eds.), Int. Conf. On Heat Exchanger Fouling and Cleaning VIII, June 14–19, 2009, Schladming, Austria, pp. 245–251. Published online. www.heatexchanger-fouling.com.

Sileri, D., 2011. Modelling of Two Phase Flow with Application to Asphaltenes Deposition and Removal in Crude-Oil Distillation Units (Ph.D. Thesis). Department of Chemical Engineering and Chemical Technology, Imperial College London, UK.

Sirota, E.B., 2005. Physical structure of asphaltenes. Energy Fuels 19 (4), 1290–1296.

Skjold-Jørgensen, S., 1984. Gas solubility calculations. II. Application to a new group contribution equation of state. Fluid Phase Equilibr. 16 (3), 317–351.

Smaïli, F., Vassiliadis, V.S., Wilson, D.I., 2001. Mitigation of fouling in refinery heat exchanger networks by optimal management of cleaning. Energy Fuels 15 (5), 1038–1056.

Smagorinsky, J., 1963. General circulation experiments with the primitive equations I. The basic Experiment. Mon. Wea. Rev. 91, 99–164.

Smith, E.M., 1979. Effectiveness-NTU relationships for tubular exchangers. Int. J. Heat. Fluid Fl. 1 (1), 43–46.

Smith, A.D., 2013. Analysis of fouling rate and propensity for eight crude oil samples in annular test section. In: Malayeri, M.R., Müller-Steinhagen, H., Watkinson, A.P. (Eds.), Proc. Int. Conf. On Heat Exchanger Fouling and Cleaning X, June 9–14, 2013, Budapest, Hungary, pp. 1–8. Published online. www.heatexchanger-fouling.com.

Soave, G., 1972. Equilibrium constants from a modified Redlich-Kwong equation of state. Chem. Eng. Sci. 27 (6), 1197–1203.

Somerscales, E.F.C., Sanatgar, H., 1989. Investigation of the deposit removal processes on a fouled surface. In: Shah, R. (Ed.), National Heat Transfer Conf. On Heat Transfer Equipment Fundamentals, Design, Applications and Operating Problems. USA. American Society of Mechanical Engineers, Philadelphia, PA, pp. 261–271.

Somerscales, E.F.C., 1990. Fouling of heat transfer surfaces: an historical review. Heat Transfer Eng 11 (1), 19–36.

Somerscales, E.F.C., 1997. Fundamentals of corrosion fouling. Exp. Therm. Fluid Sci. 14 (4), 335–355.

Somerscales, E.F.C., 1999. Mitigation of corrosion fouling. In: Bott, T.R., Melo, L.F., Panchal, C.B., Somerscales, E.F.C. (Eds.), Proc. Int. Conf. On Understanding Heat Exchanger Fouling and its Mitigation, May 1997, Castelvecchio Pascoli, Italy. Begell House, New York, pp. 23–53.

Sprague, D.E., Roy, K., 1990. Statistical determination of the performance and coking rate of fired heaters. Chem. Eng. Prog. 86 (8), 14–20.

Sprague, M., Herrara, P., Krzywicki, A., 2005. Fouling behaviour of pyrolysis gasoline over carbon steel and stainless steel alloys. In: Müller-Steinhagen, H., Malayeri, M.R., Watkinson, A.P. (Eds.), 6th Int. Conf. On Heat Exchanger Fouling and Cleaning — Challenges and Opportunities, Volume RP2 of ECI Symposium Series, Kloster Irsee, Germany, June 5–10, 2005. Engineering Conferences International Digital Archives, pp. 37–46.

Speight, J.G., 1999. The Chemistry and Technology of Petroleum, third ed. Marcel Dekker, New York.

Srinivasan, M., Watkinson, A.P., 2005. Fouling of some Canadian crude oils. Heat Transfer Eng. 26 (1), 7–14.

Stark, J.L., Asomaning, S., 2003. Crude oil blending effects on asphaltene stability in refinery fouling. Pet. Sci. Technol. 21 (3), 569–579.

Stephenson, T., Kubis, A., Derakhshesh, M., Hazelton, M., Holt, C., Eaton, P., Newman, B., Hoff, A., Gray, M., Mitlin, D., 2011. Corrosion fouling of 316 stainless steel and pure iron by hot oil. Energy Fuels 25 (10), 4540–4551.

Sun, S.-Y., Lu, Y.-D., Yan, C.-Q., 1993. Optimization in calculation of shell–tube heat exchanger. Int. Commun. Heat Mass 20 (5), 675–685.

Sun, X.G., 2005. The investigation of chemical structure of coal macerals via transmitted-light FT-IR microspectroscopy. Spectrochim Acta A 62 (1–3), 557–564.

Sundaram, K.M., Froment, G.F., 1979. Kinetics of coke deposition in the thermal cracking of propane. Chem. Eng. Sci. 34 (5), 635–644.

Suphanit, B., 1999. Design of complex distillation system (Ph.D. Thesis). UMIST, University of Manchester, UK.

Svendsen, J.A., 1993. Mathematical-modeling of wax deposition in oil pipeline systems. AIChE J. 39 (8), 1377–1388.

Szewczyk, V., Béhar, E., 1999. Compositional model for predicting asphaltenes flocculation. Fluid Phase Equilibr. 158–160, 459–469.

Szklo, A., Schaeffer, R., 2007. Fuel specification, energy consumption and CO_2 emission in oil refineries. Energy 32 (7), 1075–1092.

Taborek, J., Aoki, T., Ritter, R.B., Palen, J.W., Knudsen, J.G., 1972. Fouling: the major unresolved problem in heat transfer. Chem. Eng. Prog. 68 (2), 59–67.

Taborek, J., 1978. Heat exchanger design. In: Hooper, F.C. (Ed.), 6th Int. Heat Transfer Conf., Volume VI, Toronto, ON, Canada. Hemisphere, Washington, pp. 269–283.

Taborek, J., 1979. Evolution of heat exchanger design techniques. Heat Transfer Eng. 1 (1), 15–29.

Taborek, J., 1996. Assessment of fouling research on the design of heat exchangers. In: Panchal, C.B. (Ed.), Fouling Mitigation of Industrial Heat-exchange Equipment, June 18–23, 1995, San Luis Obispo, CA, USA. Begell House, New York, pp. 27–39.

Taborek, J., 2008a. Shell-and-tube heat exchangers: single-phase flow. In: Hewitt, G.F. (Ed.), Heat Exchanger Design Handbook. Begell House, New York.

Taborek, J., 2008b. Survey of shell-side flow correlations. In: Hewitt, G.F. (Ed.), Heat Exchanger Design Handbook. Begell House, New York.

Thackery, P.A., 1979. The Cost of Fouling in Heat Exchanger Plant. In: Proc. Conf. Fouling - Science or Art? University of Surrey, Guildford. Institution of Corrosion Science and Technology and Institution of Chemical Engineers, London.

Takatsuka, T., Kajiyama, R., Hashimoto, H., Matsuo, I., Hanazawa, T., 1989. A tubular fouling model for residue cracking furnaces. J. Chem. Eng. Jpn. 22 (2), 149–154.

Takemoto, T., Crittenden, B.D., Kolaczkowski, S.T., 1999. Interpretation of fouling data in industrial shell and tube heat exchangers. Transic. Part A 77 (8), 769–778.

Tamachkiarowa, A., Flemming, H.C., 1999. Optical fiber sensor for biofouling monitoring. In: Bott, T.R., Melo, L.F., Panchal, C.B., Somerscales, E.F.C. (Eds.), Proc. Int. Conf. On Understanding Heat Exchanger Fouling and its Mitigation, May 1997, Castelvecchio Pascoli, Italy. Begell House, New York, pp. 333–348.

Tamakloe, E.K., Polley, G.T., Picón-Núñez, M., 2013. Design of Compabloc exchangers to mitigate refinery fouling. Appl. Therm. Eng. 60 (1–2), 441–448.

Tan, S.P., Adidahrma, H., Radosz, M., 2008. Recent advances and applications of statistical associating fluid theory. Ind. Eng. Chem. Res. 47 (21), 8063–8082.

Tavares, V., Costa, A., Queiroz, E., Pessoa, F., Liporace, F., Oliveira, S., 2009. Thermo-hydraulic simulation of heat exchanger networks subjected to fouling. In: de Brito Alves, R., Oller do Nascimento, C., Chalbaud Biscaia Jr., E. (Eds.), 10th Int. Symposium on Process Systems Engineering (PSE), August 16–20, 2009, Salvador de Bahia, Brazil. Elsevier, Amsterdam, pp. 591–596.

Tay, F.H., Kazarian, S.G., 2009. Study of petroleum heat-exchanger deposits with ATR-FTIR spectroscopic imaging. Energy Fuels 23 (8), 4059–4067.

Tayal, M.C., Fu, Y., Diwekar, U.M., 1998. Optimal design of heat exchangers: a genetic algorithm framework. Ind. Eng. Chem. Res. 38 (2), 456–467.

Taylor, W.F., Wallace, T.J., 1967. Kinetics of deposit formation from hydrocarbon fuels at high temperatures. General features of the process. Ind. Eng. Chem. Prod. Res. Dev. 6 (4), 258–262.

Taylor, W.F., Wallace, T.J., 1968. Kinetics of deposit formation from hydrocarbons. Effect of trace sulfur compounds. Ind. Eng. Chem. Prod. Res. Dev. 7 (3), 198–202.

Tayyabkhan, M.T., 1962. Heat exchanger design by computer. Ind. Eng. Chem. 54 (10), 25–29.

TEMA, 1941. Standards of the Tubular Exchangers Manufacturers Association. Technical report. Tubular Exchangers Manufacturers Association.

Ten Broeck, H., 1944. Economic selection of exchanger sizes. Ind. Eng. Chem. 36 (1), 64–67.

Tihic, A., Kontogeorgis, G.M., von Solms, N., Michelsen, M.L., 2006. Applications of the simplified perturbed-chain SAFT equation of state using an extended parameter table. Fluid Phase Equilibr. 248 (1), 29–43.

Ting, P.D., Hirasaki, G.J., Chapman, W.G., 2003. Modeling of asphaltene phase behavior with the SAFT equation of state. Petrol. Sci. Technol. 21 (3), 647–661.

Tornqvist, A., 2007. A survey of Alfa Laval's experience of fouling prevention in plate heat exchanger systems. In: Müller-Steinhagen, H., Malayeri, M.R., Watkinson, A.P. (Eds.), Int. Conf. On Heat Exchanger Fouling and Cleaning VIII, June 14–19, 2009, Schladming, Austria, pp. 344–345. Published online. www.heatexchanger-fouling.com.

Tuladhar, T.R., Paterson, W.R., Wilson, D.I., Macleod, N., 2000. Development of a novel non-contact proximity gauge for thickness measurement of soft deposits and its application in fouling studies. Can. J. Chem. Eng. 78 (5), 925–947.

Tuladhar, T.R., Paterson, W.R., Wilson, D.I., 2002. Investigation of alkaline cleaning-in-place of whey protein deposits using dynamic gauging. Food Bioprod. Process. 80 (C3), 199–214.

Umeda, T., Itoh, J., Shiroko, K., 1978. Heat exchange system synthesis. Chem. Eng. Prog. 74 (7), 70–76.

Ungerer, P., Rigby, D., Leblanc, B., Yiannourakou, M., 2014. Sensitivity of the aggregation behaviour of asphaltenes to molecular weight and structure using molecular dynamics. Mol. Simul. 40 (1–3), 115–122.

Vakili-Nezhaad, G.R., Modarress, H., Mansoori, G.A., 2001. Continuous Thermodynamics of petroleum fluid fractions. Chem. Eng. Process. 40, 431–435.

Valderrama, J.O., 2003. The state of the cubic equations of state. Ind. Eng. Chem. Res. 42 (8), 1603–1618.

van der Waals, J.D., 1873. On the Continuity of the Gaseous and Liquid States (Ph.D. Thesis). Universiteit Leiden, Netherlands.

Van Nostrand, W.L.J., Leach, S.H., Haluska, J.L., 1981. Economic penalties associated with the fouling of refinery heat transfer equipment. In: Somerscales, E.F.C., Knudsen, J.G. (Eds.), Fouling of Heat Transfer Equipment, August 1979, Troy, New York. Hemisphere, Washington, pp. 619–643.

Vargas, F.M., Gonzalez, D.L., Creek, J.L., Wang, J., Buckley, J., Hirasaki, G.J., Chapman, W.G., 2009a. Development of a general method for modeling asphaltene stability. Energy Fuels 23, 1147–1154.

Vargas, F.M., Gonzalez, D.L., Hirasaki, G.J., Chapman, W.G., 2009b. Modeling asphaltene phase behavior in crude oil systems using the perturbed chain form of the statistical associating fluid theory (PC-SAFT) equation of state. Energy Fuels 23 (3), 1140–1146.

Venditti, S., Berrueco, C., Alvarez, P., Morgan, T.J., Millan-Agorio, M., Herod, A.A., Kandiyoti, R., 2009a. Deposition of solids from a petroleum heptane-soluble fraction upon heating in a micro-bomb reactor. In: Müller-Steinhagen, H., Malayeri, M.R., Watkinson, A.P. (Eds.), Int. Conf. On Heat Exchanger Fouling and Cleaning VIII, June 14–19, 2009, Schladming, Austria, pp. 86–93. Published online. www.heatexchanger-fouling.com.

Venditti, S., Berrueco, C., Alvarez, P., Morgan, T.J., Millan-Agorio, M., Herod, A.A., Kandiyoti, R., 2009b. Developing characterisation methods for foulants deposited in refinery heat exchangers. In: Müller-Steinhagen, H., Malayeri, M.R., Watkinson, A.P. (Eds.), Int. Conf. On Heat Exchanger Fouling and Cleaning VIII, June 14–19, 2009, Schladming, Austria, pp. 44–51. Published online. www.heatexchanger-fouling.com.

Venditti, S., 2010. Characterization of Petroleum Fouling with Respect to Operational parameters (Ph.D. Thesis). Department of Chemical Engineering and Chemical Technology, Imperial College London, UK.

Victorov, A.I., Firoozabadi, A., 1996. Thermodynamic micellization model of asphaltene precipitation from petroleum fluids. AIChE J. 42 (6), 1753–1764.

Visser, J., 1988a. Adhesion and removal of particles I. In: Melo, L.F., Bott, T.R., Bernardo, C. (Eds.), Fouling Science and Technology, NATO Science Series: E, May 18–30, 1987, Alvor, Portugal. Kluwer Academic, Dordrecht, pp. 87–104.

Visser, J., 1988b. Adhesion and removal of particles II. In: Melo, L.F., Bott, T.R., Bernardo, C. (Eds.), Fouling Science and Technology, NATO Science Series: E, May 18–30, 1987, Alvor, Portugal. Kluwer Academic, Dordrecht, pp. 104–123.

Visser, H., Jeurnink, T.J.M., 1997. General aspects of fouling and cleaning. In: Visser, H. (Ed.), Bulletin of the International Dairy Federation, 328. IDF, Brussels, Belgium, pp. 7–31.

Vogel, J.L., Turek, E.A., Metcalfe, R.S., Bergman, D.F., Morris, R.W., 1983. Applications of equations of state to calculate reservoir fluid properties. Fluid Phase Equilibr. 14, 103–116.

von Solms, N., Michelsen, M.L., Kontogeorgis, G.M., 2003. Computational and physical performance of a modified PC-SAFT equation of state for highly asymmetric and associating mixtures. Ind. Eng. Chem. Res. 42 (5), 1098–1105.

Walker, R.A., Bott, T.R., 1973. Effect of roughness on heat exchanger tubes. Chem. Eng. 271, 151.

Wang, J.X., Buckley, J.S., 2001. A two-component solubility model of the onset of asphaltene flocculation in crude oils. Energy Fuels 15 (5), 1004–1012.

Wang, W., Watkinson, A.P., 2011. Iron sulphide and coke fouling from sour oils: a review and initial experiments. In: Mayaleri, M.R., Müller-Steinhagen, H., Watkinson, A.P. (Eds.), Proc. Int. Conf. On Heat Exchanger Fouling and Cleaning IX, June 5–10, 2011, Crete, Greece, pp. 23–30. Published online. www.heatexchanger-fouling.com.

Wang, W., Watkinson, A.P., 2013. Deposition from a sour heavy oil under incipient coking conditions: effects of surface material and temperature. In: Malayeri, M.R., Müller-Steinhagen, H., Watkinson, A.P. (Eds.), Proc. Int. Conf. On Heat Exchanger Fouling and Cleaning X, June 9–14, 2013, Budapest, Hungary, pp. 9–17. Published online. www.heatexchanger-fouling.com.

Wang, Q., Chen, Q., Chen, G., Zeng, M., 2009. Numerical investigation on combined multiple shell–pass shell–and–tube heat exchanger with continuous helical baes. Int. J. Heat Mass Transfer 52 (5-6), 1214–1222.

Waters, A.J., Akinradewo, C.G., Lamb, D., 2009. Fouling: implementation of a crude preheat train performance monitoring application at the Irving oil refinery. In: Müller-Steinhagen, H., Malayeri, M.R., Watkinson, A.P. (Eds.), Int. Conf. On Heat Exchanger Fouling and Cleaning VIII, June 14–19, 2009, Schladming, Austria, pp. 33–38. Published online. www.heatexchanger-fouling.com.

Watkinson, A.P., Epstein, N., 1970. Particulate fouling of sensible heat exchangers. In: Proc. 4th Int. Heat Transfer Conf., Vol. 1, Paper HE1.6, Paris, France. Elsevier, Amsterdam.

Watkinson, A.P., 1988. Critical Review of Organic Fluid Fouling. Technical report. Argonne National Laboratory.

Watkinson, A.P., Wilson, D.I., 1997. Chemical reaction fouling: a review. Exp. Therm. Fluid Sci. 14 (4), 361–374.

Watkinson, A.P., Navaneetha-Sundaram, B., Posarac, D., 2000. Fouling of a sweet crude oil under inert and oxygenated conditions. Energy Fuels 14 (1), 64–69.

Watkinson, A.P., 2005. Deposition from crude oils in heat exchangers. In: Müller-Steinhagen, H., Malayeri, M.R., Watkinson, A.P. (Eds.), 6th Int. Conf. On Heat Exchanger Fouling and Cleaning — Challenges and Opportunities, Volume RP2 of ECI Symposium Series, June 5–10, 2005, Kloster Irsee, Germany. Engineering Conferences International Digital Archives, pp. 37–46.

Watkinson, A.P., 2007. Deposition from crude oils in heat exchangers. Heat Transfer Eng. 28 (3), 177–184.

Watkinson, A.P., 2008. Seminar on Crude Oil Fouling in Refineries: Current Understanding and Mitigation Methods. April, 10–11. Jesus College, Cambridge, UK.

Wertheim, M., 1984a. Fluids with highly directional attractive forces. 1. Statistical Thermodynamics. J. Stat. Phys. 35 (1–2), 19–34.

Wertheim, M., 1984b. Fluids with highly directional attractive forces. 2. Thermodynamic perturbation-theory and integral-equations. J. Stat. Phys. 35 (1–2), 35–47.

Wertheim, M., 1986a. Fluids with highly directional attractive forces. 3. Multiple attractive sites. J. Stat. Phys. 42 (3–4), 459–476.

Wertheim, M., 1986b. Fluids with highly directional attractive forces. 4. Equilibrium polymerization. J. Stat. Phys. 42 (3–4), 477–492.

Westbrook, G., 1961. Use this method to size each stage for best operation. Hydrocarb. Process. Pet. Refin. 40 (9), 201–206.

Wiehe, I.A., Kennedy, R.J., 2000. The oil compatibility model and crude oil incompatibility. Energy Fuels 14 (1), 56–59.

Wiehe, I.A., Kennedy, R.J., Dickakian, G., 2001. Fouling of nearly incompatible oils. Energy Fuels 15 (5), 1057–1058.

Wiehe, I.A., 2001a. Fouling of heat exchangers by blends of incompatible oils. In: Bott, T.R., Watkinson, A.P., Panchal, C.B. (Eds.), Proc. Int. Conf. Mitigation of Heat Exchanger Fouling and its Economic and Environmental Implications, July 18–23, 1999, Banff, Canada. Begell House, New York, pp. 198–205.

Wiehe, I.A., 2001b. Mitigating fouling by self-incompatible or by nearly incompatible petroleum crude oils. In: Müller-Steinhagen, H. (Ed.), Heat Exchanger Fouling—fundamental Approaches and Technical Solutions. July 8–13, 2001, Davos, Switzerland. PP PUBLICO Publications, Germany, pp. 259–262.

Wiehe, I.A., 2012. Asphaltene solubility and fluid compatibility. Energy Fuels 26 (7), 4004–4016.

Wilhelmsson, B., 2005. Consider spiral heat exchangers for fouling applications. Hydrocarb. Process. July, 81–83.

William, B., Teja, A.S., 1987. Continuous thermodynamics of phase equilibria using a multivariate distribution function and an equation of state. AIChE J. 32 (12), 2067–2078.

Wilson, G.M., 1964. A new expression for the excess free energy of mixing. J. Am. Chem. Soc. 86 (2), 127–130.

Wilson, D.I., Polley, G.T., Pugh, S.J., 2002. Mitigation of crude oil preheat train fouling by design. Heat Transfer Eng. 23 (1), 24–37.

Wilson, D.I., Vassiliadis, V.S., 1997. Mitigation of refinery fouling by management of cleaning. In: Bott, T. (Ed.), Understanding Heat Exchanger Fouling and its Mitigation. Begell House, New York, Lucca, Italy, pp. 299–306.

Wilson, D.I., Watkinson, A.P., 1995. Mitigation of autoxidation reaction fouling. In: Panchal, C.B. (Ed.), Fouling Mitigation of Industrial Heat-Exchange Equipment, June 18–23, 1995, San Luis Obispo, CA. USA. Begell House, New York, pp. 461–472.

Wilson, D.I., Watkinson, A.P., 1996. A study of autoxidation reaction fouling in heat exchangers. Can. J. Chem. Eng. 74 (2), 236–246.

Wilson, D.I., Vassiliadis, V.S., Smaïli, F., 2001. Mitigating fouling at the plant scale by strategic cleaning: potential, pitfalls and neEds. In: Müller-Steinhagen, H. (Ed.), Heat Exchanger Fouling - Fundamental Approaches and Technical Solutions. July 8–13, 2001, Davos, Switzerland. PP PUBLICO Publications, Germany, pp. 325–332.

Wilson, D.I., Polley, G.T., 2001. Mitigation of refinery preheat train fouling by nested optimisation. In: Advances in Refinery Fouling Mitigation Session #46. AIChE, Houston, TX, USA, pp. 287–294.

Wilson, D.I., 2005. Challenges in cleaning: recent developments and future prospects. Heat. Transfer Eng. 26 (1), 51–59.

Wilson, D.I., Ishiyama, E., Paterson, W., Watkinson, A.P., 2009. Ageing: looking back and looking forward. In: Müller-Steinhagen, H., Malayeri, M.R., Watkinson, A.P. (Eds.), Int. Conf. On Heat Exchanger Fouling and Cleaning VIII, June 14–19, 2009, Schladming, Austria, pp. 221–230. Published online. www.heatexchanger-fouling.com.

Withers, P.M., 1996. Ultrasonic, acoustic and optical techniques for the non-invasive detection of fouling in food processing equipment. Trends Food Sci. Tech. 7 (9), 293–298.

Worrell, E., Galitsky, C., 2005. Energy efficiency Improvement and Cost Saving Opportunities for Petroleum Refineries – an ENERGY STAR Guide for Energy and Plant Managers. U.S. Environmental Protection Agency. Technical report.

Wu, J., Prausnitz, J.M., Firoozabadi, A., 1998. Molecular-thermodynamic framework for asphaltene-oil equilibria. AIChE J. 44 (5), 1188–1199.

Wu, J., Prausnitz, J.M., Firoozabadi, A., 2000. Molecular thermodynamics of asphaltene precipitation in reservoir fluids. AIChE J. 46 (1), 197–209.

Xie, K., Karan, K., 2005. Kinetics and thermodynamics of asphaltene adsorption on metal surfaces: a preliminary study. Energy Fuels 19 (4), 1252–1260.

Yahyaabadi, R., 2005. Solving hydraulic problems in crude preheat train networks. Hydrocarb. Process. 84 (11), 89–95.

Yaglom, A.M., Kader, B.A., 1974. Heat and mass transfer between a rough wall and turbulent fluid flow at high Reynolds and Peclet numbers. J. Fluid Mech. 62 (3), 601–623.

Yang, M., Young, A., Crittenden, B.D., 2009. Use of CFD to correlate crude oil fouling against surface temperature and surface shear stress in a stirred fouling apparatus, Proc. Eurotherm Conference on Fouling and Cleaning in Heat Exchangers. Austria 272–280. Schladming.

Yang, M., Young, A., Crittenden, B.D., 2009a. Modelling of fouling induction periods. In: Müller-Steinhagen, H., Malayeri, M.R., Watkinson, A.P. (Eds.), Int. Conf. On Heat Exchanger Fouling and Cleaning VIII, June 14–19, 2009, Schladming, Austria, pp. 69–75. Published online. www.heatexchanger-fouling.com.

Yang, M., Young, A., Crittenden, B.D., 2009b. Use of CFD to correlate crude oil fouling against surface temperature and surface shear stress in a stirred fouling apparatus. In: Müller-Steinhagen, H., Malayeri, M.R., Watkinson, A.P. (Eds.), Int. Conf. On Heat Exchanger Fouling and Cleaning VIII, June 14–19, 2009, Schladming, Austria, pp. 272–280. Published online. www.heatexchanger-fouling.com.

Yang, M., O'Meara, A., Crittenden, B.D., 2011. Determination of crude oil fouling thresholds. In: Mayaleri, M.R., Müller-Steinhagen, H., Watkinson, A.P. (Eds.), Proc. Int. Conf. On Heat Exchanger Fouling and Cleaning IX, June 5–10, 2011, Crete, Greece, pp. 1–6. Published online. www.heatexchanger-fouling.com.

Yang, M., Crittenden, B.D., 2012. Fouling thresholds in bare tubes and tubes fitted with inserts. Appl. Energy 89 (1), 67–73.

Yang, M., Young, A., Niyetkaliyev, A., Crittenden, B.D., 2012. Modelling fouling induction periods. Int. J. Therm. Sci. 51, 175–183.

Yang, J., Matar, O.K., Hewitt, G.F., Zheng, W., Manchanda, P., 2014. Modelling of fundamental transfer processes in crude-oil fouling. In: Proc. Of the 15th International Heat Transfer Conference, Paper IHTC15-9595, Kyoto, Japan.

Yao, S.P., Zhang, J.R., Jin, K.L., 1997. Organism relics or kerogens in oils as oil-source rock correlation indicator. Sci. China Ser. D 40 (3), 253–258.

Yarborough, L., 1979. Application of a generalized equation of state to petroleum reservoir fluids. ACS Adv. Chem. Ser. 182, 385–439.

Ye, J.R., Qing, H.R., Bend, S.L., Gu, H.R., 2007. Petroleum systems in the offshore Xihu Basin on the continental shelf of the East China Sea. AAPG Bull 91 (8), 1167–1188.

Yeap, B.L., Wilson, D.I., Polley, G.T., Pugh, S.J., 2001a. Incorporation of fouling models into the design of heat transfer networks. In: Müller-Steinhagen, H. (Ed.), 4th Int. Conf. On Heat Exchanger Fouling - Fundamental Approaches and Technical Solutions, July 8–13, Davos, Switzerland. PP PUBLICO Publications, Germany, pp. 107–110.

Yeap, B.L., Hewitt, G.F., Wilson, D.I., Pugh, S.J., 2001b. Simulating the fouling behaviour in shell–and–tube heat exchangers. In: Müller-Steinhagen, H. (Ed.), 4th Int. Conf. On Heat Exchanger Fouling - Fundamental Approaches and Technical Solutions, July 8-13, Davos, Switzerland. PP PUBLICO Publications, Germany, pp. 275–282.

Yeap, B.L., 2003. Designing Heat Exchanger Networks to Mitigate Fouling (Ph.D. Thesis). Department of Chemical Engineering and Biotechnology, Cambridge University, UK.

Yeap, B.L., Wilson, D.I., Polley, G.T., Pugh, S.J., 2004. Mitigation of crude oil refinery heat exchanger fouling through retrofits based on thermo–hydraulic fouling models. Chem. Eng. Res. Des. 82 (A1), 53–71.

Yeap, B.L., Wilson, D.I., Polley, G.T., Pugh, S.J., 2005. Retrofitting crude oil refinery heat exchanger networks to minimize fouling while maximizing heat recovery. Heat Transfer Eng.. 26 (1), 23–34.

Yiantsios, S.G., Karabelas, A.J., 1994. Fouling of tube surfaces: modeling of removal kinetics. AIChE J. 40 (11), 1804–1813.

Yiantsios, S.G., Andritsos, N., Karabelas, A.J., 1995. Modeling heat exchanger fouling: current status, problems and prospects. In: Panchal, C.B. (Ed.), Fouling Mitigation of Industrial Heat–Exchange Equipment, June 18–23, 1995, San Luis Obispo, CA, USA. Begell House, New York, pp. 337–351.

Yeap, B., Wilson, D.I., Polley, G.T., Pugh, S.J., 2004. Mitigation of crude oil refinery heat exchanger fouling through retrofits based on thermo-hydraulic fouling models. Chem. Eng. Res. Des. 82 (1), 53–71.

Yee, T.F., Grossmann, I.E., 1990. Simultaneous optimization models for heat integration–II. Heat exchanger network synthesis. Comput. Chem. Eng. 14 (10), 1165–1184.

Yen, T.F., Erdman, J.G., 1962. Investigation of the structure of petroleum asphaltenes and related substances by infrared analysis. Petrol. Chem. 7 (1), 5–18.

Young, A., Venditti, S., Yang, M., Crittenden, B.D., 2009. Characterisation of crude oils and their fouling deposits. In: Müller-Steinhagen, H., Malayeri, M.R., Watkinson, A.P. (Eds.), Int. Conf. On Heat Exchanger Fouling and Cleaning VIII, June 14–19, 2009, Schladming, Austria, pp. 17–26. Published online. www.heatexchanger-fouling.com.

Young, A., Venditti, S., Berrueco, C., Yang, M., Waters, A., Davies, H., Hill, S., Millan, M., Crittenden, B., 2011. Characterization of crude oils and their fouling deposits using a batch stirred cell system. Heat Transfer Eng. 32 (3–4), 216–227.

Yuan, X., Pibouleau, L., Domenech, S., 1989. Experiments in process synthesis via mixed-integer programming. Chem. Eng. Process. 25 (2), 99–116.

Zabiri, H., Radhakrishnan, V.R., Ramasamy, M., Ramli, N.M., Do Thanh, V., Wah, C.S., 2007. Development of heat exchanger fouling model and preventive maintenance diagnostic tool. Chem. Product Process Model. 2 (2), 1–10.

Zhang, S., Xu, B., Herod, A.A., Kandiyoti, R., 1996. Hydrocracking reactivities of primary coal extracts prepared in a flowing-solvent reactor. Energy Fuels 10 (3), 733–742.

Zhang, L.Y., Lawrence, S., Xu, Z.H., Masliyah, J.H., 2003a. Studies of Athabasca asphaltene Langmuir films at air-water interface. J. Colloid Interf. Sci. 264 (1), 128–140.

Zhang, L.Y., Xu, Z.H., Mashyah, J.H., 2003b. Langmuir and Langmuir-Blodgett films of mixed asphaltene and a demulsifier. Langmuir 19 (23), 9730–9741.

Zhang, L.Y., Lopetinsky, R., Xu, Z.H., Masliyah, J.H., 2005. Asphaltene monolayers at a toluene/water interface. Energy Fuels 19 (4), 1330–1336.

Zhao, Q., Müller-Steinhagen, H., 2001. Influence of surface properties on heat exchanger fouling. In: Bott, T.R., Watkinson, A.P., Panchal, C.B. (Eds.), Proc. Int. Conf. Mitigation of Heat Exchanger Fouling and its Economic and Environmental Implications, July 18–23, 1999, Banff, Canada. Begell House, New York, pp. 217–228.

Zudkevitch, D., Joffe, E., 1970. Correlation and prediction of vapour-liquid equilibria with the Redlich-Kwong equation of state. AIChE J. 16 (1), 112–119.

INDEX

Note: Page numbers followed by "f" and "t" indicate, figures and tables respectively.

G

H

I

O

P

R

S

CPI Antony Rowe
Eastbourne, UK
December 11, 2014